Engineering Project Management

Third Edition

Edited by

N. J. Smith

Professor of Transport Infrastructure and Project Management
School of Civil Engineering
University of Leeds

Blackwell
Publishing

© 1995, 2002 and 2008 Blackwell Publishing Ltd

Blackwell Publishing editorial offices:
Blackwell Publishing Ltd, 9600 Garsington Road, Oxford OX4 2DQ, UK
  Tel: +44 (0)1865 776868
Blackwell Publishing Inc., 350 Main Street, Malden, MA 02148-5020, USA
  Tel: +1 781 388 8250
Blackwell Publishing Asia Pty Ltd, 550 Swanston Street, Carlton, Victoria 3053, Australia
  Tel: +61 (0)3 8359 1011

First edition published 1995 by Blackwell Science Ltd
Second edition published 2002 by Blackwell Science Ltd
Third edition published 2008 by Blackwell Publishing Ltd

2   2009

ISBN: 978-1-4051-6802-1

Library of Congress Cataloging-in-Publication Data
  Engineering project management / edited by Nigel J. Smith. – 3rd ed.
    p. cm.
  Includes bibliographical references and index.
  ISBN: 978-1-4051-6802-1 (pbk. : alk. paper)
  1. Engineering–Management. 2. Construction industry–Management. I. Smith, Nigel J.
  TA190.E547 2007
  658.4'04–dc22
                      2007014016

A catalogue record for this title is available from the British Library

Set in 10/12 Sabon
by Newgen Imaging Systems (P) Ltd., Chennai, India
Printed and bound in Singapore
by Fabulous Printers Pte Ltd

The publisher's policy is to use permanent paper from mills that operate a sustainable forestry policy, and which has been manufactured from pulp processed using acid-free and elementary chlorine-free practices. Furthermore, the publisher ensures that the text paper and cover board used have met acceptable environmental accreditation standards.

For further information on Blackwell Publishing, visit our website:
www.blackwellpublishing.com

Material on PRINCE2™ is Crown copyright and is reproduced with the permission of the Stationery Office.

Contents

Preface

In many sectors of industry the significance of good project management in delivering projects in accordance with predetermined objectives has been established. Industrialists and engineering institutions have called for the inclusion of a significant proportion of project management in higher-level degrees, something realised by Finniston in his review of the *Future of Engineering* in 1980. Since the publication of the first edition of this book in 1995 a number of significant developments have taken place. A British Standard for Project Management, BS 6079, has been published, and the UK-based Association for Project Management has produced a fundamental guide to processes and practice entitled *Body of Knowledge* and drafted a standard contract for employing project managers. There has also been a marked increase in the teaching and delivery of university programmes and in continuing professional development (CPD) courses for project management.

Many organisations in the engineering, financial, business, process and other sectors are appointing people as project managers, some with a very narrow, brief and precise role, whereas others have a more strategic, managerial and multidisciplinary function. This third edition builds upon the successes of the first two editions in providing a clear picture of the aim of project management based upon best practice and some consideration of its continuing evolution. The improvements to this edition have been driven by the changes to the practice of project management and by the helpful comments made by book reviewers and readers since 1995.

Changes in the management of major projects have resulted in more joint ventures, project partnering, special project vehicles and other forms of collaborative working, which are reflected in the updated and extended text, covering procurement, stakeholders and collaborative provision. The new edition includes new chapters on quality, public–private partnerships and a detailed and authoritative chapter on the PRINCE2™ project management methodology (PRINCE2™ is a Trade Mark of the Office of Government Commerce). The book is not aimed at any particular sector of engineering but relates to the management of any major technical project.

Newly appointed project managers and students of project management at the MEng, MBA and MSc level will find the enhanced text and references beneficial. The book is concerned with the practice and theory of project management, particularly in relation to multidiscipinary engineering projects, large and small, in the UK and overseas.

Acknowledgements

I am particularly grateful to my co-authors and fellow contributors to this book. I am especially indebted to those who have participated in all three editions, namely, Dr Denise Bower, Dr Tony Merna and Mr Ian Vickridge. I am also grateful to the new contributor for this edition, Mr Mark Gannon, and I thank again the contributors to the earlier editions, particularly my colleagues at the University of Leeds.

The editor and the authors would like to express their appreciation to Sally Mortimer for managing the existing text and artwork from the previous editions. I would also like to thank Sally for formatting, checking and revising each of the many draft versions of every chapter. Nevertheless, the responsibility for any errors remains entirely my own.

N. J. Smith

Author Biographies

EDITOR: N. J. Smith, BSc, MSc, PhD, CEng, FICE, MAPM
Nigel is professor of transport infrastructure and project management and the head of school in the School of Civil Engineering at the University of Leeds. After graduating from the University of Birmingham, he spent 17 years in industry mainly working on major transportation infrastructure projects. Since returning to university life, his research interests have included project management, procurement methods and, in particular, privately financed concessions, risk management and the management of maintenance. He has seen a rapid expansion in activity over the last 5 years. From a small base, 'project' management research in the school is now active in all aspects of 'engineering' management. Research funding has been attracted from a range of national and international sources and from collaborating organisation in industry. Recent work includes studies of competitiveness in airport design, PRIME Contracting, public–private partnerships in Europe, value management and light rail transit schemes.

Denise Bower, BEng, PhD, M.ASCE, ILTM
Denise is a professor in project management and the deputy head of the School of Civil Engineering, University of Leeds. Her recent work includes the evaluation of procurement strategies, assessment of corporate strategy, the development of organisational partnering guidelines and the evaluation of the success criteria for a number of partnering arrangements and recommendations of contract strategies for overseas projects. She is also the programme leader of the MSc Engineering Project Management, which attracts international students from a wide range of backgrounds, and is directing doctoral studies in the area of procurement and contracts; her particular area of research interest is in the optimisation of the procurement of contracted services. She is the author and joint author of many books and publications, including *The Management of Procurement, Engineering Project Management, Dispute Resolution for Infra-Structure Projects* and *Managing Risk in Construction Projects*.

M. J. Gannon, BSc, MSc, PhD, MBA, MAPM, CMILT
Mark is a PRINCE2™ practitioner and a senior lecturer in the Management and Professional Department at London's Metropolitan University, where he lectures on project management, strategic management and managing the business environment to postgraduate students. He has over 20 years of project development and management experience and has worked on some of the largest rail private

finance initiative/public–private partnership (PFI/PPP) projects undertaken in the UK and has published 27 papers.

Peter Harpum
Peter spent 10 years managing major projects in the oil, shipping and construction industries. Peter now works as a project management consultant advising blue-chip companies in various industrial sectors. He holds a masters degree in project management and is working on a doctoral thesis on design management theory.

Steven Male
Steve held the Balfour Beatty Chair in Building Engineering and Construction Management in the School of Civil Engineering, University of Leeds. His research and teaching interests include strategic management in construction, supply-chain management, value management and value engineering. He has extensive industrial consultancy experience and has undertaken a range of studies with construction organisations, blue-chip organisations and local and central government clients.

Anthony Merna, BA, MPhil, PhD, CEng, MICE, MAQA, MAPM
Tony is the senior partner at Oriel Group Practice, a multidisciplinary consulting organisation based in Manchester. He is also a lecturer at the Centre for Research in the Management of Projects at UMIST. He has been actively involved in the management of infrastructure projects and has wide experience of project management and consultancy in both developed and developing countries.

Krisen Moodley, BSc, MSc, AIArb
Krisen is a lecturer in construction management in the School of Civil Engineering, University of Leeds. After graduating from the University of Natal, his initial employment was as a quantity surveyor with Davis Langdon Farrow Laing in Southern Africa, before his first academic appointment at Heriot-Watt University. He spent 4 years at Heriot-Watt before joining Leeds in 1994. His research interests are concerned with the strategic business relationships between organisations and their projects. His recent books include *Engineering, Business and Professional Ethics*; *Corporate Communications in Construction* and *Construction Business Development*; *Meeting New Challenges, Seeking Opportunity*, and chapter contributions to *Construction Reports 1944–1998*, and *Commercial Management: Defining the Discipline*.

Ian Vickridge, BSc (Eng), MSc, CEng, MICE, MCIWEM
Ian runs his own civil engineering consultancy and is also a visiting senior lecturer of civil engineering at UMIST. He has over 30 years' experience in the construction industry, gained in Canada, Hong Kong, Singapore, China and Saudi Arabia, as well as the UK. He is the executive secretary of the UK Society for Trenchless Technology and a reviewer of candidates for membership of the

Institution of Civil Engineers. He has published extensively on a variety of topics related to construction, environmental management and project management.

David Wright, MA, CIChemE, ACIArb

David left Oxford with a degree in jurisprudence and spent 30 years in industry. He gained experience in the automotive industry, the electronic industry, the defence industry and the chemical engineering and process industry. He was commercial manager of Polibur Engineering Ltd. In the mechanical engineering sector, he was the European legal manager to the Mather & Platt Group. He is now a consultant on matters of contract and commercial law.

List of Abbreviations

ABS	Assembly breakdown structure
ACWP	Actual cost of work performed
ADB	Asian Development Bank
ADR	Alternative dispute resolution
AfDB	African Development Bank
APM	Association for Project Management
BAC	Budget (baseline) at completion
BCWP	Budgeted cost of work performed
BCWS	Budgeted cost of work scheduled
BOD	Build, operate, deliver
BOL	Build, operate, lease
BOO	Build, own, operate
BOOST	Build, own, operate, subsidise, transfer
BOOT	Build, own, operate, transfer
BoQ	Bill of quantities
BOT	Build, operate, transfer
BP	Basis points
BPR	Business process re-engineering
BRT	Build, rent, transfer
BTO	Build, transfer, operate
CBA	Cost–benefit analysis
CCTA	Central Computing and Telecommunication Agency
CII	Construction Industry Insitute (Texas)
CPD	Continuing professional development
CPI	Cost performance index
CRINE	Cost reduction in the new era
CS	Controlling stage
CV	Cost variance
DBOM	Design, build, operate, maintain
DBOT	Design, build, operate, transfer
DCMF	Design, construct, manage and finance
DEO	Defence Estates Organisation
DETR	Department of the Environment, Transport and the Regions
DFA	Design for assembly
DfID	Department of International Development
DFM	Design for manufacturing

DP	Directing a project
DSM	Dependency structure matrix
DTI	Department of Trade and Industry
EBRD	European Bank for Reconstruction and Development
ECC	Engineering and construction contract
ECGD	Export Credit Guarantee Department
ECI	European Construction Institute
EIA	Environmental impact assessment
EIB	European Investment Bank
EIS	Environmental impact statement
EMS	Environmental management system
EPC	Engineer, procure, construct
EPIC	Engineer, procure, install, commission
EQI	Environmental quality index
ERP	Enterprise resource planning
EU	European Union
EVA	Earned value analysis
FAST	Functional Analysis Systems Technique
FBOOT	Finance–build–own–operate–transfer
FDA	Food and Drug Administration
FIDIC	Fédération Internationale des Ingénieurs Conseils (Lausanne)
GDR	Global depository receipt
GUI	Graphical user interface
HMPS	Her Majesty's Prison Service
HSE	Health and Safety Executive
ICT	Information and communication technology
IFC	International Finance Corporation
IP	Initiating project
IPT	Integrated project team
IRR	Interest rate risk
IT	Information Technology
LIBOR	London Interbank Offered Rate
MARR	Minimum acceptable rate of return
MBO	Management buy-out
MCA	Medicines Control Agency
MoD	Ministry of Defence
MPD	Managing project delivery
NEC3	New Engineering Contract
NEPA	National Environmental Protection Agency
NGO	Non-governmental organisation
NIF	Note issuance facility
NPV	Net present value
OBS	Organisational breakdown structure
OECD	Organisation for Economic Cooperation and Development
OGC	Office for Government and Commerce

PBP	Product-based planning
PBS	Product breakdown structure
PC	Procure, construct
PCM	Project cycle management
PEP	Project execution plan
PERT	Programme Evaluation and Review Technique
PFD	Product flow diagram
PFI	Private finance initiative
PIC	Procure, install, commission
PID	Project initiation document
PIM	Personal information manager
PL	Planning
PMI	Project Management Institute
PPP	Public–private partnership
PRINCE2™	PRoject IN Controlled Environments 2
PROMPT	Project Resource Organisation Management Planning Technique
PSBR	Public sector borrowing requirement
QA	Quality assurance
QC	Quality control
QFD	Quality function deployment
QM	Quality management
QMS	Quality management system
QP	Quality planning
QST	Quality system team
RC	Relational contracting
RE	Reliability engineering
RUF	Revolving underwriting facility
SB	Stage boundaries
SCA	Structured concession agreement
SCM	Supply-chain management
SPI	Schedule performance index
SPV	Special project vehicle
SU	Starting up a project
SV	Schedule variance
TCM	Travel-cost method
TCN	Third country nationals
TQM	Total quality management
TUPE	Transfer of undertaking from previous employer
USGF	US Gulf Factor
VA	Value analysis
VE	Value engineering

VM	Value management
VP	Value planning
VR	Value reviewing
WBS	Work breakdown structure
WMG	Warwick Manufacturing Group
WTA	Willingness to accept
WTP	Willingness to pay

Chapter 1
Projects and Project Management

This chapter describes the various aspects of project management, from what a project is through the various stages of a project to the key requirements for project success.

1.1 The function of project management

Managing projects is one of the oldest and most respected accomplishments of the human race. One stands in awe of the achievements of the builders of the pyramids, the architects of ancient cities, the masons and craftsmen of great cathedrals and mosques; of the might of labour behind the Great Wall of China and other wonders of the world. Today's projects, too, command attention. People are fascinated by the sight of the Americans landing on the moon, impressed by the latest Olympic Games infrastructure and enthused by the launch of the iPOD. As a new project is commissioned, as a major building rises, as a new computer system comes online or as a spectacular entertainment unfolds, a new generation of observers is inspired.

All of these endeavours are projects, like the many thousands of similar task-orientated activities, and yet the skills employed in managing projects, whether major ones such as these or more commonplace ones, are not well known other than to the specialists concerned. The contribution that knowledge of managing projects can make to management at large is greatly underrated and generally poorly known. For years, project management was derided as a low-tech, low-value, questionable activity. Only recently has it been recognised as a central management discipline. Major industrial companies now use project management as their principal management style. 'Management by projects' has become a powerful way to integrate organisational functions and motivate groups to achieve higher levels of performance and productivity.

1.2 Projects

A project can be any new structure, plant, process, system or software, large or small, or the replacement, refurbishing, renewal or removal of an existing one. It is a one-off investment. In recent times projects have had to meet the demands

1

of increasing complexity in terms of technical challenge, product sophistication and organisational change.

One project may be much the same as a previous one and different from it only in detail to suit a change in market or a new site. The differences may extend to some novelty in the product, in the system of production or in the equipment and structures forming a system. Every new design of car, aircraft, ship, refrigerator, computer, crane, steel mill, refinery, production line, sewer, road, bridge, dock, dam, power station, control system, building or software package is a project. So are many smaller examples, and a package of work for any such project can, in turn, be a subsidiary project.

Projects thus vary in scale and complexity from small improvements to products to large capital investments. The common use of word 'project' for all of them is logical because every one is:

- an investment of resources for an objective;
- a cause of irreversible change;
- novel to some degree;
- concerned with the future;
- related to an expected result.

A project is an investment of resources to produce goods or services – it costs money. The normal criterion for investing in a proposed project is therefore that the goods or services produced are more valuable than the predicted cost of the project.

To get value from the investment, a project usually has a defined date for completion. As a result, the work for a project is a period of intense engineering and other activities but which is short in its duration relative to the subsequent working life of the investment.

A number of definitions of the term *project* have been proposed, some of which are presented below.

- The Project Management Institute (PMI), USA, defines a project as 'a temporary endeavour undertaken to create a unique product or service'.
- The UK Association of Project Managers defines a project as 'a discrete undertaking with defined objectives often including time, cost and quality (performance) goals'.
- The British Standards Institute (BS 6079) defines a project as 'a unique set of co-ordinated activities, with definite starting and finishing points, undertaken by an individual or organisation to meet specific objectives with defined schedule, cost and performance parameters'.

From the aforementioned definitions it may be concluded that a project has the following characteristics:

- temporary, having a start and finish;
- unique in some way;

- specific objectives;
- the cause and means of change;
- risk and uncertainty;
- the commitment of resources: human, material and financial.

1.3 Project management

The definition of project management stems from the definition of a project and implies some form of control over the planned process of explicit change.

- The PMI defines project management as 'the art of directing and coordinating human and material resources through the life of a project by using modern management techniques to achieve predetermined goals of scope, cost, time, quality and participant satisfaction'.
- The UK Association of Project Managers defines project management as 'the planning, organization, monitoring and control of all aspects of a project and the motivation of all involved to achieve project objectives safely and within agreed time, cost and performance criteria'.
- The British Standards Institute (BS 6079) defines project management as 'the planning, monitoring and control of all aspects of a project and the motivation of all those involved to achieve the project objectives on time and to cost, quality and performance'.

The common theme is that project management is the management of change, but explicitly planned change, such that from the initial concept the change is directed towards the unique creation of a functioning system. In contrast, general or operations management also involves the management of change, but their purpose is to minimise and control the effects of change in an already constructed system. Therefore, project management directs all the elements that are necessary to reach the objective and those that will hinder the development. It should not be forgotten that projects are managed with and through people.

Project management is needed to look ahead at the needs and risks, communicate the plans and priorities, anticipate problems, assess progress and trends, get quality and value for money, and change the plans if needed to achieve objectives.

The needs for project management are dependent upon the relative size, complexity, urgency, importance and novelty of a project. The needs are also greater where projects are interdependent, particularly those competing for the same resources.

Each project has a beginning and an end, and hence it is said to have a life cycle. A typical life cycle is defined by Wearne (Figure 1.1) as the nature and scale of activity changing at each stage. Each stage marks a change in the nature, complexity and speed of the activities and the resources employed as a project proceeds. Widely different terminologies for the various aspects of project management are used in different industries, but they can all be related to this diagram.

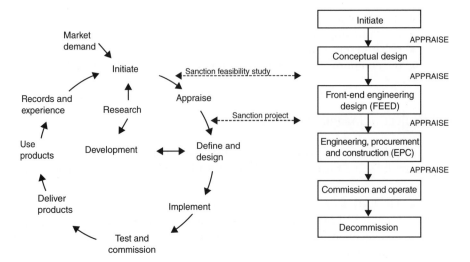

Figure 1.1   Project life cycle.

The duration of the stages vary from project to project, sometimes with delays between one and the next. They can also overlap. Figure 1.1 shows the sequence of starting them. It is not meant to show that one must be completed before the next is started. The objective of the sequence should be to produce a useful result so that the purpose of each stage should be to enable the next to proceed.

1.4 Project initiation

As shown in Figure 1.1, a project is likely to be initiated when whoever is its promoter predicts that there will be a demand for the goods or services the project might produce. The ideas for the project should then draw on records and experience of previous projects and the results from research indicating new possibilities. These sources of information ought to be brought together at this stage of a project.

Usually at this stage there will be alternative ideas or schemes that seem likely to meet the demand. Proceeding further requires the promoter to authorise the use of some resources to investigate these ideas and the potential demand for the project. The term 'sanction' is used here to mean the decision to incur the cost of these investigations. The cycle then proceeds to appraise the ideas to compare their predicted cost with predicted value.

'Feasibility' study

The appraisal stage is often referred to as the feasibility stage or as the feasibility study. Appraisal actually involves two key decisions – first a decision on the viability of the project and, if this is positive, a decision on the most feasible option for the project. The outputs from this phase can only be probabilistic, as they are based on predictions of demand and of costs whose reliability varies according to the quality of the information used, the novelty of the proposals and the amount and quality of the resources available to investigate the risks that could affect the project and its useful life. The key feature of this phase is the decision as to whether or not the project is viable, that is, high risk levels do not necessarily mean that the project will not go ahead but rather that a higher rate of return is required.

Repetition of the work up to this point is often needed after the first appraisal, as the results of the feasibility study may show that more information is needed on the possible demand or the conclusions of the appraisal may be disappointing and revised ideas more likely to meet a demand are called for. More expenditure has to be sanctioned to do so. Repetition of the work may also be needed because the information used to predict the demand for the project would have changed during this time. Feasibility studies may therefore have to be repeated several times.

Concluding this work may take time. Its outcome is quite specific: the sanctioning or the rejection of a proposed project. If a project is selected, the activities change from assessing whether it should proceed to deciding how best it should be realised and to specifying *what* needs to be done.

Design, development and research

Design ideas are usually the start of possible projects, and alternatives are investigated before estimating costs and evaluating whether to proceed any further. The main design stage of deciding how to use materials to realise projects usually follows evaluation and selection, as indicated in Figure 1.1. The decisions made during design determine almost entirely the quality and cost and therefore the success of a project. Scale and specialisation increase rapidly as it proceeds.

Development in the cycle refers to the experimental and analytical work to test the means of achieving a predicted performance. Research ascertains properties and potential performance. The two are distinct in their objectives. Design and development share one objective – that of making ideas succeed. Their relationship is therefore important, as indicated in Figure 1.1.

Most of the design and supporting development work for a project usually follows the decision to proceed. Decisions may be taken in sub-stages so as to investigate novel problems and review predictions of cost and value before continuing with a greater investment of resources.

Project implementation

Then follows the largest scale of activities and the variety of physical work typically needed to implement a project, particularly the manufacture of equipment for it and construction work.

Most companies and public bodies who promote new capital expenditure projects employ contractors and subcontractors from this stage to supply equipment or carry out construction. For internal projects within firms there is the equivalent internal process of placing orders to authorise expenditure on labour and materials.

Different sections of a project can proceed at different speeds in design and consequent stages, but all must come together to test and commission the resulting facility. The project has then reached its productive stage. It should then be meeting the specified objectives of the project.

The problems in meeting objectives vary from project to project. They vary in content and in the extent to which experience can be adapted from previous projects so as to avoid creating new problems. The criteria for appraisal also vary from industry to industry, but common to all projects is the need to achieve a sequence of decisions and activities, as indicated in Figure 1.1.

Figure 1.1 is a model of what may be typical of the sequence of work for one project. Projects are rarely carried out in isolation from others. At the start, alternative projects may be under consideration, and in the appraisal stage they may compete for selection. Those selected are then likely to share design resources with others – which may be otherwise unconnected – because of the potential advantages of sharing expertise and other resources, but they will therefore be in competition with them for the use of these resources through all the subsequent stages in the cycle.

A project is thus likely to be cross-linked with others at every stage shown in Figure 1.1. These links enable people and firms to specialise in a stage or sub-part of the work for many projects. The consequence is often that any one project depends upon the work of several departments or firms each of which is likely to be engaged on a variety of projects for a variety of customers. In all of these organisations there may therefore be conflicts in utilising resources to meet the competing needs of a number of projects, and each promoter investing in a project may have problems in achieving the sequence of activities that best suits his or her interests.

1.5 Project risks

Projects are investments of resources with a distinct increase in the level of investment, as the project passes from concept to implementation. This is demonstrated by the 'typical investment curve' as shown in Figure 1.2. From the graph, deviations from the base investment profile are identified (i.e. risks).

A – Increased income, for example, better than expected sales price/volume, or lower operational costs.

B – Delayed completion: project is completed late, but net revenue is as forecast.

C – Reduced net revenue, for example, worse sales price/volume, or higher operational costs.

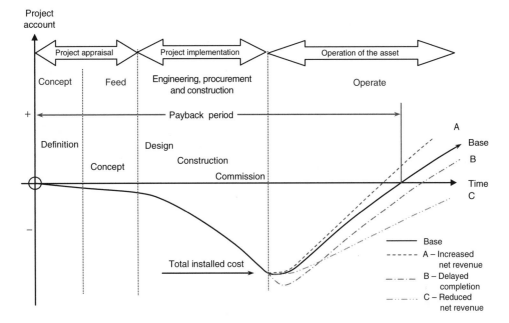

Figure 1.2   Typical investment curve.

    All projects are subject to risk and uncertainty. When you purchase goods from a retailer you are able to view them before purchase to ensure that they meet requirements. In other words, you are able to view the finished product prior to making your investment. Unfortunately, this situation is not possible in projects where the promoter is required to make investment prior to receipt of the product. Accordingly, projects are subject to uncertainty and consequent risk during the project delivery process. Such risks may be generated from factors external to the project (e.g. political change, market demand, etc.) or internally from the project activities (e.g. the effect of weather delays or unforeseen conditions). The nature of risk is that it can have both positive and negative effects on the project, that is, there are said to be upside and downside risks.

1.6   Project objectives

    It is at the front end of the project cycle that the greatest opportunity exists for influencing the project outcome. This principle is illustrated in Figure 1.3. The curves indicate that it is during the definition and concept stages that the greatest opportunity exists to reduce cost or to add value to the project. This opportunity diminishes as the project passes through sanction to implementation because as more decisions are taken the project becomes more closely defined. Conversely, the costs of introducing change are magnified, and hence the best time to explore options and make changes is at the concept stage and certainly not during implementation.

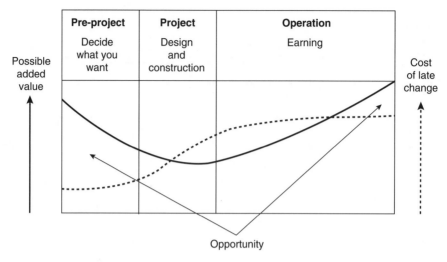

Figure 1.3　Opportunity to add value.

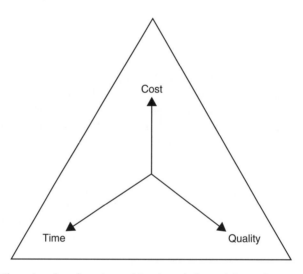

Figure 1.4　The triangle of project objectives (adapted from Barnes and Wearne, 1993).

Projects are implemented to meet the objectives of the promoter and the project stakeholders. The term *stakeholders* is used here to mean those groups or individuals who have a vested interest in the project but may or may not be investors in it. Accordingly, it is important that the project objectives are clearly defined at the outset and the relative importance of the objectives clearly established. Primary objectives are usually measured in terms of time, cost and quality, their interrelationship being shown in Figure 1.4. The use of an equilateral triangle in this context is significant, because, while it may be possible to meet one or two of the primary objectives, meeting all three is almost impossible. The positioning of the project in relation to the primary objectives is a matter of preference; where early completion is required, time is dominant, as might be the case in the launch of a new product where it is necessary to obtain market penetration first. That is not to say that the project can be completed at any cost nor does it mean that the quality can be compromised, but that time is of the essence. Where minimum cost development is required, such as for a community housing project, quality and time may have to be 'sacrificed'. Where ultimate quality is required, as, for example, in high-technology projects, cost and time may be secondary issues.

The relative importance of each objective must be given careful consideration because decisions throughout the project will be based on the balance between them. Inadequate definition and poor communication of objectives are common causes of failure in projects. Alignment meetings should be held with all key staff to ensure that all decisions are optimised in terms of the project objective.

1.7 Project success

Evidence indicates that success of projects now and in the future may depend upon the following criteria.

Definition of project objectives

The greatest lesson of project management is that the first task is to establish, define and communicate clear objectives for every project.

Risks

To succeed, the promoter's team should then assess the uncertainties of meeting the project objectives. If the risks are not identified, success cannot be achieved. The volume of events taking the team by surprise will be just too great for them to have any chance of meeting the objectives.

Early decisions

Many project successes demonstrate the value of completing much of the design and agreeing on a project execution plan before commitment to the costly work of manufacturing hardware or constructing facilities on a site.

Project planning

The form and amount of planning have to be just right. Not enough and a project is doomed to collapse from the unexpected. Too detailed plans will quickly become out of date and will be ignored.

Time and money

Planning when to do work and estimating what the resources required for it will cost must be considered together, except in emergencies.

Emergencies and urgency

A project is urgent if the value of completing it faster than normal is greater than the extra cost of the faster working. The designation of 'emergency' should be limited to work where the cost is no restraint on using any resources to work as fast as physically possible – for instance, rescue operations to save a life. An emergency is rare.

A committed project team

Dispersed project teams correlate with failure, concentration correlates to success. The committed project team should be located where the main risks have to be managed. Separation of people causes misunderstanding of objectives, communication errors and poor use of expertise and ideas. The people contributing part-time or full-time to a project should feel that they are committed to a team. The team should be assembled in time to assess and plan their work and their system of communications. Consultants, suppliers, contractors and others who are to provide goods and services should likewise be appointed in time to mobilise their resources, train and brief staff and assess and plan their work.

Representation in decisions

Success requires the downstream parties to be involved in deciding how to achieve the objectives of projects and, sometimes, in setting the objectives themselves. Human systems do not work well if the people who make the initial decisions do not involve those who will be affected later.

Communications

The nature of the work for a project changes month by month. So do the communications needed. Their volume and importance of communications can be huge. Many projects have failed because communications were poorly organised. A system of communication needs to be planned and monitored; otherwise information is received too late or goes to the wrong place for decisions, and becomes mere records of little value to control. The records then concentrate on allocation of blame for problems rather than on stimulating decisions to control them. The results of informal communications also need to be known and corrected, as bad news often travels inaccurately.

Promoter and the leader

Every project, large or small, needs a real promoter, a project champion who is committed to its success. Power over the resources needed to deliver a project must be given to one person who is expected to use it to avoid as well as to manage problems. In the rest of this book we call this person 'the project manager'. It may not be a separate job, depending upon the size and remoteness of the project. In turn, every sub-project should have its project manager with power over the resources it needs.

Delegation of authority

Inadequate delegation of authority has caused the failure of many projects, particularly where decisions have been restrained by requirements of approval by people remote to a problem who have delayed actions and so caused crises, extra costs and loss of respect and confidence in the management. Many projects have failed because authority for parts of a large project was delegated to people who did not have the ability and experience to make the decisions delegated to them. Good delegation requires prior checking that the recipients of authority are equipped to make the required decisions and then monitoring how effectively they make their decisions. This does not mean making their decisions for them.

Changes to responsibilities, project scope and plans

Some crises and the resulting quick changes to plans are unavoidable during many projects. Drive for problem solving is then very valuable, but failure to think through the decisions about problems can cause greater problems and loss of confidence in project leaders.

Control

If the plan for a project is good, the circumstances it assumes materialise and the plan is well communicated, few control decisions and actions are required. Much more is needed if circumstances change or people do not know the plan or understand and accept it. Control is no substitute for planning. It can waste potentially productive time in reporting and explaining events too late to influence them.

Reasons for decisions

In project management, every decision leads to the next one and depends upon the one before. The reasons for decisions have to be understood above, below, before and after so as to guide the subsequent decisions. Without this, divergence from the objectives is almost inevitable – and failure is its other name. Failure to give reasons with decisions and to check that they are accurately understood by their recipients can cause divergent and inconsistent actions. Skill and patience in communication are particularly needed in the rapidly changing relationships typical of the final stages of large projects.

Using past experience

Success is more likely if technical and project experience from previous projects is drawn upon deliberately and from wherever it is available. Perhaps the frequent failure to do this is another consequence of projects appearing to be unique. It is often easier to say 'this one is different' than to take the trouble to draw experience from the ones before. All projects have similarities and differences. The ability to transfer experience forward by making the comparisons is one of the hallmarks of a mature applied science.

Contract strategy

Contract terms should be designed to motivate all parties to try to achieve the objectives of the project and to provide a basis for project management. Contract responsibilities and communications must be clear, and not antagonistic. The terms of contracts should allocate the risks appropriately between customers, suppliers, contractors and subcontractors.

Adapting to external changes

Market conditions, customers' wishes and other circumstances change and technical problems appear as a project proceeds. Project managers have to be adaptable to these changes and yet be prepared to deter the avoidable ones.

Induction, team building and counselling

Success in projects requires people to be brought into a team effectively and rapidly using a deliberate process of induction. Success requires team work to be developed and sustained professionally. It requires people to counsel each other across all levels of the organisation, to review performance, to improve, to move sideways when circumstances require and to respond to difficulties.

Training

Project management demands intelligence, judgement, energy and persistence. Training cannot create these qualities or substitute for them, but it can greatly help people to learn from their own and other people's experience. A completed large project can require retraining of general management to understand and obtain the full benefit from its effect on corporate operations.

Towards perfect projects

The chapters of this book describe the techniques and systems that can be used to apply these lessons of experience. All of them should be considered, but some are chosen as priorities depending upon a situation and its problems.

All improvements cost effort and money. Cost is often given as a reason not to make a change. If so, the organisation should also estimate the cost of not removing a problem.

Further reading

Barnes, N. M. L. and Wearne, S. H. (1993) *Control of Engineering Projects*, Thomas Telford Ltd.

British Standards Institute (2002) *BS 6079 – 1:2002, Guide to Project Management*, BSI.

Hussain, R. and Wearne, S. H. (2005) Problems and needs of project management in the process and other industries, *Chemical Engineering, Research and Design*, 83, 372.

Morris, P. W. G. (1997) *The Management of Projects*, second edition, Thomas Telford Ltd.

Project Management Institute (2004) *A Guide to the Project Management Body of Knowledge*, third edition, PMI.

Smith, N. J. (2006) *Managing Risk in Construction Projects*, second edition, Blackwell Publishing.

The Association of Project Managers (2006) *Body of Knowledge*, fifth edition, APM Publishing.

Turner, R. and Simister, S. J. (2000) *The Handbook of Project Management*, third edition, Gower.

Wearne, S. H. (1989) *Control of Engineering Projects*, second edition, Thomas Telford Ltd.

Chapter 2
Value Management

This chapter includes the basic terminology and procedures associated with value management (VM). It then considers the role of VM within the content of a project.

2.1 Introduction

Over the past decade, there has been a trend towards applying value techniques at ever earlier stages in a project's life cycle. The term 'value management' has become a blanket term that covers all value techniques whether they entail value planning (VP), value engineering (VE) or value analysis (VA).

VM is used by electronics, general engineering, aerospace, automotive, construction and increasingly by service industries. VM techniques have also been successfully applied on all types of construction, from buildings to offshore oil and gas platforms, and for all types of clients from private industry to governmental organisations.

2.2 Definitions

There are no universally accepted definitions, and a number of different definitions have arisen to describe the same approach or stage of application.
According to the Institution of Civil Engineers' guide.

Value Management addresses the value process during the concept, definition, implementation and operation phases of a project. It encompasses a set of systematic and logical procedures and techniques to enhance project value through the life of the facility.

Value management (VM) is the title given to the full range of value techniques available; these include the following.

Value planning (VP) is the title given to value techniques applied during the concept or 'planning' phases of a project. VP is used during the development of the brief to ensure that value is planned into the whole project from its inception.

This is achieved by addressing and ranking stakeholders' requirements in order of importance.

Value engineering (VE) is the title given to value techniques applied during the design or 'engineering' phases of a project. VE investigates, analyses, compares and selects amongst the various options those that will meet the value requirements of stakeholders.

Value analysis (VA), or **value reviewing (VR),** is the title given to value techniques applied retrospectively to completed projects to 'analyse' or to audit the project's performance, and compare a completed, or nearly completed, design or project against predetermined expectations. VA studies are those conducted during the post-construction period and may be part of a post-occupancy evaluation exercise. In addition, the term VA may be applied to the analysis of non-construction-related procedures and process, such as studies of organisational structure, or procurement procedures.

The typical terms for VM studies at different stages of a project, including the various studies undertaken, are illustrated in the Figure 2.1. VP and VE are applied mainly in the concept and definition phases and generally end when the design is complete and the construction starts. However, VE can also be very effectively applied during construction to address problems or opportunities that

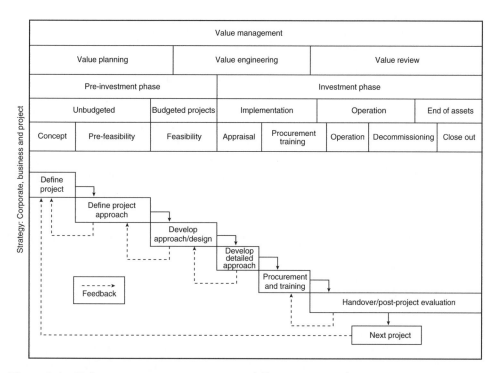

Figure 2.1   Value management structures at different project phases.

may arise. The latter typically derive from feedback from the site relating to specific conditions, performance and methods.

Maximum value is obtained from a required level of quality as least cost, the highest level of quality for a given cost or from an optimum compromise between the two. Therefore, VM is the management of a process to obtain maximum value on a scale determined by the client. VM is about enhancing value and not only about cutting cost (which is often a by-product). The philosophy of VM centres on the identification of the requirements.

VM involves functional analysis, life-cycle costing, operating in multidisciplinary groups using the job plan and creativity techniques and establishing comparative cost in relation to function.

Functional analysis is a technique designed to help the appraisal of value by a careful analysis of function (i.e. the fundamental reason why the project element or component exists or is being designed). It explores function by asking the initial question 'what does it do?' and how these functions are achieved.

The process is designed to identify alternative, more valuable and/or cost-effective ways to achieve the key functional requirements. Functional analysis is more applicable to the analysis of the detailed design of specific components (or elements) of a project. As in the project identification stage the function cannot be defined so clearly, it is less applicable to the identification stage.

A job plan is a logical and sequential approach to problem solving, which involves the identification and appraisal of a range of options, broken down into its constituent steps and used as the basis of the VM approach. The requirement is defined, different options for resolving the requirement are identified, options are evaluated and the options with the greatest potential are selected (HM Treasury Central Unit on Procurement, 1996).

The seven key steps of a job plan are the following.

(1) Orientation: identification of what has to be achieved and what are the key project requirements, priorities and desirable characteristics.
(2) Information: gathering of relevant data about needs, wants, values, costs, risks, time scale and other project constraints.
(3) Speculation: generation of alternative options for the achievement of client needs within stated requirements.
(4) Evaluation: evaluation of the alternative options identified in the speculation stage.
(5) Development: development of the most promising options and their more detailed appraisal.
(6) Recommendation for action.
(7) Implementation and feedback: examination of how the recommendations were implemented to provide lessons for future projects.

As speculation is crucial in the job plan, the quality of ideas generated determines the worth of the approach. Workshops are held where people work together to identify alternative options using idea-generating techniques such as brainstorming.

Life-cycle costing is an essential part of VM. It is a structured approach used to address all elements of cost of ownership based on the anticipated life span of a project. In the case of a building, these broad categories can be used for life-cycle costing.

(1) Investment costs: site costs, design fees (architect, quantity surveyor, engineer), legal fees, building costs, tax allowances (capital equipment allowances, capital gains, corporation tax) and development grants.
(2) Energy costs: heating, lighting, air-conditioning, lifts, etc.
(3) Non-energy operation and maintenance costs: these include letting fees, maintenance (cleaning and servicing), repair (unplanned replacement of components), caretaker, security and doormen, and insurance rates.
(4) Replacement of components (planned replacement of components).
(5) Residual or terminal credits: it is necessary to separate the value of the building from the value of the land when determining these credits in the context of a building. Generally, buildings depreciate until they become economically or structurally redundant, whereas land appreciates in value.

A **VM plan** is drawn up that should be flexible and regularly reviewed and updated as the project progresses. This plan should establish the following:

- a series of meetings and interviews;
- a series of reviews;
- who should attend;
- the purpose and timing of the reviews.

2.3 Why and when to apply VM

Projects suffer from poor definition because of inadequate time and thought given at the earliest stages. This results in cost and time overruns, claims, long-term user dissatisfaction or excessive operating costs. In addition, any project should be initiated only after a careful analysis of need. Therefore, one of the major tasks of stakeholders is to identify at the earliest possible stage the need for, and scope of, any project.

Another basic reason to use VM is that there are almost always elements involved in a project that contribute to poor value. Some of them are listed as follows:

- inadequate available time;
- habitual thinking/tradition;
- conservatism and inertia;
- attitudes and influences of stakeholders;
- lack of or poor communications;
- lack of coordination between the designer and operator;
- lack of relationship between design and construction methods;

- outdated standards or specifications;
- absence of state of the art technology;
- honest false beliefs/honest misconceptions;
- prejudicial thinking;
- lack of needed experts;
- lack of ideas;
- unnecessarily restrictive design criteria;
- restricted design fee;
- temporary decisions that become permanent;
- scope of changes for missing items;
- lack of needed information.

VM is primarily about enhancing value and not about cutting cost (although this is often a by-product). VM aims to maximise project value within time and cost constraints, without detriment to function, performance, reliability and quality. However, it should be recognised that improving project value sometimes initially requires extra expenditure. The key differences between VM and cost reduction are that the former is the following: positive, focused on value rather than cost and seeking to achieve an optimal balance between time, cost and quality; structured, auditable and accountable; and multidisciplinary, seeking to maximise the creative potential of all departmental and project participants working together.

Properly organised and executed, VM will help stakeholders achieve value for money (the desired balance between cost and function that delivers the optimum solution for the stakeholders) for their projects by ensuring that:

- the need for a project is always verified and supported by data;
- project objectives are openly discussed and clearly identified;
- key decisions are rational, explicit and accountable;
- the design evolves within an agreed framework of project objectives;
- alternative options are always considered;
- outline design proposals are carefully evaluated and selected on the basis of defined performance criteria.

Indeed, VM depends fundamentally on whether or not stakeholders can agree on project objectives from the start.

The key to VM is to involve all the appropriate stakeholders in structured team thinking so that the needs of the four main parties to a project can be accommodated where possible (HM Treasury, 1996).

The single most critical element in a VM programme is top level support (Norton and McElligott, 1995).

VM can also provide other important benefits:

- improved communication and team working;
- a shared understanding among key participants;

- better-quality project definition;
- increased innovation;
- the elimination of unnecessary cost.

It is apparent that all stakeholders should participate in the process. The only differences are connected with the level and stage of the involvement, and with the party responsible for holding the procedures.

It is vital that all stakeholders (investors, end-users and others with a real interest in the project outcome, such as project team, the owner, constructors, designers and specialist suppliers) must be involved in the process especially during VP and VE stages. In addition, while on larger or more complex projects an independent/external value manager is needed, as well as an external team with the relevant design and technical expertise, on smaller ones VM might be undertaken by the project sponsor's professional adviser, project manager or construction manager. In some cases an external professional must undertake this role. However, when establishing a structure for dealing with value for money on projects, there may be a need for expert assistance, particularly at the 'review' stages.

2.4 How to apply VM

There is no single correct approach to VM; although most projects vary, there are a number of stages common to projects themselves. Some of these stages overlap depending on the type of project. The project sponsor should ensure that a value management plan is drawn up and incorporated into an early draft of the project execution plan (PEP). This plan should establish:

- a series of meetings and interviews;
- a series of reviews;
- who should attend;
- the purpose and timing of the reviews.

It should not be a rigid schedule but a flexible plan, regularly reviewed and updated as the project progresses. It is essential that the project sponsor and the value manager prepare for reviews by deciding on the objectives and outputs required, the key participants and what will be required of them at different stages.

The precise format and timing of reviews will vary according to particular circumstances and timetable. Too many – the process may be disrupted and delayed, especially at the feasibility stage. Too few – opportunities for improving definition and the effectiveness of the proposals may be lost. Each of these reviews also provides an opportunity to undertake concurrent risk assessments. To exploit the benefits of VM while avoiding unnecessary disruption, there are at least seven obvious 'opportunity points' for reviews which arise on the majority of projects.

- During the concept stage, to help identify the need for a project, its key objectives and constraints.
- During the pre-feasibility stage, to evaluate the broad project approach/outline design.
- During the scheme design (feasibility stage), to evaluate developing design proposals.
- During detailed design (appraisal stage), to review and evaluate key design decisions as design progresses/concurrent studies.
- During construction/implementation, to reduce costs or improve buildability, functionality.
- During commissioning/operation, to improve possible malfunctions or deficiencies.
- During decommissioning/end of assets, to learn lessons for future projects.

2.5 Reviews

All reviews should be structured to follow the job plan. Issues of buildability, safety, operation and maintenance should be considered during all VM reviews and evaluation options.

The first review should:

(1) list all objectives identified by stakeholders;
(2) establish an objectives hierarchy by ranking the objectives in order of priority. It is important to stress that the aim is to produce a priority listing, not simply to drop lesser priorities. Reducing the list runs the risk of having to reintroduce priorities at a later stage with all the associated detrimental impacts on cost, time and quality. VM aims to eradicate the need for late changes – it should not encourage them;
(3) identify broad approaches to achieving objectives by brainstorming;
(4) appraise the feasibility of options: reject/abandon, delay/postpone;
(5) identify potentially valuable options;
(6) consider and preferably recommend the most promising option for further development.

The first review should result in:

(1) confirmation that a project is needed;
(2) a description of the project, that is, what has to be done to satisfy the objectives and priorities;
(3) a statement of the primary objective;
(4) a 'hierarchy' of project priorities;

(5) a favoured options for further development;
(6) a decision to proceed;
(7) a decision to reject/abandon or postpone/delay, if necessary.

This balanced statement of need, objectives and priorities, agreed by all stakeholders, helps the project sponsor produce the project brief.

The aim of the second review is to:

(1) review the validity of the objectives hierarchy with stakeholders and agree modifications;
(2) evaluate the feasibility of options identified;
(3) examine the most promising option to see if it can be improved further;
(4) develop an agreed recommendation about the most valuable option that can form the basis of an agreed project brief;
(5) produce a programme for developing the project.

The second review should result in:

(1) a clear statement of the processes to be provided and/or accommodated;
(2) a preferred outline design proposal;
(3) the basis of a case for the continuation of design development.

During the third phase/feasibility stage, some 10–30% of the design work is completed. That is why VE techniques are also part of this stage, which should:

(1) review project requirements and the objectives hierarchy agreed at the last review;
(2) check that the key design decisions taken since the last review remain relevant to the objectives hierarchy and priorities;
(3) review key decisions against the project brief by brainstorming to identify ways of improving design proposals outlined to date and to identify options;
(4) evaluate options in order to identify the most valuable one.
(5) develop the most valuable option to enhance value focusing on and resolving any perceived problems;
(6) agree on a statement of the option to be taken forward and agree on a plan for the continued development of the design.

The third review should result in:

(1) a thorough evaluation of the sketch design;
(2) clear recommendations for the finalisation of the sketch design;
(3) the basis of a submission for final approval to implement, abandon or postpone the project.

VE is aimed at finding the engineering, architectural, technical solution to help translate the VP-selected scheme design into a detailed design which provides best value; analysing, evaluating and recommending constructor's proposals, addressing problems that may emerge during construction.

By reviewing design proposals in this way, the value team will seek answers to the following to help determine value and function.

What is it (the purpose of the project or element)?
What does it do?
What does it cost?
How valuable is it?
What else could do the job?
What will that cost?

The fourth review should result in the following:

(1) promoting a continuous VM approach throughout the design process;
(2) finalising the original and proposed design and the basis for the changes, according to the findings of the previous review;
(3) describing the value proposals and explaining the advantages and disadvantages of each in terms of estimated savings, capital, operating and life-cycle costs and improvements in reliability, maintenance or operation;
(4) predicting the potential costs and savings and the redesign fee and time associated with the recommended changes;
(5) laying out the timetable for owner decisions, implementation costs, procedures and any problems (such as delays) that may reduce benefits.

The fifth review should result in the following:

(1) promoting a continuous VM approach throughout the construction process;
(2) assessing and evaluating the contractors' change proposals;
(3) investigating and verifying the feasibility of significant changes and the cost savings claimed, as well as the programme implications of including them;
(4) making forward-looking, practical recommendations for improvements which can be implemented immediately;
(5) checking that the risks to the project are being managed.

The sixth review should result in the following:

(1) measuring the success of the project in achieving its planned objectives;
(2) identifying the reasons for any problems that have arisen;
(3) determining what remedial actions should be taken;
(4) considering whether the objectives of the users/customers have been met. If those objectives changed, or were anticipated to change during the course of the project, to assess whether those changes were accommodated;
(5) ensuring that any outstanding work, including defects, has been remedied;

(6) recording the lessons that have been learnt to improve performance on subsequent or continuing projects.

Practitioners and users of VM must obtain feedback on its success, as feedback influences results by raising questions.

- Have good ideas emerged?
- Were any adopted?
- Were they implemented?
- Did the expected value improvement result?
- If not, why?

Factors that will be assessed include stakeholders' judgement, involvement, support, application, dedication, foot-dragging, approval process, systems appropriateness, use, effectiveness and management of the change process.

Actions that can emerge from the feedback include the following:

- change of personnel;
- change of approach;
- change of system;
- rerun of the exercise.

2.6 Procedures and techniques

There are many procedures and techniques available within VM for the value team to use as they see fit, whether applied formerly or intuitively. Typical techniques and procedures include:

- information gathering;
- cost analysis;
- life-cycle costing;
- Pareto's law;
- basic and secondary functions;
- cost and worth;
- FAST (Functional Analysis Systems Technique) diagramming;
- creative thinking and brainstorming;
- criteria weighting;
- value tree;
- checklists/attribute listing;
- analysis and ranking of alternatives.

2.7 Benefits of VM

Properly organised and executed VM provides a structured basis for both appraisal and development of a project, and results in many benefits to a project.

The following is a very brief list of such benefits:

- provides a forum for all concerned parties in a project development;
- develops a shared understanding among key participants;
- provides an authoritative review of the entire project, not just a few elements;
- identifies project constraints, issues and problems that might not otherwise have been identified;
- identifies and prioritise the key objectives of a project;
- improves the quality of definition;
- identifies and evaluates the means of meeting needs and objectives;
- deals with the life-cycle, not just initial, costs;
- usually results in remedying project deficiencies and omissions and superfluous items;
- ensures all aspects of the design are the most effective for their purpose;
- identifies and eliminates unnecessary costs;
- provides a mean to identify and incorporate project enhancements;
- provides a priority framework against which future potential changes can be judged;
- crystallises an organisation's brief priorities;
- maintains a strategic focus on the organisation's needs during development and implementation of a project;
- provides management with the information it needs to make informed decisions;
- permits a large return on a minimal investment;
- promotes innovation.

2.8 Summary

The key features of the value process and the application of VM to it have been described and the importance of value planning, teamwork and perseverance emphasised. The incentives and benefits to all stakeholders have been identified and discussed. These are underpinned by three particular aspects: the independence of the value manager to clearly establish the stakeholders' value criteria, planned application of team brainstorming and the inclusion of appropriate enabling clauses in contracts and agreements.

The factors needed to ensure success of VM include:

- systematic approach;
- integrated team environment;
- establishment of value criteria;
- focusing on the function;
- facilitation of creativity as a separate stage;
- consideration of a project on a life-cycle cost basis;
- collaborative and non-confrontational working environment;
- generation of records and audit trail.

VM must have comprehensive top management understanding and support and an enthusiastic, sustained and innovative approach.

References

HM Treasury Central Unit on Procurement (1996) Guidance Note 54, Value Management.

Institution of Civil Engineers (1996) *Creating Value in Engineering Projects*, Thomas Telford Ltd.

Norton, B. R. and McElligott, W. C. (1995) *Value Management in Construction: A Practical Guide*, Macmillan Press Limited.

Further reading

Connaughton, J. N. and Green, S. D. (1996) *Value Management in Construction: A Clients Guide*, CIRIA.

Kelly, J., Male, S. and Graham, D. (2004) *Value Management of Construction Projects*, Blackwell Publishing.

Webb, A. (1996) *Managing Innovative Projects*, Thompson Learning.

Chapter 3
Cash Flow, Project Appraisal and Risk Management

The primary purpose of this chapter is to describe project cash flow and then to outline the principles of project appraisal and risk management. Project appraisal, sometimes referred to as feasibility study, is an important stage in the evolution of a project. More accurately, it consists of two consecutive decisions: the project viability decision and then to identify the most feasible project option. It is important to consider alternatives, and identify and assess risks, at a time when data is uncertain or unavailable. This chapter outlines the stages of a project and describes in detail risk management techniques.

3.1 Cash flow

Projects and contracts are commercial ventures. Both the promoter of a project and the contractor invest money and take financial risks to achieve some desired benefit or return. Project and contract management is concerned with the control of both investment and risk with the aim of achieving this return. Economic or financial evaluation of cash flows is the primary basis for decisions relating to the choice, magnitude and pattern of investment.

The use of economic evaluation is not restricted to high-level management decisions such as whether to develop an oilfield or acquire or sell an entire business activity. At the level of project or contract management there are many decisions amenable to economic analysis and there is an alternative way to utilise money.

To quantify both the demand for money to meet the project or contract costs and the pattern of income it will generate, it is necessary to predict the cash flow. A cash flow is a financial model of the project or contract and even in its simplest basic form will provide vital information for the manager. Cash flow concerns the flow of money in and out of the account per time period. Income is positive and expenditure negative: the net cash flow is therefore the difference between cash in and cash out.

The model should be built up in the following stages to aid thorough understanding of the details of the investment as the basic cash flow is adjusted and progressively refined. Adherence to this structure, shown diagrammatically in Figure 3.1, will aid perception of all the implications of the investment.

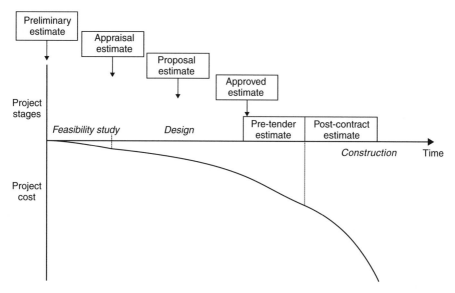

Figure 3.1   Estimates at project stage.

In many cases the desired decision may be made without completing all the stages.

(1) Compile the base-case cash flow simply by adding the costs and revenues to each activity on a bar chart programme that extends over the entire life cycle of the project or contract.
(2) Refine the base-case cash flow to take account of delays between incurring a commitment and paying or receiving the money.
(3) Calculate the resulting cost and benefit together with the investment required.
(4) Consider the implications of risk and uncertainty.
(5) If necessary, examine the implications of inflation.

In particular, it is recommended that inflation should always be considered separately when all other aspects of the investment are understood. Additional assumptions concerning future rates of inflation then have to be made, which increase the uncertainty. It is easy to be confused or misled if inflation is introduced too early in the financial analysis.

3.2 Categories of charge

There are three basic types of cost or revenue charges that make up the cash flow.

(1) Fixed charges incurred at a point in time.
(2) Quantity-proportional charges related to quantity of work completed, output or deliveries of materials. Only material costs and product revenues are strictly in this category.

(3) Time-related charges – predominantly the cost of resources. These are increasingly important, particularly in resource-dominated jobs. People and resources are normally employed or hired by the week or month.

Any flow of money can be defined in one of these categories and the realism of the cash-flow prediction will depend greatly on the correct definitions of charges. If in doubt, ask yourself how the bill will be paid.

3.3 Compiling the base-case cash flow

The most likely estimates of cost and revenue are added to each activity in a realistic programme. Initially keep the programme simple by splitting the work into a small number of major activities.

Set the base, or reference, date for the estimate, normally the date of the first flow of money, and divide the contract or project life into time periods appropriate to the accuracy required. Economists may compute annual cash flows in the early stages of project evaluation, but it is normal to consider monthly cash flows when appraising or managing the project while it is advisable to calculate weekly cash flows for short-duration contracts.

The base-cash flow is compiled using the costs and exchange rates as they were at the base date. Distinguish between fixed, quantity-proportional and time-related charges, and when necessary convert all cash flows to one currency. Assume zero inflation.

It is recommended that the cash-flow patterns be sketched on each activity in the programme prior to computing the period cash flows and constructing the cumulative curves. These should be completed before proceeding to add the refinements introduced in later sections. The process is illustrated in the following example and Figures 3.2 and 3.3.

Example

The contract for the manufacture and installation of a machine has been defined in Figure 3.2 as five activities on a bar chart, that is, design, fabrication, installation and commissioning of the machine plus construction of the foundation. The 26-week programme is known to be tight and the contract includes modest liquidated damages.

The costs associated with the completion of each activity are calculated and the resulting cash-flow patterns sketched onto the bar chart utilising the appropriate cost categories.

- It is assumed that a design team is allocated to this job for 6 weeks. They generate a time-related charge of pounds/week, which is shown to accumulate linearly. If the activity takes longer than predicted, then the cost will extend (and subsequent activities may be delayed).

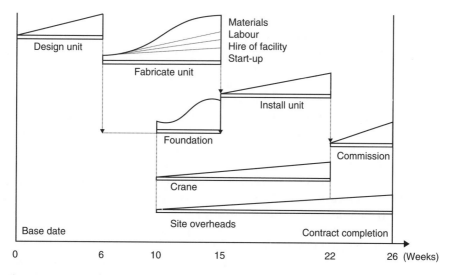

Figure 3.2　Development of a cash flow model on the contract programme for the manufacture and installation of a machine.

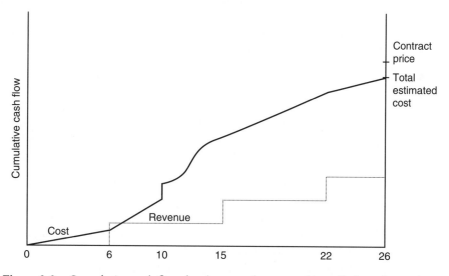

Figure 3.3　Cumulative cash flow for the manufacture and installation of a machine.

- Fabrication of the unit incurs several different costs: a fixed cost for setting up the job, time-related charges for labour and the use of the facility and material charges as they are scheduled to be delivered, all as sketched against this activity on the bar chart. By categorising costs the consequence of a delay or of an increase in material costs can quickly be deduced.
- Similar cost patterns are estimated for the other activities.

- General costs associated with several activities have been isolated into 'indirect' cost centres or 'hammocks'. It is assumed that the same crane will be retained on site for construction of the foundation and installation of the unit, also that overheads are incurred as a weekly charge while work is proceeding on site. These time-related hammock costs are an important mechanism in the development of this simple 'time-and-money' model. They should be viewed as elastic and they are, of course, sensitive to changes in programme. The author strongly advises that all general resources should be separated as hammocks in any model.

The total weekly costs are then accumulated to give the cumulative cost curve, as shown in Figure 3.3. This is the best prediction of the progressive commitment of expenditure on this contract.

The pattern of income is then added to that diagram as shown by the broken line. In this case payments are staged – on completion of design, fabrication and installation – with the largest payment at the end of the contract. This revenue curve predicts how money will be earned, and the area between the cost and revenue curves is an indication of the investment required from the contractor.

Even in this very simple form the base cash-flow model offers a useful management tool as the implication of any change is easy to visualise. In this example, adherence to the schedule is of great importance as any delay in completion will rapidly increase interest payments on the contractor's investment, and liquidated damages may also be incurred.

Contract cash flow

It is easy to see from Figures 3.2 and 3.3 that the investment would be greatly affected by any change in the payment characteristics or any circumstances that delayed either expenditure or receipts. The contractor's cash flow can therefore be said to be 'sensitive' to such changes. Remember that cash flow in a contract that is of relatively short duration is sensitive to the following:

(1) the type of contract, for example, admeasurement or cost-reimbursable;
(2) the contractual payment characteristics imposed by that contract, including mobilisation fees and retentions;
(3) the delay between incurring costs and paying the bills;
(4) disruption of the work plan;
(5) the speed and extent of reimbursement of variations and claims.

Inflation is only likely to be significant if the contract duration exceeds 12 months and the underlying rate of cost escalation is 5% or more.

The investment and confidence in the cash-flow forecasts will also be affected by uncertainty surrounding any aspect of the contract. A fully quantified risk analysis of the type applied later in this chapter to a project cash flow of long duration will rarely be necessary for a small contract, but risks must not be ignored. The implications of each variable should be considered and appropriate contingencies allocated in the programme and estimate.

Refining the contract cash flow

The simple basic model will assist decision making, but for accounting and investment predictions it will be necessary to adjust the base cash flow to take into account supplier credit and the lag in payment of earned revenue. The actual cash flow will be later than shown in the basic model. In most cases the delay in payment of revenue (e.g. in accordance with the terms of the contract 'payment will be made within x weeks of submission of the certificate') will be greater than the lag in expenditure as most workers expect to be paid weekly!

The refined cash-flow predictions linked to a specified rate of interest will generate figures of the investment required to complete the contract. This becomes an element of the contractor's cost and must be included when determining the contract price.

3.4 Project cash flow

There is some similarity between the contractor's balance of account for the aforementioned contract and that for the promoter developing a new facility or project. In each case, a period of investment is followed by a period of surplus in the account. The time scale of the project is, however, much longer, and consequently, this investment is unlikely to be sensitive to delays of a few weeks in payments. The sensitive factors are here likely to be the following:

(1) market forecasts of the quantity and price of the product;
(2) staging the developing, that is, the project is constructed in either one or a number of separate stages;
(3) operational efficiency and reliability;
(4) delay in commissioning;
(5) inflation.

In this case, adjustments to the base cash flow for the lag in payments are more likely to be introduced in the operational phase of the project, perhaps to allow for the delay between producing the product and selling it. Emphasis must here be given to the analysis of risk.

3.5 Profitability indicators

Although it is sometimes possible to compare different investments by reference to their cash-flow diagrams, it is advisable to employ and tabulate several of the various profitability indicators that quantify the investment when choosing between alternatives. These include profit, maximum capital lock-up and payback period amongst others.

Profit – the arithmetical difference between total payments and total receipts – is an obvious case. Maximum profit could be the criterion for the choice of one of several alternative plans, all of which satisfy the time and resource requirements. The profit figure does not, however, give any measure of the investment required or of the effect of time on the flow of money that constitutes that investment.

Similarly, maximum capital lock-up and payback period, which quantify, respectively, the maximum demand for capital and the time taken from the start of the contract or project for the investor's account to move into credit, could be used. By reference to Figure 3.4, jointly these three figures define the key ordinates on the cumulative net cash-flow curve; none of them, however, takes into account the time-value of money. This means that it is better to receive £1 today rather than in 12 months' time, because of the use of the money in the intervening period.

The process used to introduce a measure of the time-value of money is called discounting. The individual-period cash flows incurred over the duration of the contract or project are converted to their equivalent values at a single point in time, normally the start of the investment. If a stream of future payments representing an investment is discounted to the present time, the present value of

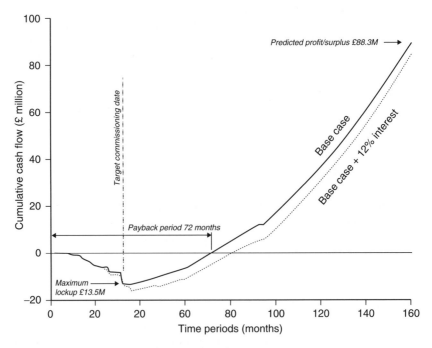

Figure 3.4 Cumulative cash flow predicted.

the investment is calculated. When both payments and receipts are discounted, the net present value is given.

The present value p of a cash flow C_n in period n is given by:

$$p = C_n \times \frac{1}{(1+r)^n}$$

where r is the discount rate per time period (expressed as a decimal).

Care must be taken to ensure that all the alternatives to be compared are evaluated on the same basis, that is, that the same discount rate, base date, and time period or interval are used. In a non-inflationary situation, the discount rate is normally selected to represent the cost of capital to the investor: the time period may be years or months for a project and months or weeks for a contract.

It is rarely necessary to calculate the discount factor $1/(1+r)^n$ as this is tabulated and is readily available in many textbooks. Calculation of the net present value is then simply a matter of multiplying the net cash flow in the first time period by the relevant factor to give the discounted cash flow and then repeating this process for each successive time period. Receipts are designated as positive and payments as negative cash flows. The net present value is the arithmetical sum of the individual discounted values.

Calculations of present value assume a discount rate and it is sometimes convenient to utilise the internal rate of return, which is defined as the discount rate that will produce zero net present value, that is, net present value is a measure of the gain from a project or contract, whereas internal rate of return measures gain relative to outlay. It must be emphasised that discounting is an expedient technique used only to aid comparison of investments. This technique for taking into account the 'time-value' of money should not be confused with provision for cost escalation, which should be considered separately.

It is probable that either net present value or internal rate of return would be used in conjunction with several of the other criteria for the selection of a project or to identify a preferred contract cash flow. For example, a positive net present value or some specified minimum value of internal rate of return may be required in addition to a minimum level of profit. A short payback period is also desirable and is given prominence in the selection of many commercial projects. The use and tabulation of several profitability indicators is strongly advised when comparing investments. The same criteria are also a convenient way of evaluating change in either contract or project. In both cases, the investor is greatly concerned with the effect of change on the investment, and all the aforementioned factors to which cash flow was said to be sensitive may be assessed in this way.

3.6 Inflation

Inflation is the decrease in the purchasing power of money arising from escalation of costs and prices. The effect is compound, cumulative and, consequently, extremely time sensitive. When base-case cash flows are escalated to predict

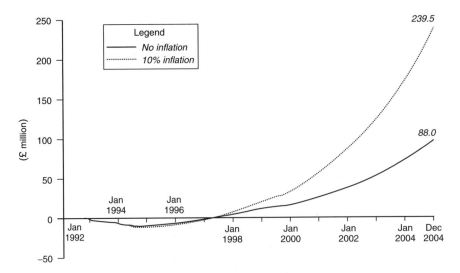

Figure 3.5   New industrial plant: cumulative cash flows.

'money of the day figures', the increases can appear dramatic – they can also be misleading!

Figure 3.5 illustrates this effect when the base-case costs and prices for new industrial plant are all increased at an inflation rate of 10% p.a. In 'money of the day' terms both the demand for capital and the predicted profit increase significantly, the former from £13.5 million to £16.2 million and the profit from £88 million to £240 million. The apparent improvement in benefit is, however, a myth as the purchasing power of money will decrease over the life of the project.

The difficulty of predicting future rates of inflation is obvious when one considers the technical, political and economic changes of the last 20 years – all of which affect commodity values, exchange rates and market confidence. Different elements of cost will also escalate at different rates, for example, changes in the cost of labour or fuel may not be reflected in the cost of imported materials or construction plant. Published national 'rates of inflation' are normally linked to a collection of domestic prices to indicate movement in 'the cost of living' and may be very different from the costs of raw materials used by your particular project.

For projects of more than 10 years' overall duration and particularly where the cash flow analysis is performed primarily for ranking purposes, escalated forecasts should be given relatively little weight and the decision be based on the non-escalated figures. In all cases the maximum escalated demand for capital should be determined. Fortunately the project manager is unlikely to be required to generate 'money of the day' figures when estimating project costs and benefits. This is more often the responsibility of economists or accountants who may simulate the implications of a range of inflation rates. Tax and/or royalty payments levied on the profits generated by the project will, of course, be calculated at 'money of the day' values. In engineering contracts the risk arising from variations in exchange rates should be clearly allocated to one of the parties. For contracts exceeding 12 months in duration it is normal to compensate the contractor by the application of a contract price adjustment formula linked to published national indices.

3.7 Initiation

The individual project, however significant and potentially beneficial to the promoting organisation, will only constitute part of corporate business. It is also likely that, in the early stages of the project cycle, several alternative projects will be competing for available resources, particularly finance. The progress of any project will therefore be subject to investment decisions by the parent organisation that may allow the project to proceed.

In most engineering projects the rate of expenditure changes dramatically as the project moves from the early stages of studies and evaluations, which consume mainly human expertise and analytical skills, to design, manufacture and construction of a physical facility. A typical investment curve is shown in Figure 3.6, which indicates that considerable cost will be incurred before any benefit accrues to the promoter from use of the completed project (or asset).

When considering the investment curve, the life cycle of the project splits into three major phases: appraisal and implementation of the project followed

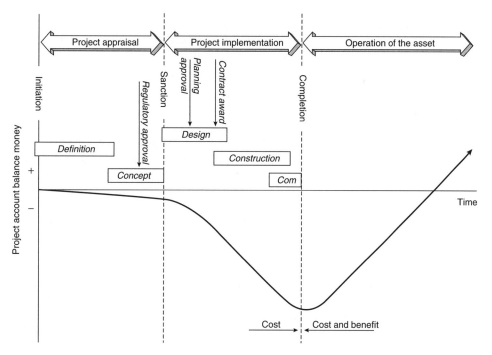

Figure 3.6   Project cash flow.

by operation of the completed facility. The precise shape of the curve will be influenced by the nature of the project, by external factors such as statutory approvals and by the project objectives. In the public sector, appraisal may extend over many years and be subjected to several intermediate decisions to proceed. In a commercial situation the need for early entry into a competitive market may outweigh all other considerations.

Two important factors emerge when studying the investment curve.

- Interest payments compounded over the entire period when the project account is in deficit (below the axis in Figure 3.6) form a significant element of project cost.
- The investor will not derive any benefit until the project is completed and is in use.

It is likely that at least two independent and formal investment decisions will be necessary – labelled 'initiation' or 'viability' and 'sanction' on this diagram. The first signifies that the ill-defined idea evolved from research and studies of demand is perceived to offer sufficient potential benefit to warrant the allocation of a specific project budget for further studies and development of the project concept. The subsequent sanction decision signifies acceptance or rejection of these detailed proposals. If positive, the organisation will then proceed with the major part of the investment in the expectation of deriving some predicted benefit when the project is completed.

All these estimates and predications, which frequently extend over periods of many years, will generate different degrees of uncertainty. The sanction decision therefore implies that the investor is prepared to take risk.

3.8 Sanction

When the project is sanctioned the investing organisation commits itself to major expenditure and assumes the associated risks. This is the key decision in the life cycle of the project. To make a well-researched decision the promoter will require the following.

Clear objectives. The promoter's objectives in pursuing this investment must be clearly stated and agreed to by senior management early in the appraisal phase, for all that follows is directed at achievement of these objectives in the most effective manner. The primary objectives of quality, time and cost may well be in conflict, and it is particularly important that the project team know whether minimum time for completion or minimum cost is the priority. These are rarely compatible, and this requirement will greatly influence both appraisal and implementation of the project.

Market intelligence. This relates to the commercial environment in which the project will be developed and later operated. It is necessary to study and predict trends in the market and the economy, anticipate technological developments and the actions of competitors.

Realistic estimates/predictions. It is easy to be over optimistic when promoting a new project. Estimates and predictions made during appraisal will extend over the whole life cycle of implementation and operation of the project. Consequently, single-figure estimates are likely to be misleading and due allowance for uncertainty and exclusions should be included.

Assessment of risk. A thorough study of the uncertainties associated with the investment will help to establish confidence in the estimate and to allocate appropriate contingencies. More importantly at this early stage of project development it will highlight areas where more information is needed and frequently generate imaginative responses to potential problems, thereby reducing risk.

Project execution plan. This should give guidance on the most effective way to implement the project and to achieve the project objectives, taking account of all constraints and risks. Ideally this plan will define the likely contract strategy and include a programme showing the timing of key decisions and award of contracts.

It is widely held that the success of the venture is greatly dependent on the effort expended during the appraisal preceding sanction. There is, however, conflict between the desire to gain more information and thereby reduce uncertainty, the need to minimise the period of investment and the knowledge that expenditure on appraisal will have to be written off if the project is not sanctioned.

Expenditure on appraisal of major engineering projects rarely exceeds 10% of the capital cost of the project. The outcome of the appraisal as defined in the concept and brief accepted at sanction will, however, freeze 80% of the cost. The opportunity to reduce cost during the subsequent implementation phase is relatively small, as shown in Figure 3.7.

3.9 Project appraisal and selection

Project appraisal is a process of investigation, review and evaluation undertaken as the project or alternative concepts of the project are defined. This study is designed to assist the promoter to reach informed and rational choices concerning the nature and scale of investment in the project and to provide the brief for subsequent implementation. The core of the process is an economic evaluation; based on a cash-flow analysis of all costs and benefits that can be valued in money terms, which contributes to a broader assessment called cost–benefit analysis. A feasibility study may form part of the appraisal.

Appraisal is likely to be a cyclic process repeated as new ideas are developed, additional information received and uncertainty reduced, until the promoter is able to make the critical decision to sanction implementation of the project and commit the investment in anticipation of the predicted return.

It is important to realise that, if the results of the appraisal are unfavourable, this is the time to defer further work or abandon the project. The consequences

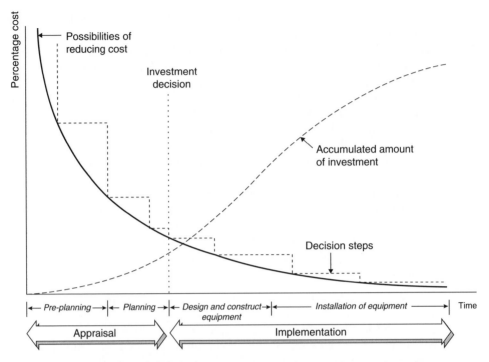

Figure 3.7 Graph of the percentage cost against time taken, showing how the important decisions for any project are made at the start of that project.

of inadequate or unrealistic appraisal can be expensive – as in the case of the Montreal Olympic Stadium – or disastrous.

Ideally all alternative concepts and ways of achieving the project objectives should be considered. The resulting proposal prepared for sanction must define the major parameters of the project: the location, the technology to be used, the size of the facility, the sources of finance and raw materials together with forecasts of the market and the predictions of the cost/benefit of the investment. There is usually an alternative way to utilise resources, especially money, and this is capable of being quantified, however roughly.

Investment decisions may be constrained by non-monetary factors such as:

- organisational policy, strategy and objectives;
- availability of resources such as manpower, management or technology.

Programme

It will be necessary to decide the best time to start the project based on the previous considerations. Normally this means as soon as possible, because no profit can be made until the project is completed. Indeed, it may be that market conditions or other commitments impose a programme deadline, for example, a customer will not buy your product unless he can have it by mid-2010, when his processing factory will be ready. In inflationary times, it is doubly important to complete a project as soon as possible because of the adverse relationship between time and money. The cost of a project will double in 7.25 years at a rate of inflation of 10%.

It will therefore be necessary to determine the duration of the appraisal design and construction phases so that:

- the operation date can be determined;
- the project costs can be determined;
- the promoter's liabilities can be assessed and checked for viability. (It may well be that the promoter's cash availability defines the speed at which the project can proceed.)

The importance of time should be recognised throughout the appraisal. Many costs are time-related and would be extended by any delay. The programme must therefore be realistic and its significance taken fully into account when determining the project objectives.

Risk and uncertainty

The greatest degree of uncertainty about the future is encountered early in the life of a new project. Decisions taken during the appraisal stage have a very large impact on final cost, duration and benefits. The extent and effects of change are frequently underestimated during this phase although these are often considerable, particularly in developing countries and remote locations. The overriding conclusion drawn from recent research is that all parties involved in construction projects would benefit greatly from reductions in uncertainty prior to financial commitment.

At the appraisal stage the engineering and project management input will normally concentrate on providing:

- realistic estimate of capital and running costs;
- realistic time scales and programmes for project implementation;
- appropriate specifications for performance standards.

At appraisal the level of project definition is likely to be low, and therefore risk response should be characterised by a broad-brush approach. It is recommended that effort should be concentrated on the following:

(1) seeking solutions that avoid/reduce risk;
(2) considering whether the extent or nature of the major risks is such that the normal transfer routes may be unavailable or particularly expensive;
(3) outlining any special treatment which may need to be considered for risk transfer, for example, for insurance or unconventional contractual arrangements;

(4) setting realistic contingencies and estimating tolerances consistent with the objective of preparing the best estimate of anticipated total project cost;
(5) identifying comparative differences in the riskiness of alternative project schemes.

Engineering/project managers may not have full responsibility for identifying the revenues and benefits from the project: this is usually shared with the marketing or development planning departments. The involvement of project managers in the planning team is recommended as the appraisal is essentially a multidisciplinary brainstorming exercise through which the promoter seeks to evaluate all alternative ways of achieving these objectives.

For many projects this assessment is not easy as not all the benefits/disbenefits may be quantifiable in monetary terms. For others it may be necessary to consider the development in the context of several different scenarios (or views of the future). In all cases the predictions are concerned with the future needs of the customer or community. They must span the overall period of development and operations of the project, which is likely to range from a minimum of 8 or 10 years for a plant manufacturing consumer products to 30 years for a power station and much longer for public works projects. Phasing of the development should always be considered.

Even at this early stage of project definition, maintenance policy and requirements should be stated as these will affect both design and cost. Special emphasis should be given to future maintenance during the appraisal of projects in developing areas. The cost of dismantling or decommissioning may also be significant but is frequently ignored.

3.10 Project evaluation

The process of economic evaluation and the extent of uncertainty associated with project development require the use of a range of financial criteria for quantification and ranking of the alternatives. These will normally include discounting techniques, but care must be taken when interpreting the results for projects of long duration.

Cost–benefit analysis

In most engineering projects, factors other than money must be taken into account. If a dam is built, we might drown a historical monument, reduce the likelihood of loss of life due to flooding, increase the growth of new industry because of the reduced risk, and so on. Cost–benefit analysis provides a logical framework for evaluating alternative courses of action when a number of factors are highly conjectural in nature. If the evaluation is confined to purely financial considerations, it fails to recognise the overall social objective of producing the greatest possible benefit for a given cost.

At its heart lies the recognition that no factor should be ignored because it is difficult or even impossible to quantify in monetary terms. Methods are available to express, for instance, the value of recreational facilities, and although it may not be possible to put a figure on the value of human life, it is surely something that cannot afford to be ignored.

The essential cost–benefit analysis is to take into account all the factors that influence either the benefits or the cost of a project. Imagination must be used when assigning monetary values to what at first sight might appear to be intangibles. Even factors to which no monetary value can be assigned must be taken into consideration. The analysis should be applied to projects of roughly similar size and patterns of cash flow. Those with the higher cost–benefit ratio will be preferred. The maximum net benefit ratio is marginally greater than the next most favoured project. The scope of the secondary benefits to be taken into account frequently depends on the viewpoint of the analyst.

It is obvious that, in comparing alternatives, each project should be designed at the minimum cost that will allow the fulfilment of objectives including the appropriate quality, level of performance and provision for safety.

Perhaps, more important, the viewpoint from which each project is assessed plays a critical part in properly assessing both the benefits and cost attributed to a project. For instance, if a private electricity board wishes to develop a hydro-electric power station, it may derive no benefit from the coincidental provision of additional public recreational facilities, which cannot therefore enter into its cost–benefit analysis. A public sector owner could quite properly include the recreational benefits in its cost–benefit analysis. Again, as far as the private developer is concerned, the cost of labour is equal to the market rate of remuneration, no matter what the unemployment level. For the public developer, however, in times of high unemployment, the economic cost of labour may be nil, as the use of labour in this project does not preclude the use of other labour for other purposes.

3.11 Engineering risk

An essential aspect of project appraisal is the determination of whether the level of risk is acceptable to the investor. This process starts with a realistic assessment of all known uncertainties associated with the data and predictions generated during appraisal. Many of the uncertainties will involve a possible range of outcomes, that is, it could be better or worse than expected. Risks arise from uncertainty and are generally interpreted as factors that have an adverse effect on the achievement of the project objectives.

It is helpful to try to categorise risk associated with projects both as a guide to identification and to facilitate the selection of the most appropriate risk management strategy. One method is to separate the more general risks that might influence a project but may be outside the control of the project parties from the risks associated with key project elements; these are referred to as global and elemental risks, respectively.

Global risks may be capable of being influenced by governments and can be subdivided into four sections; political, legal, commercial and environmental risks. Political risks would include events such as public inquiry, approvals, regulation of competition and exclusivity, whereas changes in statute law, regulations and directives would all be legal risks. Commercial risks can include the wider aspects of demand and supply, recession and boom, social acceptability and consumer resistance. Environmental risks are easier to identify, but as global risks they are more specifically to do with changes in standards, in external pressure and in environmental consents.

Elemental risks are those associated with elements of the project, namely, implementation risks and operation risks and, for some projects, financial and revenue risks. These risks are more likely to be controllable or manageable by project parties. Typical examples of these risks are given in Tables 3.1–3.4.

Table 3.1   Implementation risks.

Risk category	Description
Physical	Natural, pestilence and disease, ground conditions, adverse weather conditions, physical obstructions
Construction	Availability of plant and resources, industrial relations, quality, workmanship, damage, construction period, delay, construction programme, construction techniques, milestones, failure to complete, type of construction contracts, cost of construction, insurances, bonds, access, insolvency
Design	Incomplete design, design life, availability of information, meeting specification and standards, changes in design during construction, design life, competition of design
Technology	New technology, provision for change in existing technology, development costs

Table 3.2   Operational risks.

Risk category	Description
Operation	Operating conditions, raw materials, supply, power, distribution of offtake, plant performance, operating plant, interruption to operation due to damage or neglect, consumables, operating methods, resources to operate new and existing facilities, type of O&M (operation and maintenance) contract, reduced output, guarantees, underestimation of operating costs, licences
Maintenance	Availability of spares, resources, sufficient time for major maintenance, compatibility with associated facilities, warranties
Training	Cost and levels of training, translations, manuals calibre and availability of personnel, training of principal personnel after transfer

Table 3.3 Financial risks.

Risk category	Description
Interest	Type of rate, fixed, floating or capped; changes in interest rate, existing rates
Payback	Loan period, fixed payments, cash flow milestones, discount rates, rate of return, scheduling of payments, financial engineering
Loan	Type and source of loan, availability of loan, cost of servicing loan, default by lender, standby loan facility, debt–equity ratio, holding period, existing debt, covenants, financial instruments
Equity	Institutional support, take-up of shares, type of equity offered
Dividends	Time and amounts of dividend payments
Currencies	Currencies of loan, ratio of local/base currencies

Table 3.4 Revenue risks.

Risk category	Description
Demand	Accuracy of demand and growth data, ability to meet increase in demand, demand over concession period, demand associated with existing facilities
Toll	Market- or contract-led revenue, shadow tolls, toll level, currencies of revenue, tariff variation formula, regulated tolls, take and/or pay payments
Developments	Changes in revenue streams from developments during concession period

3.12 Risk management

The logical process of risk management may be defined in four steps.

- Identification of risks/uncertainties.
- Analysis of the implications (individual and collective).
- Response to minimise risk.
- Management of residual risk.

Risk management can be considered as an essential part of the continuous and structured project planning cycle. Risk management:

- requires an acceptance that uncertainty exists;
- generates a structured response to risk in terms of alternative plans, solutions and contingencies;
- is a thinking process requiring imagination and ingenuity;
- generates a realistic (and sometimes different) attitude in project staff by preparing them for risk events rather than being taken by surprise when they arise.

If uncertainty is managed realistically, the process will:

- improve project planning by prompting 'what if' questions;
- generate imaginative responses;
- give greater confidence in estimates;
- encourage provision of appropriate contingencies and consideration of how they should be managed.

Risk management should impose a discipline on those contributing to the project, both internally and on customers and contractors. By predicting the consequences of a delayed decision, failure to meet a deadline, or a changed requirement, appropriate incentives/penalties can be devised. The use of range estimates will generate a flexible plan in which the allocation of resources and the use of contingencies are regulated.

Risk reduction

Risk reduction involves measures to mitigate the likely impact of risk and may include:

- obtaining additional information;
- performing additional tests/simulations;
- allocating additional resources;
- improving communication and managing organisational interfaces;
- allocating risk to the party best able to control or manage the risk.

Market risk may frequently be reduced by staging the development of the project. All these will incur additional cost in the early stages of project development.

The role of people

The consequences of the aforementioned risks may be aggravated by the inadequate performance of individuals and organisations contributing to the project.

Control is exercised by and through people. As the project manager will need to delegate, he or she must have confidence in the members of the project or contract team and, ideally, should be involved in their selection.

Involve staff in risk management so as to utilise their ideas and to generate motivation and commitment. The roles, constraints and procedures must be clear, concise and understood by everyone with responsibility.

3.13 Risk and uncertainty management

Recently, much attention has been given to the prospect of considering both the risks and the opportunities associated with a particular project, sometimes known as uncertainty management. Both the evaluation of risk and of opportunity is affected by uncertainty. However, it is important to note that these are different processes requiring a different mindset and different data, and cannot be easily integrated at a detailed level for any project.

In the UK the Turnbull Report requires all companies to include in their annual statement of accounts a guide showing how risks and opportunities are monitored and managed. This has tended to lead companies to produce simple risk and opportunity matrices with various levels of impact. This has its own dangers and should be in addition to detailed risk management and value management studies being undertaken as and when required.

No standard approach has been developed for the application of uncertainty and/or complexity management, but this is another example of the continuing evolution of project management theory and practice, as described in Chapter 21.

Further reading

Perry, J. G. and Thompson, P. A. (1992) *Engineering Construction Risks: A Guide to Project Risk Analysis and Risk Management*, Thomas Telford Ltd.

Rogers, M. (2001) *Engineering Project Appraisal*, Blackwell Publishing.

Smith, N. J. (2006) *Managing Risk in Construction*, second edition, Blackwell Publishing.

Chapter 4
Quality Management in Projects

Quality management (QM) is paramount to all businesses and projects. It is no longer acceptable for a project manager to consider meeting only time and budget as measures for a project's success. QM has a purpose, that is, to ensure that projects meet specification and customer requirements. QM must be structured to meet the requirements of any business or project. This requires organisations to identify their goals – what are they trying to achieve – and identifying where they are now and where they want to be. When involved in quality planning (QP), it is important to ensure that the processes involved can be reviewed and that the documentation forming a quality management system (QMS) is relevant to the business or project in hand. This chapter provides a brief history of QM, the major elements and functions of a QMS and total quality management (TQM), international quality standards, a brief description of the costs associated with the implementation and operation of a QMS, a typical quality production format and how to improve and integrate a project quality.

4.1 History of quality management

The approach to QM has gradually developed to its present-day position. During the 1920s separate inspection departments appeared in many companies. Industrial workers often lacked skills to identify inferior products. During the Second World War, as production increased and simultaneously the availability of labour decreased, inspection work had to be more efficient and statistical quality control (QC) became an increasingly effective tool.

Quality assurance (QA) came to the forefront of business in the UK in the 1980s although the Institute of Quality Assurance was founded in 1919 in the UK. In the USA the origins of quality stem from work carried out by Frederick W. Taylor, generally regarded as the Father of Scientific Management, in the early part of the twentieth century. The Japanese attributed much of their success to Deming and Juran, who introduced QM in Japan in 1946.

More sophisticated products increased the possibilities for mistakes and reliability engineering (RE) became important. Special methods to ensure reliability were developed in the 1950s. RE was given a big boost by the development of space technology, at first concentrating on the electronic sector but later developed to include mechanical products. In the 1960s total QC was beginning to emerge. It was soon realised that, to produce high-quality products at competitive prices, a quality system operating throughout all stages of production was required.

In the 1970s product safety became an issue. As highly complex and expensive plants with a great potential risk element were being constructed, such as nuclear power stations, the regulatory authorities and others began insisting on procedures that not only ensured satisfactory quality but also provided evidence of a safe and controlled operation. QA was seen as the answer to this problem. QA ensures that all planned actions deemed necessary to provide confidence to all stakeholders to a project satisfy given requirements. Suppliers, distributors and others providing products and projects were beginning to be held responsible for any damages caused by the product to persons or property – product liability. To ensure safety and to limit consequences of any damage, additional factors were incorporated into quality activities.

Arising from this and inspired by Japanese successes, primarily in the manufacturing field, the need for continuous quality improvements became an accepted norm in all manufacturing organisations. Human resources were also taken into account, as it became increasingly important to make use of all available skills within a company to prevent mistakes and to increase motivation. People became important to operational efficiency.

Today quality is a management concern of vital importance to organisations – the need for quality requires all stakeholders to be aware of the importance of quality. Each new development in the field of QM has been a base for, and an influence to, the next. Today, all past experience is being used to develop further the quality activities in order to achieve efficiency. The pursuit of quality is a dynamic process.

More and more business organisations require a QM, in most cases achieved through a QMS as a condition of tender for any contract entered into. The signs are that this is increasing and those businesses that ignore quality are going to lose business opportunities.

4.2 Definitions

There are many definitions of quality, a few of which are listed below.

- The ability to meet market and customer expectations, needs and requirements.
- The supplying of goods that do not come back to customers who do.
- To be in conformance with user requirements (Crosby, 1979).
- Fitness for use (Juran, 1950).

For the purposes of this chapter the following definition of quality, compatible with the requirements of project management, is suggested:

the ability to manage a project and provide the product or service in conformance with the user requirements on time and to budget, and where possible maximise profits without affecting quality.

The following definitions of terms used in QM are proposed.

- **Continuous improvement.** The ongoing cycle of plan, implement, evaluate and review of every aspect of the business
- **Customer.** Any individual, group or organisation that may receive a product or service from any other individual, group or organisation.
- **Mission statement.** An expression of intent concerning the organisation's business and how this will be managed and operated with respect to both employees and customers.
- **Quality.** The totality of features and characteristics of a product or service that bear on its responsibility to satisfy stated or implied needs.
- **Quality management system.** The organisational structure procedures, processes and resources needed to implement QM.
- **Quality policy.** The overall quality intentions and directives of an organisation as regards quality and as formally defined by top management.
- **Value.** The desired balance between cost and function that delivers the optimum solution for the customer.

The major elements of QM are:

- confidence;
- control;
- consistency;
- cost-effectiveness;
- commitment;
- communication.

The preceding are essential elements of any project that is described as an undertaking which has a beginning and an end and is carried out to meet established goals within cost, schedule and quality objectives (Haynes, 1990). These elements of QM should be considered both internally and externally and not in isolation.

4.3 Quality planning

QP involves the design of a very detailed plan to audit and maintain the QMS. Harrington and Mathers (1997) suggest a 15-step plan, which includes forming and training a quality system team (QST), defining the support, structure and formats, identifying the major quality processes, development of a quality manual and procedural documentation, development of communication, training, implementation and support plans and establishment of a tracking and review system.

A QP is a document setting out the specific quality practices, resources and activities relevant to a particular process, service or project (Oakland, 1992).

Ashford (1989) advises that quality plans should define:

(1) the quality objects to be attained;
(2) the specific allocation of responsibility and authority during the different phases of the project;
(3) the specific procedures, methods and work instructions to be applied;
(4) suitable testing, inspection examination and audit programmes at appropriate stages;
(5) a method for changes and modifications in a quality plan as projects proceed;
(6) other measures necessary to meet objectives.

To be of value, the first issue of a quality plan must be made before the commencement of work on site. It is also essential that it should be a document that will be read, valued and used by those in control of the work. Their preparation should be commenced at tender stage as part of the normal routine of project planning. Quality plans should be as succinct as possible and discussed with all those involved in its implementation.

4.4 Quality management

Dale *et al.* (2003) suggest that there are four stages of QM: inspection, QC, QA and TQM. Inspection and QC are retrospective; they act in a detection mode aiming to find problems that have occurred. QA and TQM aim to reduce and ultimately avoid problems occurring, hence bringing about improvement.

QC involves the operational techniques and activities that are used to fulfil the requirements for quality and is often regarded as an extension of inspection. Inspection is a detection activity such as measuring, examining, testing or gauging one or more characteristics of an entity and comparing these results with specified requirements so as to establish whether conformity is achieved for each characteristic.

QA considers all those planned and systematic actions necessary to provide adequate confidence that a product or service will satisfy given requirements for quality (prevention).

4.5 Total quality management

The fundamentals of TQM are the following:

- customer-focused (internal and external);
- examines 'work processes' not 'individual performance';
- applied to all work processes and all staff;
- monitoring, measurement and reporting;
- continual improvements.

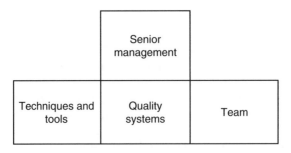

Figure 4.1   Total quality management.

The author suggests there are four main elements to TQM: management commitment, teamwork, techniques and the quality system (Figure 4.1). The main ingredients of TQM are as follows:

- it provides quality that meets the projects requirements;
- quality is not an alternative to productivity but a means of achieving it;
- every activity of a project contributes to the total quality of the project;
- TQM is a means of achieving project success;
- managing quality involves systems, techniques and individuals;
- TQM is a way of managing a project.

TQM according to BS EN ISO 8402 (1995) is a management approach of an organisation, centred on quality, based on the participation of all members and aiming at long-term success through customer satisfaction, and benefits to all members of the organisation and to society. In most cases TQM follows the implementation of QA, although this is not a prerequisite.

The advantages of adopting TQM in project management include:

- quality in meeting the project specification saves money;
- TQM alleviates the quality costs that result from poor quality;
- costs are reduced by preventing poor quality;
- capacity is increased by improving quality;
- TQM provides a basis for teamwork and techniques to interact;
- TQM helps recognise the need to balance risk, benefit and cost;
- TQM allows for changes in the project;
- TQM provides a basis for recording all activities in the project.

In devising and implementing TQM for an organisation it may be useful to ask first if the managers have the necessary authority, capability and time to carry it through. A disciplined and systematic approach to continuous improvement may be established in a quality council, whose members are the senior management team. The use of the team approach to problem solving has many advantages over allowing individuals to work separately.

- A greater variety of complex problems may be tackled by pooling the expertise and resources.
- Problems are exposed to a greater diversity of knowledge, skill and expertise and are solved more efficiently.
- The approach is more satisfying to team members, and boosts morale and ownership through participation in problem solving and decision making.
- Problems that cross departmental or functional boundaries can be dealt with more easily, and potential/actual conflicts are more likely to be identified and solved.
- The recommendations are more likely to be implemented than individual suggestions, as the quality of the decision making in good teams is high.
- Management commitment/support (leadership) is high.

These are achieved through management commitment, teamwork, techniques and the quality system.

4.6 Quality management systems

A QMS connects all the activities influencing the project's quality. It incorporates all stages of construction or manufacture from initial identification of needs to the operation of the product and completion of the project. By focusing on the early feasibility and design stages, quality and costs may be identified. To achieve the desired quality without unnecessary costs, a project manager must pursue an efficient system to coordinate the project activities affecting quality.

The QMS should ensure that:

- quality products and services always meet the expressed or implied requirements of the customer;
- company management knows that quality is achieved in a planned and systematic way;
- the customer feels confident about the quality of goods or services supplied and the method by which they are achieved.

The QMS must always be adjusted to suit the project's operation and final product. It must be designed so that the emphasis is put on preventive actions, at the same time allowing the project manager to correct any mistakes that do occur during the project life cycle. In most projects control documentation provides the data required to improve the system and hence the projects quality.

The QMS should be based on the following four activities.

- Planning.
- Execution.

- Checking.
- Action.

To achieve a satisfactory result, all activities and tasks affecting the quality of a project require planning. Execution should be based on necessary expertise and resources, and results must be checked. Checking must be followed by action. Defective products or sections of work must be removed. Information gained must be analysed and recorded so as to prevent the same defects from appearing again. Continuous upgrading must take place.

W. Edwards Deming, one of the founders of TQM and the quality movement developed the Deming circle (Deming, 1986). The Deming circle represents the problem analysis process and the quality improvement cycle and provides focus on defect correction as well as defect prevention. Figure 4.2 illustrates the Deming circle, often referred to as the Plan-Do-Check-Act (PDCA) cycle, which illustrates the interrelationship between planning, execution, checking and action.

Certain common rules are necessary to coordinate and direct a project's operation to achieve the quality objectives. These must be made known to all members of the company and the project team, and include the following:

- quality policy;
- quality principles;
- quality routines;
- working instructions.

Quality policy: This is the main guide to a company's approach to quality matters. It clarifies the overall principles for the company's attitude towards, and handling of, QM. The quality policy must not make any statement that will not have the backing and resources necessary that the company will not support and cannot resource. This means all key personnel must be committed to the

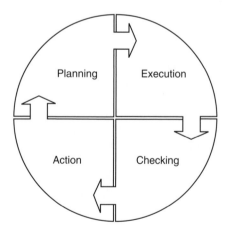

Figure 4.2   Deming circle (Jonson 1988).

success of a project. It is essential there is buy-in from senior management (all the way to the top), the staff, the shareholders and the customers.

Quality principles: These give more specialised guidelines as to the specific project phases. For example, they may deal with various stages of construction and how to implement improvements

Quality routines: These define patterns and sequences of actions to be taken and coordinate the processes important to quality matters. They distribute the responsibility between different functions, those being departments within the company and members of the project team.

Working instructions: These give detailed information as to the execution of various stages of construction related to the project.

Executing involves educating, training and informing relevant people of what is going to happen and who is doing what and implementing the necessary changes. Checking involves checking the effects of implementation to see that the change is happening as required.

It is important to note that QMSs are dynamic and must be regularly audited and improved to meet changes in the type of work carried out by an organisation. Often, redundant processes and documentation remain locked in a QMS even though an organisation is no longer involved in the type of work described. QMSs are only as good as the organisation managing them.

Most organisations use one of two systems, centralised and de-centralised systems for managing quality.

Centralised systems lay stress on the practice of QC or the operational techniques and activities that are used to fulfil requirements for quality. These obviously vary from one application to another. However, they will almost certainly include taking samples at various stages, comparing them with specified requirements and rejecting items that do not comply. Under a centralised system, these items will be the responsibility of a QC department with its own management hierarchy independent of production departments. In a large organisation the QC department may include technical experts such as engineers, physicists, metallurgists as well as inspectors and testing operatives. They will have access to the project to take samples as required and operate the laboratories and testing facilities.

Such centralised organisations can be very powerful. They provide both control and assurance (QC/QA). They have the advantage that the authority and independence of the quality manager and his staff are clearly established and they can be effectively insulated from commercial pressures that might compromise their judgement. The disadvantage is that QC departments tend to grow into separate empires working in parallel with but in isolation from those responsible for production. While they can be very effective in rejecting defective work, their isolation precludes them from participating in the planning and organisation of procedures for the prevention of defects. Furthermore, in attempting to achieve and maintain the technical initiative they are often prone to duplicate, if not outdo, the skills and qualifications of the production teams. Centralised systems therefore tend to be both expensive and prone to friction.

The difference between centralised and de-centralised systems is that in the latter the responsibility for controlling quality is placed firmly on the shoulders of those actually doing the work. This follows the principle that production management has a duty to make things that comply with specification, a duty which it should not be permitted to relinquish or to share with others.

In a de-centralised system, production managers are required to develop plans, procedures and routines for inspection and testing that will ensure that work is done properly. Inspection and testing is then carried out mainly by staff within the production hierarchy in accordance with strictly defined and documented QC programmes. With such arrangements, it requires little imagination to visualise that occasions will arise when an inspector's integrity will become stretched by the knowledge that his superiors will be displeased by a rejection, which will disrupt production schedules. Protection against such pressure is provided by an independent QA manager who has powers to approve or reject quality plans and to supervise their implementation by surveillance, random checks, examination of documentation and formal audit.

4.7 Quality manual

In order to ensure personnel are made aware of the methods of working and their responsibilities, it is essential that policy, principles and routines are documented. This is done in the form of a quality manual. In some cases an existing company quality manual is used as the basis for a project specific manual or plan. The aim of the quality manual is to aid the company and project team in directing their QM. It is becoming increasingly common for customers to demand proof from suppliers of a documented quality system; the manual will also fulfil this function.

The quality manual should provide an adequate description of the QMS while serving as a permanent reference in the implementation and maintenance of that system. The quality manual is also used as a means to educate new employees by identifying the routines and responsibilities of all other employees within the company or specific to a project.

The company management and the project manager should periodically review the quality actions. A tool for this is quality audits, sometimes called quality reviews, to make them more politically acceptable. Quality audits are similar to other in-house audits but concentrate on quality aspects. A quality audit is a systematic and objective analysis conducted by a third party on a periodic basis. It can be directed towards goals (goal audit), system (system audit), its structure and practicality (operation audit). It may sometimes involve only results, for example, the finished product (result audit). In some industries regulatory audits are carried out by regulatory bodies against relevant regulations for the manufacture and supply of products. National regulatory bodies, such as the Medicines Control Agency (MCA) in the UK and Food and Drug Administration (FDA) in the US are statutorily responsible for carrying out such audits in the pharmaceutical industry.

4.8 International standards for quality

As of December 2003 the ISO 9000:1994 series of standards became obsolete. This means that a large number of organisations needed to take steps to realign their QMS to the requirements of the new standard ISO 9000:2000.

ISO 9001:1994 is a quality system model for QA in design development, production, inspection and testing and installation and servicing. ISO 9002 is a quality system model for QA in production, inspection and testing and installation and servicing. ISO 9003 is a model for QA in inspection and testing only. ISO 9003 is applicable to organisations setting up their management systems to control their activities without any documented system to cover in-process control. They only need demonstrate to the client the quality of the product or service by a final test. ISO 9003 has only 16 quality elements and many of these are of reduced scope.

ISO 9000:2000 is far more process based than ISO 9000:1994, which was primarily based on procedures. ISO 9001:2000 sets new and different standards. It demands that you continuously assess the processes you operate and investigate how to improve them. You are also asked to anticipate what might go wrong and plan ways to prevent problems before they arise. This is of paramount importance in the management of projects as time, money and quality are the main criteria for success.

The type of quality system adopted by organisations depends on the organisational system in place. Quality can be improved by the use of management techniques at different stages of a project's life cycle. A major influence on quality and quality systems has been the introduction of business process re-engineering, which can often result in major changes to an organisation's activities and thus the processes it goes through to ensure quality is achieved.

For an organisation to function effectively and efficiently, it has to identify and manage numerous linked activities. When an activity uses resources, they must be managed to ensure the transfer of inputs to outputs.

Figure 4.3 illustrates the model of a process-based QMS outlined in a QMS guideline for performance improvements – EN ISO 9001-2000, Second Edition 2000-12-15. This figure shows that interested parties play a significant role in defining requirements as inputs. Monitoring the satisfaction of interested parties requires the evaluation of information relating to the perception of interested parties as to whether the organisation has met their requirements.

When used in a QMS, this approach emphasises the importance of the following:

- understanding and fulfilling the requirements;
- the need to consider processes in terms of added value;
- obtaining results of process performance and effectiveness;
- continual improvement of processes based on objective measurement.

This process-based system is far more flexible than the procedures-based system, particularly as it allows added value to be considered on every process.

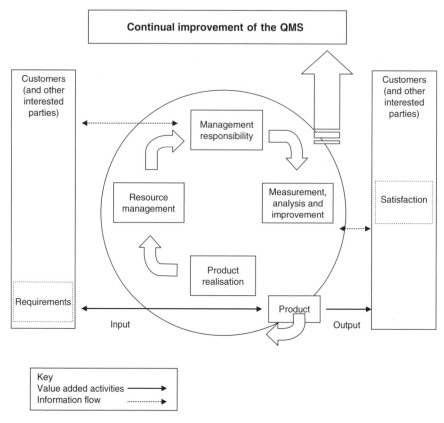

Figure 4.3   Model of process-based QMS (IS0 9000:2000).

Practising project managers will most likely be familiar with the aforementioned standards and also with ISO 10006 (1997). This standard provides guidance on quality system elements, concepts and practices for which the implementation is important to, and has an impact on, the achievement of quality in project management and supplements the guidance given in ISO 9004-1-1994. This standard is applicable to projects of varying complexity, small or large, of short or long duration, in different environments and irrespective of the kind of project product (including hardware, software, processed material, service and combinations thereof). This may necessitate some tailoring of the guidance to suit a particular project.

4.9 Implementing a project quality system

Implementing a project-specific quality system is similar to implementing any other quality system and involves the following:

• determine the size and characteristics of the project, project team and its locations;

- identify the status of existing documentation and determine whether or not existing documentation is compatible with the proposed project activities;
- determine the range of project activities within the specific project;
- identify the level of activities within the project and the method of work allocation;
- determine the responsibilities and commitment of the project management team;
- evaluate what customers really want and then map the processes relevant to improving customers satisfaction.

A greater awareness has emerged of the importance of quality as a means of competing. Quality has become essential in improving profitability and strengthening a company's position in the market place.

The advantages to the sponsor and project team of developing and introducing a quality system to a project are the following:

- profits would be increased through better management control systems;
- quality levels would be uniform throughout the project;
- safer deliveries and storage of materials would be ensured;
- costs should also be reduced through higher productivity, fewer defects and changes;
- capital utilisation can be improved through faster and safer throughput;
- clients would be satisfied.

4.10 Types of quality-related costs

Quality costs more, but on the other hand, lack of quality costs even more. Today's society depends heavily on reliable products and services. The consequences of failure are costly, both on society as a whole and the project, incurring additional expense and inconvenience.

Crosby (1979) suggests that improved quality is cheaper in the medium term than poor quality. The costs of improving quality are:

costs of assurance procedures;
costs of control procedures;
costs of dealing with failures, rework, scrap and repairs.

The need to identify the costs of quality is not new, although the practice is not widespread. Attempts to budget for various types of project quality-related costs are rare, and planning activities to identify, measure and control these costs is even rarer. The project manager is primarily concerned with controlling time and costs, and with the inclusion of a quality system, the project manger should be in a position to control time, costs and quality.

Quality costs can be described as one of two types: operating costs and failure costs (or losses).

Operating costs are those incurred by a project in order to attain and ensure specified quality levels; these include the following:

(1) Prevention costs of efforts to prevent failures, which includes
  - Design reviews.
  - Quality and reliability training.
  - Vendor QP.
  - Audits.
  - Construction/manufacturing and installation prevention activities.
  - Product or project qualification.
  - Quality engineering.
(2) Appraisal costs of testing, inspection and examination to assess whether specified quality is being maintained, which includes
  - Test and inspection.
  - Maintenance and calibration.
  - Equipment depreciation tests.
  - Production line quality engineering.
  - Installation, testing and commissioning.

Failure costs (or losses) include internal failure costs resulting from the failure of a project or service to meet performance specification prior to delivery, such as product or project service, warranties, returns direct costs and allowances or product recall costs or liability costs. These include:

- design changes;
- vendor rejects;
- rework;
- scrap and material renovation;
- warranty;
- commissioning failure.

An important factor when considering introducing a QMS is the benefits in terms of cost savings. If a QMS costs x to implement and the savings are only y, where y is less than x, and no new orders are received, then the benefits of the QMS must be reconsidered. The development of a QMS must be managed as though it is itself a project to ensure benefits accrue.

4.11 Introducing a quality-cost system

Only by knowing where costs are incurred and their order of magnitude can project managers monitor and control them. Quality costs must be collected and recorded separately; otherwise they become absorbed and concealed in numerous overheads. Regular financial reports are vital if there is to be management visibility in quality-related costs. Quality costs should be regularly reported to and monitored by the project manager and can be related to other costs, such as the project cost curve, cost–benefit ratio or individual project package costs (BS 6143, 1992).

It is necessary to evaluate the adequacy and effectiveness of the QMS. Additional areas requiring attention may be unique to the project considered and require the establishment of quality and cost objectives. This may require the following:

- calculation of those costs that are directly attributable to the project's quality function;
- identifying the costs that are not directly the responsibility of the project's quality function;
- identifying internal failure costs for which budgets were allocated;
- identifying internal failure costs for unplanned failures;
- identifying the cost of failures after the change of ownership.

Sources of quality-cost data will be varied in most projects. For data to be collected consistently, detailed cost sheets are required to address each of the project quality-cost types. Costs associated with the development, implementation and running of a QMS must be balanced by the benefits accrued.

4.12 Example of a typical quality and production format

The QMS is connected to all activities influencing a product's quality. It incorporates all stages of production from initial identification of needs to when the end product is on the market fulfilling customer requirements and expectations. By focusing on the early production stages, quality and costs may be most easily influenced (Jonson, 1988).

The production process can typically be divided into the following:

- market research;
- product development and design of product or project specification;
- procurement;
- manufacturing and checking;
- distribution and sales;
- installation;
- feedback.

In addition, a stage for after-sales service and maintenance may be required.

Where quality is concerned, it is highly likely that it will involve people from different areas of an organisation.

Market research and the standards and specifications of market requirements are important for quality. Expectations, needs and requirements of the selected market must initially be collected and utilised in the right way. This means identifying the following:

- the need for and use of a product or service;
- the market sector to concentrate on and to assess the likely demand. It is important to decide on grade, quality, price and timing of the product launch;
- the customer requirements from existing contracts and data through customer inquiries and contracts and use of market surveys. Unexpressed needs and prejudices should also be revealed.

Market needs and requirements may be identified through market surveys. Studies of the competition are also useful. A product idea must be related to market needs. The collection of ideas and the evaluation of these are therefore an important part of quality work, as well as the assessment of the ability of the company to realise its ideas both technically and financially. When evaluating ideas a company must make use of all its knowledge and past experience of previous products and services. Presumed market needs without factual research cannot be allowed to influence decisions.

Product requirements are stated in a requirement specification, which will form the basis for further developments. In the requirement specification, customer requirements and expectations are transferred to the first set of specifications. Apart from the customer requirements, legal requirements relating to safety and the environment should be included in addition to those inherent in the company's own quality policy.

The product has to be designed and specified. The requirements stated are transcribed onto technical specifications showing raw material parts and manufacturing methods and processes, giving clear definite information as regards purchase, manufacturing and QC. Accepted tolerances are indicated. Through the development and testing of different designs, a basis evolves for the final specification.

It is essential that requirements important to the use of the product are defined and that the risks of misuse are avoided. Stress levels and acceptable environmental conditions, for example, should be established through testing. The results of these tests are necessary not only for the design work but also for the planning of after-sales service and maintenance.

At set intervals the design work reviews should take place and involve staff from various functions in the company. At these reviews possible problems should be identified and precautions defined to ensure that the final product meets the customer requirements. Through systematic review of various designs, the experience and know-how of not only sales and manufacturing staff but also of purchasing, service and quality staff can be utilised and the design established so that unnecessary work and costs are prevented in the later stages of development.

The number of design reviews required will depend on the complexity of the product, its usage, applied technology and similarity to earlier products. The design should also be subjected to tests to ensure that it meets the standards of the requirements of the specification. The number of tests should relate to the risks involved. It is important that the final product design and specification is fully documented and contains all essential information.

A design may be altered, requiring documents to be changed. It is important that these changes are done in the correct manner and at the correct time and also on the relevant physical products. Above all, it is important to ensure the product still meets the stated requirements even after changes.

The manufacturing planning stage is an important link between the design and manufacturing stages. Even if the design has already taken into account the capability of the manufacturing process, the product specification must be transferred to the instructions for manufacturing. The instructions should give a clear step-by-step guide to the manufacturing team, indicating what raw materials to use and quality requirements.

Product quality should be ensured during the manufacturing process. It is, therefore, essential to know how the process affects the quality and end product and to be able to use this information for in-process control.

In all manufacturing processes there are variances among the products produced. The size of spread will depend on which process has been selected or the condition of the equipment. The capability of the manufacturing line will indicate how large or small the spread will be. The capability index shows the ratio of the spread in relation to the tolerances allowed. This index is important when selecting machines to use, when to check and how often.

Just as important as the ability to manufacture the product is the ability to check products against requirements at the right time. Inspection points and instructions must be determined in close co-operation with designers and manufacturing engineers.

During the planning stage it must be proved that the manufacturing process, with its built-in inspection points, is able to produce goods that meet the requirements of the product specification. There should always be a continuous effort to improve the quality of a finished product.

Purchases must be planned and controlled. In all purchases, quality should be taken into account. The lowest price may in the long run prove to be very costly.

A project or business depends on the quality produced by its suppliers. An important part of quality activities must therefore be an evaluation of suppliers and their ability to meet standards of quality, delivery times and total cost. The suppliers must be able to ensure the quality of their products and that deliveries will be on time. It is often costly and time consuming for buyers to check the quality of goods bought. To ensure a satisfactory utilisation of capital and rapid throughput, it is prudent to demand information from various suppliers about their QC and to insist that relevant final checks are made, so that the consignment can go straight onto the production line.

In most cases receiving inspections have to be made. These should be planned in the same way as other inspections. A close relationship should be established with suppliers chosen by the company, and there should be good exchange of information regarding both requirements and results.

As the end product will finally emerge from the manufacturing stage, it is important to ensure that the manufacturing is run under controlled conditions exactly as planned. Preventive maintenance of machinery and equipment to be used is needed to ensure the capability assumed in the planning. Necessary checks should be performed and the costs and the influence that inferior quality may have on the output must be taken into account. Inspection should be directed to in-process control and not just to final inspection. There may have to be:

- first piece inspection;
- operator inspection;
- automatic inspection;
- inspection stations in the process;
- patrol inspections when inspectors monitor specific activities.

The raw material, as well as all parts and components entering the manufacturing line, should be in perfect condition. Raw material should be stored appropriately. Storage facilities should be such that material and products pending use are satisfactorily protected. Special attention must be paid to material with a shelf life. Material parts should be transported, stored and handled and protected with care to avoid damage. It should be possible to identify the treatments and checks of the raw material, parts and components throughout the manufacturing process. They should have markings to avoid mix-ups and incorrect usage. When traceability is required, documented information must be kept to identify materials and methods used.

The skills of the operators influence the manufacturing process and final quality. Through operator inspection information is quickly fed back. Corrections can be made immediately and the costs of any mistakes can be limited. The classic mistake is to blame the operator for defects and deficiencies. Usually they are not the operator's fault but are due to incorrect and insufficient information. This is why work instructions, specifications and drawings are so important to the manufacturing team.

Alterations in the manufacturing process should not be haphazard. There must be rules on how to make necessary changes. They should also be recorded, and it should be ensured that the changes do not have a negative effect on the product quality.

To obtain reliable results from checks made during the various production stages, the measuring and test equipment must be accurate and show true values. The equipment should be calibrated regularly, appropriately marked for identification and stored correctly.

It is essential to know how discrepancies should be handled and how corrective actions should be applied to avoid recurrence. Corrective actions should apply to the causes of the problem and their removal. Where there are failures, substandard products should be segregated and marked, and decisions made about what is to be done.

How a product is introduced to the market can influence the way it is perceived in terms of quality. Ideally a product should not be presented to the whole of the market at once. It is preferable to make controlled field tests, introducing the product to a limited section of its potential market where its acceptability and performance can be easily monitored and any corrective actions taken at a relatively low cost.

This approach can ensure that, if and when a full launch is undertaken, the product is at the appropriate quality standard. Instruction manuals covering assembly, installation, use, maintenance and repairs are important to quality. Problems can be avoided if such information is accurate, complete and easily understood.

Similarly, advertising influences the customer's perception of a product's quality. Advertisements should be truthful and realistic and build the correct expectations.

The packaging is part of the product and must be designed for adequate protection. It can also be used for information and marking.

To launch a product correctly, a good customer service and distribution system has to be established. The service and maintenance organisation must be adequately equipped and people must be trained. There should be a sufficient supply of spare parts and equipment. Product guarantees are often included and must be technically and administratively catered for.

The quality of goods is, as mentioned earlier, its ability to respond to the needs and requirements of the market. Therefore, it is not until the product reaches the market, when customers' reactions are felt, that the company can finally determine how successful its quality system is. It is important to follow up market reactions and customer perception. Silent customers are not always happy customers. By recording customer views the company will be able to further improve its products and services.

Market reactions should therefore be reported, referred to and used. All information relating to quality should be analysed and form the basis for appropriate changes in the product itself or in the way the company is operating.

The costs associated with maintaining quality must be reflected in the benefits accrued. If a customer requires a product or service to a certain specification, then the costs of producing the necessary system to permit this must be balanced against the profitability of the project.

Initially a brainstorming session is held to determine any attainable improvement in a project activity. Typically the activities associated with brainstorming or thought showering as it is now called is shown below.

- A time limit.
- A clear statement of the problem at hand.
- A method for capturing ideas such as a flip chart.
- Leave the ideas somewhere visible and let them incubate.
- Observe the principle that no idea is a bad idea.
- Suspend judgement.
- Everyone should put evaluation to one side.
- Participants should freewheel by letting go of their normal inhibitions and let themselves dream and drift around a problem.
- Go for quantity by encouraging lots of ideas regardless of quality (evaluation can come later).
- Cross-fertilise ideas by picking up group ideas and developing them.

In effect, the project activity is discussed with other members of the project team. A number of good ideas are identified and proposals made on how the problem may be solved. Finally the least expensive solution is implemented.

4.13 Improving project quality

The need to improve from a quality point of view can be divided into two parts:

- Implement improvement programmes.
- Meet the standards demanded by the customer regarding the project's quality system.

The objective in the first case is to improve the operation in terms of reducing costs and to gain other competitive advantages (Lascelles and Dale, 1993). In the second case, the objective is to establish a QMS based on customer demands for the operation. In most cases the two systems frequently overlap, but the work routine differs somewhat.

Quality is dynamic and improvement needs to be continuous. Never accept a QMS as gospel. The quality function should be the organisation's focal point of the integration of the business interests of customers and suppliers into the internal dynamics of the organisation. Its role is to encourage and facilitate quality improvement, monitor and evaluate progress, promote the quality chains, plan, manage, audit and review systems, plan and provide quality training, counselling and consultancy, and give advice to management. In larger organisations a quality director will contribute to the prevention strategy (Harrington and Mathers, 1997). Smaller organisations may appoint a member of the management team to this task. An external adviser is usually required.

To encourage involvement by all staff in project quality activities, it is essential that the company set goals and encourage and recognise efforts by staff to reach these goals. Every employee must have the appropriate training.

Quality circles are often used as a method of refining the local and project working conditions and improving productivity. It is based on two factors.

- Staff can often see problems that are not evident to their managers.
- The best people to fix a problem are those who stand to benefit from its solution.

There are only two ground rules that must be adhered to if a quality circle is to be successful.

- Implement totally within its own resources – usually at zero cost.
- Tackle one problem at a time, usually the one that yields the highest benefit.

The way quality circles work is by taking a small but representative team from one department. An activity is discussed and the task of defining the problem undertaken. In the case of project-specific quality circles the project manager will organise such discussions.

Initially, a brainstorming session is held to determine any attainable improvement in a project activity. Secondly, the project activity is discussed with other members of the project team. Thirdly, a number of good ideas are identified and proposals made on how the problem may be solved. Finally, the least expensive solution is implemented within the project as a whole.

In most projects a QMS is developed as the project proceeds. Many predetermined ideas are often unsuitable to solve particular problems during the project life cycle. To ensure problems are addressed and catered for as they occur, the project manager can utilise the experience of the project team in quality circles.

Although many of the preceding techniques were originally developed outside the sphere of quality their adoption at specific intervals in a project's life cycle can provide valuable benefits both to the customer and the organisation itself.

4.14 Integrating quality into common business practices

Quality will lead to success only if it stops functioning as a separate entity. It has to be integrated into common business practices and change them from within. The discrepancies between business objectives and quality objectives, still widely present, are a major problem. Especially in regressive times, well-thought-out QMSs, which were often built up with the intention of improving overall results, might collapse if not fully integrated into the whole business system. A real integration of quality into business therefore holds the highest priority.

Current TQM strategies aimed at optimising and improving the quality–cost–time triangle are necessary but seemingly no longer sufficient. The whole organisation should be able to continually review itself, its procedures, project strategies, organisational hierarchy, competencies and capabilities. Managers should learn how to improve and renew the organisation, using quality tools. Management quality is the best companion on the road to business excellence. Quality concepts should pervade the entire organisational system and its processes.

Learning to manage a PDCA cycle becomes vital. It can be the first and most significant move towards the integration of quality concepts into common corporate business practices. The adoption of best practice through the QMS will alleviate many of the risks formerly accepted as part of the business environment.

4.15 Summary

The type of QMS adopted by organisations and the way it is managed depends on the organisational system in place.

The benefits of QM are numerous in terms of meeting specification, meeting customer requirements and reducing corrective action. QM, however, cannot stand still. Changes in business activities and processes and the introduction of new technology means that QMSs must reflect an organisation's working practices.

All projects are unique and require the development of QMSs specific to a project. In many cases, existing QMSs can be used as the basis of a project-specific system with the addition of control documentation covering activities not previously encountered.

Many project-specific systems are developed as a project proceeds, thereby requiring continuous updating of the system. Although a system specific to a project may be required, the same principles for monitoring project activities will be adopted as for a system carrying out the repetitive manufacture of a product.

A QMS is a management system used by a project team to ensure the project meets the quality required. TQM needs to involve all members of an organisation as well as external elements such as suppliers and customers as and when necessary.

The costs associated with maintaining quality must be reflected in the benefits accrued. If a customer requires a product or service to a certain specification, then the costs of producing the necessary system to permit this must be balanced against the profitability of the project.

References

Ashford, J. L. (1989) *The Management of Quality in Construction*, E & F.N. Spon.

BS EN ISO 8402 (1995) *Quality Management and Quality Assurance – Vocabulary*, British Standards Institute.

Crosby, P. B. (1979) *Quality is Free*, McGraw-Hill.

Dale, B. G., Lascelles, D. M. and Boaden, R. J. (2003) Levels of total quality management adoption. In B. G. Dale (ed.) *Managing Quality*, fourth edition, Blackwell Publishing.

Deming, W. E. (1986) *Out of the Crisis*, MIT Press.

Harrington, J. H. and Mathers, D. D. (1997) *Chapter 4, Phase II ISO 9000 and Beyond: From Compliance to Performance Improvement*, McGraw-Hill.

Haynes, M. E. (1990) *Project Management: From Idea to Implementation*, Kogan Page.

ISO 9001 (2000) *Quality Management Standard*, International Standards Organisation.

Jonson, K. (1988) *Quality Systems – Quality a Challenge for Everyone*, Studentlitteratur.

Juran, J. M. (1950) *Quality Control Handbook*, McGraw-Hill (reprinted 1982).

Lascelles, D. M. and Dale, B. G. (1993) *Total Quality Commitment: The Japanese approach. The Road to Quality*, IFS Ltd.

Oakland, J. S. (1992) *Total Quality Management: The Route to Improving Performance*, second edition, Butterworth Heinemann.

Further reading

Feigenbaum, A. V. (1991) *Total Quality Control*, third edition, McGraw-Hill.

ISO (2003) *ISO 10006: Quality Management Systems – Guidelines for Quality Management in Projects*, International Standards Organisation.

Juran, J. M. and Godfrey A. B. (1998) *Juran's Quality Handbook*, fifth edition, McGraw-Hill.

Chapter 5
Environmental Management

In the past the main tools for project appraisal were cost–benefit analysis (CBA) and cost-effectiveness analysis, both logical and quantitative methods for identifying whether or not a project was worth implementing. Many publicly funded projects, such as water supply systems, were assumed to have such overwhelming benefits that only the costs were assessed to determine which of the various alternative methods of achieving the project objectives was the most cost-effective. It is now recognised that all projects will result in some unquantifiable costs and benefits, and therefore today these appraisal methods are supplemented by environmental impact analysis and assessment. By its nature, environmental impact assessment (EIA) requires a much more qualitative approach to project appraisal than CBA, although this is gradually changing as new methods for valuing environmental impacts emerge and develop.

This chapter covers the main elements of EIA as it is currently used for the appraisal of projects in many countries, and introduces the basic elements of an environmental management system (EMS). Developments in the evaluation and 'monetising' of environmental impacts are also reviewed and discussed.

5.1 Environmental impact

Before discussing environmental impact, it is first necessary to define what is meant by environment, as it can mean a variety of things to different people – from the light and heat inside a building to the outdoor environment of natural woodlands, moors, rivers and seas, to the ozone layer surrounding the planet. In the context of the environmental impact of projects, and the discipline of EIA, the term *environment* is taken in its widest context to include all the physical, chemical, biological and socio-economic factors that influence individuals or communities. It not only includes the air, water and land, and all living species of plants, animals, birds, insects and microorganisms, but also man-made artefacts and structures, and factors of importance to the social, cultural and economic aspects of human existence.

In this context, all projects have an effect or impact on the environment – indeed it could be argued that unless they did, there would be no point in implementing projects at all. Some of these may be seen as positive impacts (benefits), while others will have a detrimental effect (costs). The purpose of EIA is to evaluate these positive and negative effects as objectively as possible and present the information in a manner that is accessible to decision makers so that it can be used in conjunction with other appraisal tools, such as CBA.

Engineering projects may have an impact on the full range of environmental features including air, water, land, ecology, sound, human aspects, economics and natural resources. Many of these impacts can be measured in terms of changes to specific quality parameters such as concentration of particulates or hydrocarbons in the air, or dissolved oxygen concentration in water. Other impacts, such as the aesthetic qualities of a landscape or a structure, or the importance of preserving an historic building, are not so easily quantified.

Some of the impacts will be directly attributable to the project – such as noise from an airport or road, the visual impact of a structure or the pollution of the air or water from a factory. These are referred to as *direct* or *primary* impacts. Other impacts may arise indirectly through the use of materials and resources required for the project. Examples include the pollution of the air from the manufacture of cement used for a project or the impact of quarrying for raw materials needed for the project. These impacts may affect areas remote from the project itself and are termed the *indirect* or *secondary* impacts. Lower-order impacts can also be identified, such as the impacts on the environment caused by the manufacture of equipment used for quarrying or cement manufacture.

The range of environmental impacts arising from a project will continue throughout the operating life of the project, and can therefore be regarded as permanent or long-term impacts. The pollution of the air caused by the operation of a thermal power station is thus an example of a long-term impact. Short-term or temporary impacts, on the other hand, are those arising from the planning, design and construction phases of the project. Typically these might include the noise and dust generated by the construction process itself or, for example, temporary changes to water table levels.

Environmental impacts may therefore be temporary and direct, temporary and secondary, permanent and direct or permanent and secondary, and as discussed later in this chapter, it is necessary to differentiate between these categories of impact within an environmental impact statement (EIS).

5.2 Environmental impact assessment

Environmental impact assessment is a logical method of examining the actions of people, and the effects of projects and policies on the environment, so as to help ensure the long-term viability of the earth as a habitable planet. EIA aims to identify and classify project impacts, and predict their effects on the natural environment and on human health and well-being. EIA also seeks to analyse and interpret this information and communicate it succinctly and clearly in the form of an EIS, which can be used by decision makers in the appraisal of projects.

The natural environment is not in a completely steady state; changes occur naturally over time, some extremely slowly but others at a much faster rate. Therefore, any study of the impacts of a project on the environment must be seen in the context of what would have happened if the project had not been implemented. The environmental impact is thus any environmental change that occurs over a specified period, and within a defined area, resulting from a particular action, compared to the situation as it would have been if the action had not been undertaken.

The report of all environmental impacts that are predicted to arise from a particular project is normally termed an EIS, and is now often required to obtain sanction for a project. In many countries this procedure has now been incorporated in the legal processes of obtaining project and planning approval, and in other situations it has been used voluntarily as part of general project preparation and evaluation.

EIA is a process rather than an activity occurring at one point in time during the project cycle. The various steps in the process are shown in Figure 5.1. The process starts with *screening*, which should be done in the early stages of a project as it is concerned with determining whether or not a detailed EIA is required or necessary.

Scoping is the second step of the EIA process and is essentially a priority-setting activity. It may be seen as a more specific form of screening, aimed at establishing the main features and scope of the subsequent environmental studies and analysis. The results of the scoping exercise provide the basis and guidelines for the next steps in the process – the baseline study and impact assessment itself.

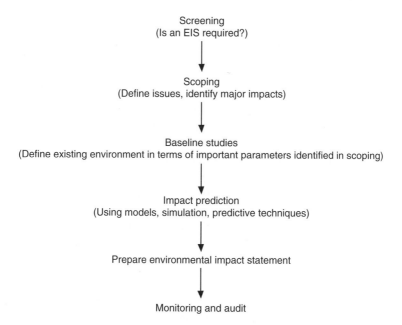

Figure 5.1   Environmental impact assessment process.

Scoping sets the requirements for the *baseline study*, which is the collection of the background information of the ecosystem and the socio-economic setting of a proposed development project. The activities involve the collection of existing information and the acquisition of new data through field examination and the collection of samples. This step can be the most expensive step of the EIA because it needs a large number of expert people to carry out field surveys and analysis.

The baseline study and the project proposals are then used as the basis for predicting how environmental parameters will change during both the construction and operational phases of the project. Environmental and social scientists will be required for this step as, often, sophisticated modelling and simulation techniques must be used.

The predictions and the baseline study are then used to prepare the report – the EIS – which is used in the appraisal and approval processes. Further details of the structure of the report and the more common methods of presenting EIA information are given in subsequent sections of this chapter.

Because of the uncertainty associated with environmental impact predictions, it is important that major environmental parameters are monitored throughout the implementation of the project. This can provide valuable information for future EIAs and generally improve the accuracy of forecasting models and methods. The process of comparing the impacts predicted in an EIA with those that actually occur after the implementation of the project is referred to as *auditing*.

5.3 Screening

Screening is the first step of the EIA process; it is the selection of those projects requiring an EIA. For example, construction of a simple building rarely requires an EIA, whereas the construction of a large coal-burning power station would need a full-scale EIA to be conducted and an EIS to be prepared, which would include the results of the required scientific and technical studies, supporting documentation, calculations and background information.

The extent to which the EIA is needed for a project is also defined in this step, as the impact of a project depends not only on its own specific characteristics but also on the nature of the environment in which it is set. Therefore, the same type of project can have different impacts, or a different intensity of some impacts, in different settings. So, for any particular project, a full-scale EIA may be required for one site but not for another.

The requirements for an EIA also largely depend on the legislative policy of the country in which the project is to be carried out. Through the screening process, only those projects that require an EIA are selected for study. By doing this, unnecessary expenditure and delay is avoided.

The screening process will depend on whether or not appropriate legislation exists, and where it does, the laws will often differentiate between those types of project that need an EIA and those that do not. The legislation may also define the extent of the study that will be needed for a particular type of project. Where there is no appropriate law relating to EIA, or if the types of project requiring EIA are not specifically defined, the project manager will have to undertake a preliminary study to ascertain the most significant impacts of the project on the environment. On the basis of the results of this study, and after studying the environmental law and assessing the political situation of the country, the project manager will have to decide whether a full-scale EIA is needed.

5.4 Environmental legislation

The concept, and eventually the practice, of EIA evolved, developed and incorporated into legislation in the USA in the 1970s. The idea was subsequently introduced into other countries, such as Canada, Australia and Japan, but it was not until 1985 that EIA was fully accepted within the European Community. At that time a European Community Council Directive 'on the assessment of the effects of certain public and private projects on the environment' was issued. This required all member states to implement the recommendations set out in that directive by July 1988.

The directive called for governments of member states to enact legislation that would make EIA mandatory for certain categories of project and recommended for other types of project. The projects where an EIA is *mandatory* include:

- crude oil refineries;
- gasification and liquefaction of coal and shales;
- thermal power stations;
- radioactive waste storage and disposal facilities;
- integrated steel and cast-iron melting works;
- asbestos extraction and processing works;
- production of asbestos products;
- integrated chemical installations;
- motorways and express roads;
- long-distance railways;
- airports with runways greater than 2100 m;
- trading ports and inland waterways and ports for traffic over 1350 tonnes;
- waste disposal for the incineration, chemical treatment or landfill of toxic and dangerous wastes.

The categories of projects where an EIA is *recommended* include the following.

Agriculture. For example, poultry and pig rearing, salmon breeding, land reclamation, water management, afforestation, restructuring of rural land holdings and use of uncultivated land for intensive agricultural purposes.

Extractive industries. Deep drilling, extraction of sand, gravel, shale, salt, phosphates, potash, coal, petroleum, natural gas, ores, and installations for the manufacture of cement.

Energy industries. Production of electricity, steam and hot water; installations for carrying gas, steam, and hot water; overhead cables for electricity; storage of fossil fuels; production and processing of nuclear fuels and radioactive waste; and hydroelectric installations.

Processing of metals. Iron and steel works, non-ferrous and precious metals installations, surface treatment and coating of metals, manufacture and assembly of motor vehicles, ship yards, aircraft manufacture and repair; and manufacture of railway equipment.

Manufacture of glass.

Chemical industries. Production of pesticides, pharmaceuticals, paint, varnishes, peroxides and elastomers; and storage of chemical and petrochemical products.

Food industries. Manufacture of oils and fats, packing and canning, dairy product manufacture, brewing, confectionery, slaughter houses, starch manufacturing, sugar factories, fish-meal and fish oil factories.

Textile, leather, wood and paper industries.

Rubber industries.

Infrastructure projects. Urban and industrial estate development, cable cars, roads, harbours, airports, flood relief works, dams, water storage facilities, tramways, underground railways, oil and gas pipelines, aqueducts, marinas, hotels, holiday villages, race and test tracks for cars and motor cycles, waste disposal installations, waste water treatment plants, sludge disposal, scrap iron storage, manufacture of explosives, engine test beds, manufacture of artificial fibres, knackers' yards.

In the UK, these requirements for EIA have been incorporated into the existing planning legislation, and the developer or initiator of a project now has to supply the relevant environmental impact information in order to obtain sanction for the project. Prior to the enactment of the legislation, planning procedures only addressed land-use considerations, but now issues of pollution control and environmental impact must also be considered.

It is worth noting that environmental impact issues must also be addressed in many overseas projects, either as a result of relevant legislation having been introduced by particular countries or, perhaps more importantly, due to the fact that bilateral funding agencies (such as the UK's DfID – Department for International Development) and multi-lateral funding agencies (such as the World Bank and the Asian Development Bank) now have requirements for EIAs built into their loan and grant approval procedures.

5.5 Scoping

The US National Environmental Protection Agency (NEPA) Council on Environmental Quality defined scoping as 'an early and open process for determining the scope of issues to be addressed and for identifying the significant issues related to a proposed action'. Scoping is therefore intended to identify the type of data to be collected, the methods and techniques to be used and the way in which the results of the EIA should be presented.

The question of what is a significant impact is not always an easy one to answer, and requires judgement, tact and an understanding of not only the technical issues but also the social, cultural and economic ones. It is likely to involve value judgements based on social criteria (aesthetic, human health and safety, recreation and effect on life styles), economic criteria (the value of resources, the effect on employment and the effect on commerce), or ethical and moral criteria (the effect on other humans, the effect on other forms of life and the effect on future generations).

The baseline study and the impact assessment are expensive and time consuming and the determination of the impacts of a proposed project or policy depends on the current state of knowledge relating to any particular aspect of the environment. Very often this depends on the latest developments in the science and technology of environmental monitoring and prediction as well as the professional experience and judgement of those who carry out the EIA. It is therefore important to ascertain what can be reasonably carried out within the allowed time and budget so that the best use can be made of available resources. This is one of the tasks of the scoping exercise. The other main task is that of determining the most important impacts that may occur, and this will vary with the environment surrounding the project. As it is the people living in or near the area of a proposed project who are most likely to have detailed local knowledge and to be most concerned about the local environment, it is extremely important to solicit local public opinion about the project and its perceived impacts on the environment. This is best done at the scoping stage, as local knowledge is a valuable source of data for the baseline studies.

There are two major steps to take in seeking public opinion, the first of which is to identify the affected population/target groups and attempt to obtain their opinions. Local political and environmental concern groups should be consulted at an early stage to avoid unnecessary confrontation at a later stage. To carry out this public consultation exercise, a schedule of meetings, with the affected population and target groups, needs to be arranged. Care should be taken to ensure that all affected and interested groups are included in the programme. At these consultation meetings, the objectives, possible impacts and activities associated with the project should be explained, and everyone should be encouraged to express their opinions on the project and its possible impacts, and to suggest mitigation measures. Sufficient time should be allowed for these meetings, and all suggestions and comments should be recorded in meeting minutes and made public. Major impacts and areas of concern may be identified from this public consultation exercise.

This opinion and data collection procedure is quite a time- and money-consuming activity, and requires patience and diplomacy. In many cases, the EIA has to be completed in a very short period of time, and so meetings with the affected people and other interested groups can become very difficult, if not impossible. In such cases, sample questionnaire surveys can be carried out and, by using statistical methods, probable results may be formulated. Where appropriate, it may be possible to carry out some of these surveys by telephone.

A method used in Canada to ensure the public's involvement is to use a panel of four to six experts who are selected to examine the environmental and related implications of a particular project. The panel is then responsible for issuing guidelines for preparing the EIS and reviewing it after completion. The panel will collect public views through written comment, workshops or public meetings before completing the guidelines for the EIS.

The advantage of scoping is that it helps to obtain advance agreement on the important issues to be considered in the following activities of the EIA and helps to use resources that are scarce efficiently to analyse the impacts by preventing unnecessary investigation into the issues/impacts that have little importance. Like good, early project planning, scoping can save enormous expense and time during the later and more costly steps of EIA. Good scoping requires planning, using competent staff and providing adequate resources.

The outcome of a scoping exercise should be a list of priority concerns and guidelines for the preparation of the EIS, including a structure for the baseline study.

5.6 Baseline study

The scoping exercise should provide information about which data are required and which are unnecessary. If this information is not provided by the scoping team, a great deal of time and money can be wasted in gathering irrelevant and unnecessary data. Unfortunately there are a large number of examples of EIA baseline studies that have included a lot of fairly irrelevant material simply because those preparing them have included all the data that were available, even if they were unnecessary. The baseline study should concentrate on gathering the important data as identified in the scoping exercise. Some of this will be available from previous studies or other sources, but it may be necessary to conduct extensive monitoring exercises to gather other needed data. Costly gathering of data that are of little or no importance must be avoided.

The baseline study should result in a description of the existing environment, and it should be remembered that this is not static; changes will occur even without the proposed project. Information should be relevant to the impacts discussed, and one should be selective in the information used to describe the baseline. For example, there is no need to spend a great deal of time and energy on data collection and description of present and future noise levels, if the project is very unlikely to generate any additional noise. The baseline description should be confined to those impacts and parameters that matter. Too often the focus of the baseline survey has been on what is available rather than on what is needed.

The outcome of the baseline study is a form of environmental inventory based on primary environmental data collection where appropriate, the opinions of preselected individuals and groups and other sources, such as the monitoring and evaluation data from similar completed projects.

As the impacts of the proposed project on the environment are forecast on the basis of the data collected during the baseline study, it is important that the study is carried out accurately and thoroughly.

5.7 Impact prediction

For each alternative, the consequences or impacts should be predicted using the most appropriate methods, which may include predictive equations, modelling and simulations techniques. The predictions should include, and differentiate between, primary, secondary, short- and long-term effects. There should be an attempt to show the effect of all project activities on a comprehensive range of environmental parameters, which should have been identified during the scoping stage. These parameters may be grouped into various categories such as biological, physical, social, economic and cultural, and wherever possible quantitative predictions should be made.

Decision makers need to address the question of the accuracy of these predictions – imprecise predictions can generally be counted on to be fairly accurate, but there is often a degree of uncertainty associated with precise forecasts. For example, if a power station is being considered, an imprecise prediction such as 'the levels of particulate concentration in the air in the immediate vicinity will increase' is likely to be very accurate. On the other hand, making a precise prediction of the value of the particulate concentration will almost certainly suffer from inaccuracies. Some assessment of the accuracy of predictions could be included.

5.8 Environmental impact statement

The EIS itself is a detailed written report on the impact of the project on the environment and should contain the following:

The need for the project, which is a description of the aims and objectives of the project, explaining clearly why the project is required.

The baseline study report, a comprehensive description of the existing environment, indicating the levels of important environmental parameters, quantitatively wherever possible.

A list and description of all reasonable and possible project alternatives, including the 'do-nothing' option. Even options that might be considered 'non-viable' should be mentioned and reasons given for their rejection. For each of the main alternatives, the following should be included.

- A clear description of the project during construction and operation, giving details of the use of land, materials and energy, and estimates of expected levels of pollution and emissions.

- A clear estimate of the environmental consequences of each alternative, including predictions of the effects of all the significant impacts on human health and welfare, water, land, air, flora, fauna, climate, landscape and cultural heritage. This section provides the scientific and technical evidence on which comparisons between the alternatives can be made. The report should clearly draw the reader's attention to any serious adverse impacts that cannot be avoided or mitigated, as well as any irreversible environmental consequences and irretrievable use of natural resources.
- The severity of each impact and a clear description of the methods used for measuring and predicting it.

The environmental consequences of each alternative should be compared, possibly in tabular or other easily readable form, as discussed later in this chapter. Many consider this to be the most important section of an EIS, and it may often be the first section to be referred to, as it provides an easily understood comparison of all alternatives and their main environmental consequences.

The statement conclusions indicating the preferred option and describing any mitigation measures that may be required to minimise environmental impacts of this option.

A non-technical summary.

An index and any necessary appendices for detailed calculations and data analysis.

5.9 Presenting EIA information

One of the difficult tasks in preparing an EIS is that of presenting the information in a way that is comprehensive, yet readily understood by decision makers. This is particularly true of comparisons of the severity of impacts of alternative proposals. A number of commonly accepted ways of doing this have emerged, and because of their critical importance within the EIS, they are often referred to as EIA 'methods'. These include a number of different forms of checklists, overlay mapping techniques, networks, multi-attribute utility theory, and matrix methods.

Checklists

Checklists are one of the oldest of the EIA methods and are still in use in many different forms. Usually checklists consist of lists (prepared by experts) of environmental features that may be affected by the project activities.

Sometimes a list of project actions that may cause impacts is also incorporated. Checklists may be a simple list of items or more complex variations that incorporate the weighting of impacts. There are five types of checklist commonly in use.

Simple lists. A list of environmental factors and development actions, with no guidance on the assessment of the impacts of the project on these factors.

The principal use of this method is to focus attention and to ensure that a particular factor is not omitted in the EIS.

Descriptive checklists. This type of checklist gives some guidance on the assessment of impacts, although it does not attempt to determine the relative importance of impacts.

Scaling checklists. These consist of a list of environmental elements or resources, accompanied by criteria for expressing the relative value of these resources. For each alternative being assessed, the appropriate criteria are selected and any impact that exceeds the defined limit is considered to be significant and should be highlighted for the decision makers' attention.

Scaling-weighting checklists. A panel of experts decides on the weight to be assigned to each environmental parameter. The idea is to assign a weighted score or scale to each possible impact to enable one impact to be compared with another, and to provide an overall environmental impact score for each alternative.

Questionnaire checklists. This type of checklist consists of a series of direct linked questions posed to a variety of professionals who are asked to respond with yes, no, or unknown.

Overlay mapping

This procedure consists of producing a set of transparent maps showing environmental characteristics (such as physical, social, ecological and aesthetic) of the proposed project area. A composite characterisation of the regional environment is produced by overlaying those maps, and different intensities of the impacts are indicated on the maps by different intensities of shading. This method is suitable for the selection of routes for new highways or electrical transmission lines, for example.

One of the main difficulties of using this method is that of superimposing all the maps in a comprehensive way, especially when the number of maps is large. However, by using computers, these problems can be overcome, and a number of software packages have been developed for this purpose.

Networks and systems diagrams

These methods were developed for the identification of secondary, tertiary and higher-order impacts from an initial impact. The methodology starts from a list of project activities to establish cause–condition–effect relationships; in the case of systems diagrams these effects are quantified in terms of energy flows. It attempts to recognise a series of impacts that may be triggered by a project. By defining a set of possible networks, the user can identify impacts by selecting the appropriate project actions.

The advantage of these approaches is that they can show the interdependency of parameters and the effects of changes in one parameter on other parameters. The limitations are that they are only really suitable for the assessment of ecological impacts, and they are expensive, time consuming and require periodic updating. Systems diagrams depend on knowledge of the ecological relationships in terms of energy flow, which is often very difficult to characterise.

Multi-attribute utility theory

Utility theory in EIA has been applied most often to site selection, especially for projects such as major power stations and waste disposal facilities. Different projects have different environmental impacts and exhibit different levels (intensity) of the same impact. These methods provide a logical basis for comparing the impacts of alternatives to aid in decision making. The methodology follows the steps indicated here:

(1) Determine the environmental parameters that may be affected and can be measured. For each parameter, use appropriate prediction methods to estimate the value of each parameter after the proposed project has been implemented.
(2) Establish the desirability or otherwise of different levels of each parameter, and formulate the utility function through systematic comparison of those different levels. The utility $U_i(x_i)$ of each parameter x_i is measured on a scale of 0 to 1, where 0 is the lowest utility (i.e. the worst possible situation) and 1 is the highest utility (the best possible situation). This step is highly subjective and relies on the values, knowledge, and experience of 'experts'.
(3) Determine a scaling value for each of the environmental parameters that will reflect the relative importance as perceived by decision makers. This is denoted by k_i. Again the scaling factors are highly subjective and value laden.
(4) The preceding steps have to be carried out for each of the alternatives separately, and the total utility, or a composite *environmental quality index* (EQI) for each of the alternatives is then calculated as shown below:

$$\text{EQI} = U(x) = 3 \sum_{i=1}^{n} k_i U_i(x_i) \qquad (5.1)$$

where k_i is scaling factor of parameter x_i; $U_i(x_i)$ is utility function of parameter x_i and $U(x)$ is multi-attribute utility function.
(5) Determine which alternative has scored the highest EQI value. This will be the least environmentally damaging and the best of the alternatives.

One of the advantages of this method is that the concepts of probability and sensitivity analysis can be incorporated into it and, using computers, a number of 'what-if' scenarios can be examined. The fact that a single number – the EQI – is produced is seen by some as an advantage, because it is easy for the decision maker to compare one alternative with another. Others see this as a major drawback, because it produces what appears to be an objective numerical comparison, but is in fact based on hidden subjective assessments of the utility factors and the scaling factors. Another criticism of the method is that it militates against public understanding and involvement by being unnecessarily technocratic and complex.

Finally, a further limitation of this methodology is that it assumes the environmental parameters are independent of each other, which is not the case, and that they are fully dependent on the probability assumptions.

Matrices

Simple interaction matrix

This form of matrix is simply a two-dimensional chart showing a checklist of project activities on one axis and a checklist of environmental parameters on the other axis. Those activities of the project that are judged by experts to have a probable impact on any component of the environment are identified by placing an X in the corresponding intersecting cell. Matrix methods were originally proposed and developed by Leopold and his colleagues of the US Geological Survey to identify the impacts on the environment for almost any type of construction project. Their matrix, now known as the Leopold matrix, consists of 100 specified actions of the project along the horizontal axis and 88 environmental parameters along the vertical axis.

A simple two-dimensional matrix is, in effect, a combination of two checklists. It incorporates a checklist of project activities on one axis and a checklist of environmental characteristics that may be affected by the actions of a project on the other axis. From the matrix, the cause–effect relationships of actions and impacts can be identified easily by marking the relevant cells. Various modifications have been made to Leopold's original lists of activities and environmental characteristics to suit particular situations.

Quantified and graded matrix

The original Leopold matrix is slightly more complex, in that a number ranging from 1 to 10 is used to express the magnitude and importance of the impacts in each cell. Thus, a grading system is used in place of the simple X.

Modification of matrices has been going on over time with the advancement of EIA. In 1980, Lohani and Thanh suggested another grading system to incorporate the relative weights of each development activity. This method can identify major activities and areas that need more attention.

The advantages of the matrix are that it can rapidly identify the cause–effect relationships between impacts and project activities and it can express the magnitude and importance in both qualitative and quantitative forms. Another major advantage of the matrix method is that it can communicate the results to the decision makers as a summary of the EIA process. It can be used for any type of project and can be modified as required.

5.10 Monitoring and auditing of environmental impacts

Monitoring

EIA is largely concerned with making predictions about the effects projects will have on the environment. Unless the actual impacts are measured and monitored during and after project implementation, and compared with those predicted, it is not possible to assess the accuracy of predictions or to develop improved methods of prediction. The monitoring of environmental impacts is thus an essential component of the EIA process.

The aim of monitoring is first to detect whether an impact has occurred or not and, if so, to determine its magnitude or severity, and second to establish whether or not it is actually the result of the project and not caused by some other factor or natural cause. To do this it is desirable to identify 'control' or 'reference' sites that should be monitored, as well as the project site. The reference site should be as similar as possible to the project site, except that the predicted project impacts are unlikely to occur. To be effective, monitoring should begin during the baseline study and continue throughout the construction and operational phases of the project.

During the scoping stage, a framework for monitoring should be established. This should outline which parameters to monitor, the required frequency of sampling, the magnitude of change of each parameter that is statistically significant and the probability levels of changes occurring naturally. The establishment of this framework is perhaps more important than the actual details relating to data measurement and collection, as there is little point in spending a great deal of time and money in gathering irrelevant or insignificant data. The monitoring process can be very expensive.

While the main objective of monitoring is to assess the validity and accuracy of predictions (auditing), it is worth noting that monitoring data can provide an early warning of possibly harmful effects in time to initiate mitigating measures and thus minimise adverse impacts. Monitoring improves general knowledge about the environmental effects of different types of projects on a variety of environment parameters. As more monitoring exercises are carried out, the data available will increase and become more reliable, thus improving future EIA studies.

Auditing

Auditing is the process of comparing the predicted impacts with those that actually occur during and after project implementation. Auditing thus requires accurate and well-planned monitoring as explained earlier.

Auditing and monitoring are steps in the EIA process that are often disregarded or not given sufficient attention. One of the reasons for this is that EIA is seen by many as a hurdle in the approval and sanction phase of a project and, once this has been granted, the environmental effects are given much less attention. If EIA is to be used as a method of reducing environmental impacts and/or mitigating them, the auditing process must be given a much higher priority than is currently the case. Auditing not only helps to provide information on the accuracy of environmental impact predictions but also highlights best practice for the preparation of future EISs.

Audit studies of EIAs produced in the past have shown considerable variation and inaccuracies in environmental parameter predictions, and a large number of impacts studied could not be audited for a variety of reasons. These included inappropriate forms of prediction, the fact that design changes had been made after the EIS had been completed, and inadequate or non-existent monitoring. In some cases less than 10% of the impacts evaluated in an EIA could be audited, and of these less than 50% were shown to be accurate. There is thus great room for improvement in the preparation and presentation of EISs, and effective monitoring and auditing are absolutely essential if these improvements are to be made.

5.11 Environmental economics

Decision makers are often faced with two documents: the mainly qualitative report on environmental effects (the EIS) and the 'objective' numerical statement of costs and benefits (the cost–benefit appraisal). Both are concerned with costs and benefits, but they are written and presented in different terms and forms. For this reason, a relatively new science of *environmental economics* has developed, which helps to coordinate these two appraisal tools and perhaps eventually to combine them.

Cost–benefit analysis

In its simplest form, the principles of CBA can be expressed by the following equation:

$$3(B_t - C_t)(1 + r)^{-t} > 0 \tag{5.2}$$

where B_t is benefits accruing at time t; C_t is costs incurred at time t; r is discount rate and t is time.

However, only those costs and benefits that can be easily quantified are normally included, and the environmental and social costs and benefits are ignored. There are two questions to address, the first being whether or not there are indeed any real costs and benefits that can be ascribed to the environment, and the second being how can they be valued in the same terms as other project costs and benefits, that is, in money terms.

If both these questions can be answered satisfactorily, the CBA equation could be rewritten as follows:

$$3(B_t - C_t \pm E_t)(1 + r)^{-t} > 0 \tag{5.3}$$

where E_t is environmental change (cost or benefit) at time t.

To address the first question, consider the situation where a community relies on catching fish, from an unpolluted river, for its income. If a factory sited upstream then discharges waste directly into the river, it is likely that the fish population in the river will be affected and, as a direct consequence, the livelihood and welfare of the fishing community downstream will decline. In this instance, there is clearly a cost associated with the pollution of the environment, that is, the loss of income suffered by the community.

This simple example illustrates that environmental 'goods' are not free. Whenever they are used, either as a resource or as a waste depository, they do incur some form of cost, whether that be in terms of loss of income, or a reduction in health or welfare, as a consequence of air pollution, high noise levels, polluted land or water, or in a number of other ways. The only reason that these costs are not always evaluated is because there is no market for environmental goods – we cannot buy clean air, or quiet surroundings, at least not directly. Nevertheless, these things do have value to people, and they increase or decrease their welfare in much the same way as other commodities, such as a new car, a washing machine or the provision of electricity, affect the quality of people's lives.

But how can environmental costs and benefits be measured in money terms – how can they be 'monetised'? Over the last few years, environmental economists have been developing new approaches to valuing the environment, and many of these techniques can now be used to incorporate environmental costs and benefits in traditional CBA.

Environmental values

Environmental economists ascribe three distinct forms of value to the environment. The first of these is what is termed *actual use value*, and this refers to those benefits directly derived by people who actually make use of the environment and gain direct benefit, such as farmers, fishermen and hikers. Polluters also derive a benefit from the environment as a consequence of being able to freely dispose of their waste.

The second form of environmental value is *option use value*, which is the value that some people put on the environment to preserve it for future use – either by themselves or by future generations. For example, they may believe that a particular area of scenic beauty is worth preserving because, even if they do not use it now, they may want to use in the future.

In addition to this, some people place a value on the environment for its mere existence, even if they see no probability that they will ever make use of that environmental asset. Some people, for example, place a value on the rainforests or the preservation of certain endangered species, even though there is no possibility of them ever seeing them. This is the *existence value* of the environment.

The *total economic value* of an environmental asset is equal to actual use value plus the option use value plus the existence value.

Methods for valuing the environment

The main methods that have been developed for 'monetising' environmental costs and benefits are the following:

- the effect on production methods;
- preventative expenditure and replacement cost;
- human capital;
- hedonic methods;
- the travel-cost method (TCM);
- contingent valuation.

The effect on production methods

In its simplest form, an example of this method might be the reduction in fish catches due to pollution of a watercourse, as described previously. However, changes in market prices for goods and services must also be considered, as these will influence the net cost or benefit. For example, if increased production saturates the market and prices then fall, the net benefit may be reduced.

The procedure for quantifying the effect on production is a two-stage process. First, the link between the environmental impact and the amount of lost production must be established; second, the monetary value associated with the changes must be calculated.

It is often very difficult to establish the physical link between the affected environment and the change in output, for example, to establish the relationship between the effect of fertilizers used in an upland catchment area of a lake and on diminishing fish stocks in the lake. There may be other causes, and it is often an extremely complex task to disentangle the various factors contributing to a loss in production, particularly when trying to separate man-made from natural effects. Determining the effect requires a 'with' and 'without' scenario to be established. This is particularly difficult, if not impossible, when project actions have already been initiated.

Despite the aforementioned difficulties, this technique is probably the most widely used environmental valuation method.

Preventative expenditure and replacement cost

This method assesses the value placed on the current environment in terms of what people will spend to preserve it or stop its degradation (defensive expenditure) or to restore the environment after it has been degraded. For example, the cost of additional fertilizers used to compensate for the loss of natural fertility in the soil arising from a particular project could be used as the environmental cost of the project. Similarly, the cost of installing double glazing in houses affected by noise from a road or airport can be regarded as the minimum cost of this form of environmental pollution arising from the project. In some cases, of course, these costs are real and actually become part of the project cost. However, even if the preventative measures are not taken, there is a cost to those affected and it should be given consideration in the appraisal of the project.

Replacement cost is often calculated by estimating the cost of a shadow or compensating project, which would be required to bring the environment to the same quality as it was before the main project. An example of this might be the drilling of boreholes and the establishment of an irrigation system for people displaced, by a dam project, from a village that had good natural irrigation to an area where crop irrigation is problematic.

Human capital

This method considers people as economic units and their income as a return on investment. It focuses on the increased incidence of disease and poor health, and the consequent cost to society in terms of increased health care costs and reduced potential for earnings by the affected individuals. The method suffers from the problem that it is often very difficult to isolate a single cause of productivity loss. For example, poor health may be caused by unclean water, poor domestic hygiene, polluted air or a wide variety of other factors.

However, if a direct causal link with a particular environmental pollutant can be proven, the environmental cost of the pollution could be calculated by assessing the loss of earnings of those people affected and the additional costs of medical care required. Although relevant statistics may be available in developed countries, it is unlikely that the method could be used in many developing countries, because of the lack of records and inadequate information.

Hedonic methods

These methods use surrogate prices for environmental 'goods' for which there are no direct markets. They seek to link environmental goods with other commodities that do have a market value. For example, the price of a three-bedroom semi-detached property will vary according to the environment in which it is located. People are prepared to pay more money for a house located in a quiet area with a pleasant view than they would be for essentially the same house located next to a noisy, dirty factory. The house market can therefore be used as a 'surrogate market' for the environment. There are of course many other factors that affect house prices, such as the proximity to schools and transportation routes; and these factors must be isolated and removed, using techniques such as multiple regression analysis if the method is to prove in any way reliable. The method thus requires large amounts of data, which are then very difficult to analyse. However, it has been used successfully to evaluate the costs of noise and air pollution on residential environments.

Travel-cost method

This method is particularly appropriate for valuing recreational areas with no or low admission charges. The value of the site is calculated as the cost incurred by the visitors in both travelling expenses and the time spent travelling. The further people travel to the site, the greater value they place on it as they will use more fuel, incur higher fares and spend more hours in transit to and from the site.

Surveys must be carried out at the site in question. Visitors are asked how far they have travelled, how long the journey took, what type of transport they used, whether they visited any other locations and how often they visit the site. The results of the survey, together with information on the total number of visitors to the site each year, are used to determine an annual total cost incurred by all visitors to the site. It is assumed that this cost must be a minimum value that is placed on the amenity.

The method is mainly used to determine the value of preserving national parks and other sites of scenic, recreational and scientific interest.

Contingent valuation method

The contingent valuation method is a technique that can be used where actual market data are lacking. The method involves asking people what their preferences are and how much they are willing to pay ('willingness to pay', or WTP) for a certain environment benefit or how much they would need to be paid ('willingness to accept', or WTA) in compensation for losing a certain environment benefit.

The technique usually involves questionnaires being sent to the concerned parties or house-to-house interviews. This technique may be used on its own or in conjunction with the other techniques described before.

The future

EIA is now a recognised and accepted practice, and is used widely in appraising projects, primarily at the sanction stage. In most countries it is now incorporated into the legislation relating to planning and approval, and many international funding agencies now require an EIS as part of their loan or grant approval procedures.

If EIA is to continue to develop and to maintain credibility, it is essential that it is seen not just as a means to obtain a permit to develop, but as an environmental management tool to be used throughout the life of the project – from inception to design, construction, operation and decommissioning. This will only come about if monitoring and auditing are conscientiously carried out and the results fed back into the EIA process.

The valuation of environmental impacts is likely to improve as more data are gathered and confidence in the methods grows. It is already possible to incorporate some environmental impacts as costs and benefits in conventional CBA. The sceptics point to inaccuracies associated with the monetary evaluation of environmental impacts, but even direct costs of design and construction cannot be predicted accurately. The way ahead is to recognise that, as with direct costs, social and environmental costs can only be estimated within certain limits of confidence, and to build these ranges, rather than discrete values, into our cost–benefit appraisal calculations.

5.12 Environmental management

Environmental management is not separate to project management; it must be seen by project managers, design engineers and constructors as an integral part of their work at all stages of the project cycle. Although some elements of EIA, such as monitoring and auditing, should be an integral part of the design, construction, operation and decommissioning phases of a project, EIA is still used primarily at the feasibility, appraisal and sanction stages of the project cycle. However, EIA is not the only 'tool' or process that can be used for environmental management, and one that has recently risen to prominence throughout the world, and in a wide range of industries, is the EMS.

Whereas EIA is an analysis and appraisal tool that enables decision makers to predict the environmental impacts of a project, an EMS is an action tool, the purpose of which is to provide managers and other decision makers with clear guidelines on how to improve an organisation's environmental performance, in every aspect of the work they do (which may include carrying out projects). An EMS is part of an organisation's overall management system, and as such it includes strategic planning, organisational structure and implementation of an organisation's environmental policy as an integrated part of the organisation's activities. In this sense, an EMS bears a lot more resemblance to a quality management system than to an EIS. The idea of the EMS was first formally recognised and given authority by the publication, in 1992, of British Standard *BS 7750 – Specification for Environmental Management Systems*. This was superseded first in 1994 by a new version of BS 7750, and again in 1996 by the international standard EN ISO 14001, part of the ISO 14000 family of documents on environmental management.

It should be stressed that, whereas EIA is now a legal requirement for certain types of project in most countries, there is no legal requirement for a company or organisation to set up an EMS. However, it is becoming more and more common for major clients to require their contractors to have an EMS and, as a consequence, contractors to require their subcontractors and suppliers to have one also. Thus, the spread of the use of the EMS is being driven by the supply chain rather than by legislation.

The ISO 14001 standard requires a company or organisation establish and maintain a system that enables it to:

- establish an environmental policy;
- identify the environmental aspects of their operations;
- determine the *significant* environmental impacts. Identify relevant laws and regulations;
- set environmental objectives and targets;
- establish a structure and programme to implement objectives;

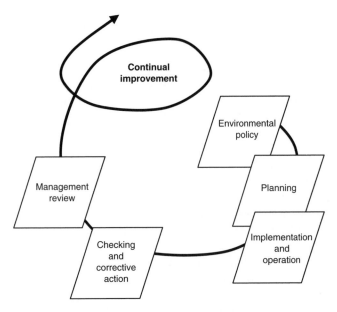

Figure 5.2   Steps necessary to set up an EMS for an organisation (after ISO 14001).

- facilitate planning, control, monitoring, review and audit;
- be capable of adapting to change.

Like a quality management system, an EMS is a management standard; it is not a performance or product standard. The underlying purpose of ISO 14001 is that companies will improve their environmental performance by implementing an EMS, but there are no standards for performance or the level of improvement. It is a process for managing company activities that impact on the environment.

Figure 5.2 illustrates the various steps necessary to set up an EMS for an organisation.

Environmental policy

The first step in implementing and EMS is to gain commitment to environmental improvement from top-level management, and this must then be made public by issuing a corporate environmental policy statement. A preliminary review of all aspects of the organisation in the light of strengths, weaknesses, risks and opportunities in relation to environmental issues and effects should then be undertaken. The review should cover all legislative and regulatory requirements and an assessment of previous non-compliance, together with an evaluation of known significant environmental impacts of the organisation's activities, and an examination of existing environmental management procedures.

This review sets the baseline for the organisation's environmental policy. To meet the ISO 14001 standard the policy must contain a commitment to at least meet all relevant environmental legislation and show a continuing commitment to improvement in environmental performance over and above that required by the legislation. It must also be made public. These are the essential requirements for meeting the standard. It should be noted that the standard does not require an organisation to meet externally set targets for environmental performance (other than minimum legal requirements) – the organisation must set these targets itself.

Planning

ISO 14001 requires an organisation to set up and maintain programmes and procedures to identify any environmental effects arising from support functions (planning, finance, sales and procurement) of the organisation as well as the main activities. These effects may be direct ones, such as the release of pollutants into environment, or indirect ones, which may include the possible misuse of the company's products by others, or the environmental impacts arising from suppliers' and subcontractors' activities. The plan should also include procedures for identifying and ensuring access to relevant environmental legislation and regulations.

A major element of planning is the setting of objectives and targets. These must be consistent with the organisation's policy, they must be documented and they need to be developed for, and communicated to, all levels of the organisation. Wherever possible they should quantify the targeted levels of improvement (e.g. reducing hazardous waste emissions by 50% of current levels within the next 12 months) in environmental performance.

Once objectives have been set, programmes to achieve them must be developed and documented. These programmes will formally assign responsibilities and specify the means and time frame by which they will be achieved.

Implementation and operation

To implement the programme, the necessary resources must be provided, and these must be fully detailed in the plans. A management representative must be appointed to ensure that the EMS is established, implemented and maintained. In addition, individuals need to be nominated for ensuring that the organisation is kept up to date with all relevant environmental legislation and for training at all levels within the organisation. To conform to the requirements of the standard, organisations must implement procedures for internal communication between the various levels and functions of the organisation and for dealing with relevant communication from external parties, such as suppliers, subcontractors, environmental groups, public organisations and the general public. The documentation for the EMS need not be a separate manual; it can be integrated with other management systems such as quality, and health and safety. The important thing is that it is easily available to all employees, and to this end, it may be in paper or electronic format as appropriate to any given situation.

Operations and activities, including maintenance, which can have a significant impact on the environment need to be identified and then controlled and monitored to ensure compliance with legislation and the organisation's EMS. The organisation should also identify any significant environmental impacts associated with goods and services that it uses, and communicate the relevant procedures and requirements to its suppliers and contractors. Methods for preventing or mitigating environmental impacts associated with emergencies should also be developed and documented.

Checking and corrective action

Characteristics of operations and activities that can have a significant impact on the environment need to be monitored and measured regularly. Records of monitoring and measurement are needed to assess performance, to demonstrate that operational controls are effective and to show that objectives and targets are being met. The results of any monitoring then need to be compared with the legislative and regulatory requirements to demonstrate compliance. For each environmentally relevant activity, the organisation should document how verification is to be carried out and the required accuracy of the measurements. Actions to be taken in the case of non-conformance should be detailed in the EMS together with any the reasons for non-conformance, and the procedural changes implemented as a result.

The EMS itself must also be subjected to systematic and regular audit to provide assurance that the system is being implemented satisfactorily and to provide information for the management review. Procedures for audit need to be established to identify the frequency of audits, which aspects of the organisation's activities will be audited, who will be responsible for the auditing and how the results will be documented and communicated. Although an internal auditor may conduct audits, an auditor, externally verified by the appropriate agency, must be used in order to gain formal accreditation to ISO 14001. Whether internal or external auditors are used, they must have the support and authority necessary to obtain the necessary information.

Management review

The EMS cycle is concluded by the management review, which is conducted by senior management to ensure that the system is operating effectively. It provides the opportunity to make any changes to policies, objectives or the EMS itself, all of which may be required owing to changes in legislation, modifications to business operations, advances in technology, results of audits or for continual improvement.

Further reading

Harrington, H. J. and Knight, A. (1999) *ISO 14000 Implementation: Upgrading Your EMS Effectively*, McGraw Hill.

Hughes, D. (1992) *Environmental Law,* second edition, Butterworth.

International Organisation for Standardisation (1996) *ISO 14001 Environmental Management Systems – Specifications and Guidance for Use,* ISO.

Roberts, H. and Robinson, G. (1998) *ISO 14001 EMS Implementation Handbook,* Butterworth Heinemann.

Turner, R. K., Pearce, D. and Bateman, I. (1994) *Environmental Economics – An Elementary Introduction,* Harvester Wheatsheaf.

Wathern, P. (ed.) (1988) *Environmental Impact Assessment – Theory and Practice,* Routledge.

Winpenny, J. T. (1991) *Values for the Environment – A Guide to Economic Appraisal,* HMSO.

Wood, C. (1995) *Environmental Impact Assessment – A Comparative Review,* Longman.

Chapter 6
Project Finance

All projects require financing. No project progresses without financial resources. However, the nature and amount of financing required during different phases of the project varies widely. In this chapter various financial instruments available for project financing are discussed.

6.1 Introduction

In most engineering projects the rate of expenditure changes dramatically as the project moves from the early stages of studies and evaluation, which consumes mainly human expertise and analytical skills, to design, manufacture and construction of a physical facility. Broadly, a project may be said to pass through three major phases:

(1) project appraisal;
(2) project implementation/construction;
(3) project operation.

A typical cash-flow curve for a project is illustrated in Figure 6.1.

The precise shape of the cash-flow curve for a particular project depends on various factors such as the time taken in setting up the project objectives,

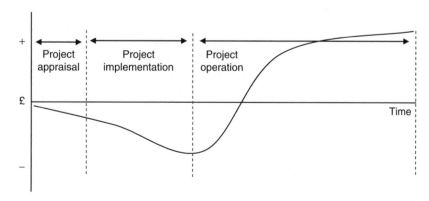

Figure 6.1 Typical project cumulative cash-flow stages.

obtaining statutory approvals, design finalisation, finalisation of the contracts, finalisation of the financing arrangement and the rate and amount of construction and operation speed. The negative cash flow, until the project breaks even, clearly indicates that a typical project needs financing from outside. The rate of spend during this period is known as the cash burn. The shape of the curve also reveals that in the initial phase of the project relatively less financing is required. As the project moves on to implementation phase there is a sudden increase in the finance requirement, which peaks at the completion stage. The rate of spending is also depicted by the steepness of the curve; the steeper the curve, the greater the need for finance to be available.

Once the project is commissioned and starts to yield revenues, the requirement of financing from outside the project becomes less and less. Finally, the project starts to generate sufficient resources for the operation and maintenance and also a surplus. However, even after the break-even point, the project may require financing for short periods, to meet the mismatch between receipts and payments.

In project financing, it is this future cash flow that becomes the basis for raising resources for investing in the project. It is the job of the finance manager of the project to package this cash flow in such a way that it meets the needs of the project and at the same time is attractive to the potential agencies and individuals willing to provide resources to the project for investment. To achieve this objective effectively, a thorough knowledge of the financial instruments and the financial markets in which they trade is essential.

6.2 Definition of project finance

The concept of project finance is widely used in business and finance in developed countries. Many developing countries are also using project finance to raise funds for their infrastructure projects, such as Malaysia, China and Indonesia. There is, however, no precise legal definition of project finance as yet. However, project finance is used to refer to a wide range of financing structures. These structures have one feature in common – the financing is not primarily dependent on the credit support of the sponsors or the value of the physical assets involved. In project financing, those providing the senior debt place a substantial degree of reliance on the performance of the project itself.

The term project finance is best defined as:

> A financing of a particular economic unit in which a lender is satisfied to look initially to the cash flows and earnings of that economic unit as the source of funds from which a loan will be repaid and to the assets of the economic unit as collateral for the loan.

Similarly, Nevitt (1983) describes project finance as:

> Financing of a stand-alone project in which the lender looks primarily to the revenue stream created by the project for repayment, at least once operations

have commenced, and to the assets of the project as collateral for the loan. The lender has limited recourse to the project sponsors.

Thus, project finance is a useful financing technique for sponsors who wish to avoid having the project debt reflected on their balance sheets or to avoid the conditions or restrictions on incurring debt contained in existing loan documents, or when the sponsors' creditworthiness or borrowing capacity is less than adequate. The cutting edge of developments in project finance is in emerging countries and to a large extent relates to private infrastructure projects as governments in these countries grapple with limited foreign exchange reserves, budgetary restraints and burgeoning infrastructure requirements.

Esty (2004) states that project finance is not any of the following:

secured debt – because the debt has recourse to corporate assets;

vendor-financed debt – because the debt has recourse to corporate assets (vendor finance is a kind of secured debt in which the manufacturer of the goods provides the debt);

subsidiary debt – not if the debt has recourse to corporate assets.

6.3 Basic features of project finance

From the preceding brief descriptions, the following basic features of project finance may be identified:

Special Project Vehicle (SPV);
non-recourse or limited recourse funding;
off-balance-sheet transaction;
sound income stream of the project as the predominant basis for financing;
a variety of financial instruments;
a variety of participants;
a variety of risks;
a variety of contractual arrangements.

Projects have to raise cash to finance their investment activities. This is normally done by issuing or selling securities. These securities, known as financial instruments, are in the form of a claim on the future cash flow of the project. At the same time, these instruments have a contingent claim on the assets of the project, which acts as a security in the event of future cash flows not materialising as expected. The nature and seniority of the claim on the cash flow and assets of the project vary with the financial instrument used. Merna and Owen (1998), describe financial instruments as the tools used by the SPV to raise money to finance the project.

Traditionally, financial instruments were either in the form of debt or equity. Developments in the financial markets and financial innovations have led to the development of various other kinds of financial instruments that share the characteristics of both debt and equity. These instruments are normally described as mezzanine finance.

Most projects are financed utilising debt as part of the financing package. In project finance, mobilising commercial debt can be quite difficult due to several reasons.

High demand, cautious lenders. Lenders face the same risks as equity investors when arranging loans to finance a project. They also fear the risk of not getting any money back in the event of default. In addition, there is a limit to how far loan pricing can be pushed. So, the lenders often have the major say in how financing is to be structured and seek to reduce the projects risks by negotiating the conditions with the promoter under whom they will participate.

Foreign lenders. In developing countries, most domestic markets cannot mobilise high volumes of long-term debt, so they turn to foreign lenders. Foreign lending involves foreign currency, thus exposing the promoter to currency risks.

The number of banks involved in project finance is still relative small, although more have entered the market in the last 5 years. Each bank has exposure limits to the project finance environment, and thus organising a syndication of lenders is complex and time consuming.

In developing countries, there are many potential non-bank lenders such as pension funds and insurance companies. However, this potential is not being met as many of these companies are publicly owned monopolies and most developing countries also require pension funds and insurers to invest mostly in government securities.

Instruments refer to those securities issued by the project that makes it liable to pay a specified amount at a particular time. Debt is senior to all other claims on the project cash flow and assets.

Ordinary equity refers to ownership interest of common stockholders in the project. On the balance sheet, equity equals total assets less all liabilities. It has the lowest rank and therefore the last claim on the assets and cash flow of the project.

Mezzanine finance occupies an intermediate position between the senior debt and the common equity. Mezzanine finance typically takes the form of subordinated debt, junior subordinated debt, and preferred stock or some combination of each.

Besides debt, equity and mezzanine finance a project may also utilise certain other types of instruments such as leasing, venture capital and aid.

Senior debt

In terms of seniority, senior debt ranks the highest among the financial instruments in terms of claims on the assets of a project. This means that in the event of a default, the lenders of senior debt have the right to claim on the assets of a project first.

Subordinated debt

In project financing, there is another type of debt called subordinated debt. Subordinated debt is subordinate to senior debt and generally only has second claim to the collateral of the project company. This means that in the event of default by the promoter, all senior debt must be met before any claim can be made by subordinated debt lender. As is it is second only to senior debt in terms of claims on the projects assets, lenders seek higher returns on subordinated debt. The interest rate on subordinated debt is usually higher than the interest rate on senior debt. For example, interest rate on senior debt may be London Interbank Offered Rate (LIBOR) + 200 basis points (BP), but interest rate on subordinated debt can be LIBOR + 400 BP. Subordinated debt is used mainly for refinancing needs or for restructuring of the finance package of a project.

Standby loans

Standby loans are often arranged with lenders by promoters to meet required draw-downs in excess of the term loans due to higher interest rate and lower-than-expected revenues in the early phase of the operation and for additional works in the construction phase.

As the financing requirement of a project depends on the future cash flow, which depends on time, another way of classifying financial instruments is temporal in nature, that is, long-term financing instruments and short-term financing instruments. Long-term financing refers to raising debt, equity and mezzanine finance, which have repayment obligation beyond 1 year. Short-term financing refers to financial instruments which normally have repayment obligation up to 1 year.

From a project-financing point of view, it is more reasonable to discuss the financial instruments according to their maturity, that is, short term and long term.

Long-term financing instruments

A project raises long-term financing primarily for long-term investment purposes. Long-term financing is needed because the asset created by the project has a gestation lag, before it starts to yield revenues. Long-term financing helps the project by deferring, partly or fully, the servicing of the securities sold until the project starts to generate revenues. A discussion of the major debt, equity and mezzanine finance instruments follows.

Debt instruments

Debt instruments refer to the raising of term loans from banks, other financial institutions (including commercial banks, merchant banks, investment banks, development agencies, pension funds and insurance companies), debentures and export credits (buyer's credit and supplier's credit).

Term loans

Term loans are negotiated between the borrower (project) and the financial institutions. In large infrastructure projects, where the funds required to be raised are very high, normally, term loan is not provided by one bank or financial institution. A group of banks and financial institutions pool their resources to provide the loans to the project. For example, in the Channel Tunnel Project, the original financing structure of this project involved 210 lending organisations, with 30% of the loan arranged through Japanese banks. The lead bank was NatWest Bank. Banks and financial institutions set their own internal exposure limits to particular types of project. This helps in spreading the risk. Generally an investment bank or a merchant bank acts as the agent or lead bank to manage the debt issue. Many banks specialise in lending to certain types of infrastructure projects that they have both technical and financial experience of.

The terms and conditions of loans vary from lender to lender and borrower to borrower. It can have fixed interest rate or floating interest rate. Repayment of the loan could be between 7 and 10 years for an oil sector project and from 16 to 18 years for a power project. It varies with the gestation lag of the project. The servicing of the loan, that is the repayment of the principal amount (amortisation) and interest payment can take various forms. It could be from equal instalments spread throughout the life of the loan to a bullet or sunset payment at the end of the period. Another alternative could be initial moratorium on repayment of the principal amount but with regular interest payment. A third option could be moratorium on both principal repayment and interest payment until the project is planned to operate. The type of loan is determined by the project characteristics and availability of the instruments. As stated earlier, the success of the project lies in projecting attractive future cash flows that are acceptable to lenders as in project finance the basic security to the lenders is the project's cash flow.

The cost of raising debt capital includes certain fees besides the interest.

Management fee, as a percentage of the loan facility for managing the debt issue, normally to be paid up front

Commitment fee, calculated on the un-drawn portion of the loan to be paid when the loan is fully drawn

Agency fee, normally an annual fee to be paid to the lead bank for acting as the agent to the issues after loan has been raised

Underwriting fee, paid up front as a percentage of the loan facility to the bank or financial institution which guarantees to contribute to the loan issue if it is not fully subscribed

Success fee, paid up front as a percentage of the total loan, once all loans have been secured

Guarantee fee, paid annually on the outstanding loan amount, if it is guaranteed against default

All, none or some of these may be present in a specific loan proposal. In certain cases the lenders submerge all the fees in the interest rate they offer. A careful analysis of the cost of loans offered from different sources is therefore required.

Still, the overall cost of raising a term loan is less as compared to any other mode of large-scale financing because the project has to negotiate and deal with only a small number of lenders of money through a lead manager of the issue. Also, in the event of default, it is easier to renegotiate a term loan as compared to any other instrument of financing.

Euro currency loans

Euro currency loans are the loans provided by banks from the unregulated and informal market for bank deposits and bank loans denominated in a currency other than that of the country where the bank initiating the transaction is located. For example, a British bank in London that lends in US dollars to an Argentinean company is in fact lending in Euro currency. Euro currency finance originated in the 1950s. The development of the Euro currency loan market coincided with the development of inter-bank market for Euro currency deposits and placings. Euro currency deposit can be defined as a deposit in a currency other than in the currency of the country in which the bank is located.

Euro currency market provides a lot of flexibility in borrowing by offering funds with fixed or floating interest rates. These loans are generally for a fixed period. When fixed-rate funds are scarce or too expensive for the borrower, the lending bank provides commitment to lend for, say, 5 years on a rolling basis, where the rates are adjusted at agreed intervals, monthly, quarterly or semi-annually, in line with the then current market rate.

Another feature of Euro currency finance is the multi-currency credit facility. This enables the borrower to draw down in any currency of his choice, although this condition is subject to the availability of funds in the currency of choice.

These funds are usually provided by banks out of their Euro currency deposits that a bank takes for a short term – for major currencies up to 1 year and in the case of US dollars and sterling up to 5 years, depending upon the amount. The loans are provided on a matching basis with deposits procured. The largest sector of the Euro currency market is deposits and loans in US dollars, and therefore it is also known as Eurodollar market. London is the principal source of Euro currency finance.

Debentures

A debenture is similar to a term loan except for the fact that the loan is divided into securities and sold through the stock market to a variety of investors. They are usually in the form of a bond, undertaking the repayment of the loan on a specified date, and with regular stated payments of interest between the date of issue and date of maturity. In the UK, The Companies Act defines the word debenture as including debenture stock and bonds. In the UK, often the expressions debenture and bonds are used interchangeably. However, in this book debenture and bonds are treated as two different financial instruments primarily because of two reasons. Firstly, bonds are normally treated subordinate to debenture in the event of liquidation of the borrowing entity, and secondly, bond markets have become much more flexible through the introduction of a variety of bond instruments. Bonds have, therefore, become very close to equity issues as far as liquidity is concerned. Liquidity refers to the ease and quickness of converting assets to cash. It is also called marketability. Bonds have therefore been classified as mezzanine finance instruments.

Export credit

Export credit is a pure form of loan provided by either the exporter of the product/equipment or by a bank in the exporting country or by any governmental agency in the exporting country set up to promote exports. These loans are guaranteed by the export credit guarantee agencies in the exporting countries, such as, ECGD in the UK, EDC in Canada, COFACE in France, HERMES in Germany, SACE in Italy, MITI in Japan and USEXIM in the USA. The guarantee fee depends largely on the political risk perception about the borrowing country. For example, during the Gulf Crisis of 1991, when India was facing an acute balance of payment problems and there was very heavy risk of default, most of these agencies raised their insurance premium to 16–18% up front as against the normal premium of 4–5% on a 5-year loan. Export credit is either in the form of supplier's credit or buyer's credit.

Supplier's credit is normally structured in two ways:

(1) If the supplier of the machinery or equipment is in a position to sell on a deferred credit basis, it enters into an agreement with the importer to supply the equipment and receive payments for it over a period of time. There is no involvement of any bank or financial institution in this structure. The supplier gets the deferred credit insured from the export credit guarantee agency of the country.
(2) Alternatively, if the supplier company is unable to extend deferred credit, it negotiates with a bank or financial institution or governmental export promotion agency to seek a loan for the export with a guarantee from the export credit guarantee agency. The exporting company takes the loan in its own name and extends deferred credit to the importer as mentioned earlier. These are called supplier's credit because the supplier of the equipment extends credit to the importing project.

Ordinary equity

As stated earlier, the process of a project proposed to be funded on a project-financing basis starts with the setting up of a particular project legal entity, known as SPV. The sponsors of the project provide the initial equity capital known as the seed equity capital:

> Pure equity is the provision of risk capital by investors to an investment opportunity and usually results in the issuance of shares to those investors. A share

may be described as an intangible bundle of rights in a company, which both indicates proprietorship and defines the contract between the shareholders.

The terms of the contract, that is, the particular rights attaching to a class of shares, are contained in the article of association of the company. Equity is the residual value of a company's assets after all outside liabilities (other than to shareholders) have been allowed for. Equity is also known as risk capital, because these funds are usually not secured and have no registered claim on any assets of the business, thus freeing these assets to be used as collateral for the loans (debt financing). Equity, however, shares in the profits of the project and any appreciation in the value of the enterprise, without limitation. The compensation for equity are the dividends (dividends are the amount of profits paid to shareholders). No dividends are paid if the business does not make profits. Dividends to the shareholders can be paid only after debt claims have been met. The return on the equity, therefore, is the first to be affected in case of financial difficulties being faced by the project entity. This means that equity investors, in the worst-case scenario, may be left with nothing if the project fails, and hence they demand greater return on their capital so as to bear a greater risk. This explains the general rule that high-risk projects use more equity while low-risk projects use higher debt.

A high proportion of equity means a low financial leverage and a high proportion of debt means a high leverage. Leverage is measured by the ratio of long-term debt to long-term debt plus equity. Leverage is also called gearing, or 'debt–equity ratio'.

High financial leverage means that relatively more debt capital has been used in the project, signifying more debt service and less funds being available for distribution to the equity holders as dividend payments. However, once the project breaks even and profit starts to grow fast, shareholders receive a higher dividend. The seed capital provided by the sponsors of the project, which is normally a very small amount as compared with the total finances raised for the project, is also known as founders or deferred shares. These are lower in status as compared with ordinary and preference shares in the event of winding up.

In non-recourse financing, the debt–equity ratio may be higher if the interest rate is high, provided lenders are satisfied with the risk structure of the project. If, however, a project is considered innovative, then more equity will be demanded by lenders and the equity will be drawn down before debt becomes available to the project.

Ordinary share capital is raised from the general public. Holding of these shares entitles dividends, provides the right of one vote per share held, and the right to a pro-rata proportion of the project's assets in the event of winding up of the project. The right to participate in the assets of the project provides the opportunity for highest return on the capital invested.

The SPV can raise equity capital from the market in various ways:

public issue;
offers for sale;

issue by tender;
private placement;
rights issue.

Cost of equity issue

The detailed procedures involved in getting a new issue (public offer, offer for sale, issue by tender and private placement) registered, issue of detailed prospectus and aggressive marketing through advertisements make it very expensive. On top of that the issuer has to pay underwriting commission, stock exchange listing fee, legal and broker's fee and capital duties. It is estimated that for a public offer the total cost comes to as high as 12.5% of the proceeds for a 5 million issue.

Mezzanine finance instruments

There are a host of financial instruments in this category. They are senior with respect to an equity issue and lower than debt. Some of them are close to a debt issue and some of them share features of an equity issue.

A bond, like any other form of indebtedness, is a fixed-income security. The holder receives a specified annual interest income and a specified amount at maturity – no more, no less (unless the company goes bankrupt). The difference between a bond and other forms of indebtedness such as term loans and secured debenture is that bonds are a subordinate form of debt as compared with term loans and secured debentures. Similar to debentures, these are issued by the borrowing entity in small increments, usually US$1000 per bond in the USA. After issue, the bond can be traded by investors on organised security exchanges.

Four variables characterise a bond: its par value, its coupon, its maturity and its market value.

Par value. Par value is also known as nominal value or face value or principal or the denomination. It is stated on the bond certificate. It is the amount of money which the holder of the bond will receive on maturity.

Coupon rate. It is a percentage of par value the issuer promises to pay the investor annually as interest income. This is normally paid semi-annually. Bonds are normally issued with attached coupons for the payment of the earned interest. The holder of the bond in such cases separates the coupons and sends it to the company for the payment of interest. This is the reason why interest in the case of bonds is known as the coupon rate.

Maturity date. The date on which the last payment on the bond is due is known as the maturity date.

Market value of bond. On the date of issue of the bonds the issuer usually tries to set the coupon rate equal to the prevailing interest rate on other bonds of similar maturity and quality. This ensures that the bond's initial market price is almost equal to its par value. After issue, the market value of the bond can differ substantially from its par value. As the stated return on the bond (its coupon) is fixed, the market value of the bond depends on the general level of interest rate prevailing in the market. As compared to the coupon rate, if the general interest rate in the market is low, then the price of the bond moves up as holding of the bond yields higher return than other forms of investment in the market.

Bond ratings

The success of a bond issue, *inter alia*, depends on its quality. There are many companies that analyse the investment qualities of publicly traded bonds. They publish their findings in the form of bond ratings. The ratings are determined by using various financial parameters of the borrowing agency, general market conditions in which the borrower operates, the political situation of the country in which the project is located and other sources of finance that have been tied up by the project. It is based, in varying degree, on the following considerations.

- The likelihood of default by the bond issuer on its timely payment of interest and repayment of principal.
- The nature of and provisions of the obligations.
- The protection afforded by the indenture in the event of default, bankruptcy, reorganisation or other arrangements under the law and the laws affecting creditors rights.

The ratings are normally depicted by letters such as A or a combination of letters and numbers such as BB. In certain financial markets public issue of bonds is not permitted if the bonds have not been rated, such as the US bond market. Rating is also important because bonds with lower ratings tend to have higher interest costs. The rating agencies keep reviewing the financial performance of the borrower, the general market situation and the political situation in the country of the borrower. Depending on the emerging situations, the ratings are revised upwards or downwards. Table 6.1 gives a summary of the common long-term ratings assigned by the two leading bond rating agencies Moody's Investor Service and Standard & Poor's.

High-grade bonds. Capacity to pay interest and principal is extremely strong.

Medium-grade bonds. Strong capacity to pay interest and repay principal, although it is somewhat more susceptible to the adverse effects of changes in circumstances and economic conditions. Both high-grade and medium-grade bonds are investment quality bonds.

Low-grade bonds. Adequate capacity to pay interest and repay principal. However, adverse economic conditions or changing circumstances are more likely to lead to a weakened capacity to pay interest and repay principal. They are regarded as predominantly speculative bonds. BB and Ba indicate the lowest degree of speculation and CC and Ca the highest degree of speculation.

Very-low-grade bonds. This rating is reserved for income bonds on which no interest is being paid. This rating is in default, and payment of interest and/or repayment of principal is in arrears.

Table 6.1   Bond ratings.

Bond ratings

Standard & Poor's	Moody's	Comments
AAA	Aaa	**High-grade bonds**
AA	Aa	Capacity to pay interest and principal is very strong
A	A	
BBB	B	**Medium-grade bonds**
BB	Ba	Strong capacity to pay interest and repay principal,
B	B	although it is somewhat more susceptible to the adverse effects of changes in circumstances and economic conditions. Both high-grade and medium-grade bonds are investment quality bonds
CCC	Caa	**Low-grade bonds**
CC	Ca	Adequate capacity to pay interest and principal, although
C	C	adverse economic conditions or changing circumstances are more likely to lead to a weakened capacity to pay interest and principal. These are regarded as mainly speculative bonds, with CC and Ca being the bonds with the highest degree of speculation
D	D	**Very-low-grade bonds**
		This rating is reserved for income bonds on which no interest is being paid
		This rating is in default, and payment of interest and/or repayment of principal is in arrears

Source: Adapted from Merna and Dubey (1998).

At times, both rating agencies use adjustments to these ratings. Standard & Poor's uses plus and minus signs: A+ is the strongest and A− the weakest. Moody's uses a 1, 2 or 3 designation, with 1 being the strongest.

Wrapped and unwrapped bonds

Wrapped bonds are guaranteed by a monoline insurer, which makes them very creditworthy. As a result of the guarantee, the bonds are rated AAA/Aaa, therefore reducing the cost of borrowing. Unwrapped bonds have no guarantor and the bond is rated on the project itself. The bond pricing will, in turn, be driven by the projects rating.

The use of bond finance, through private placement, usually depends on the size of the project finance required. The Office of Government Commerce (2002) suggests that in the UK bond finance tends to be used in projects requiring in excess of £90 million. For projects between £60 and £70 million, bond finance

Table 6.2 Characteristics of bond and bank financing (Office of Government Commerce, 2002).

Financial characteristic	Bank financing	Bond financing
Source of Funds	Direct from banks	Bond investors
Arrangement of funds	Negotiations between bank and lender	Via bond arranger
Certainty of funds	After agreement: Certain	Less certainty: Only known if funding is forthcoming when the bond goes on sale
Maturity repayments	Up to 30 years	Up to 38 years
Flexibility	High: Early payments can be made, and refinancing is possible	Very little. No room for negotiation on interest and capital repayment
Receipt of funds	Staged: Works on a draw down process	Whole: After the bond is sold
Assessment of project risk	Banks assess risks	Bond arranger assesses risks
Costs	Interest of the funds borrowed, and a commitment fee for funds yet to be drawn down	Interest to the bond investors, a fee to the bond arranger, and an insurance fee (Optional)
Ongoing project scrutiny	Significant: Possible step in clauses	Very little: Bond investors have little influence on the project once it is funded
Optimum size	No optimum size	Approximately £100–400 million
Opportunities for refinancing	Yes, if project risks become less than those assumed in the initial financing	Unlikely: Bond terms tend to be fixed for the life of the project

needs to be assessed in greater detail by monoline insurers to determine whether such finance can be cost-effective due to the costs associated with raising bond finance. Monoline insurers, for example, seek a return of 1–2% of the total bond finance raised to cover identified risks.

Table 6.2 illustrates the characteristics of bank and bond financing.

At the time of writing this chapter, UK, EU and US interest rates were 3.75%, 2.0% and 1.0%, respectively. These low interest rates have meant that investors have sought debt rather than bonds as debt is the cheapest form of lending.

Other forms of bonds

Bonds discussed so far are known as plain vanilla bonds. The other types of bonds are junk bonds, floating rate bonds, deep discount bonds (zeros), income bonds, Euro bonds and revenue bonds.

Other financing instruments

Depository receipts

A depository receipt is a negotiable certificate representing a project's equity or debt. Depository receipts are created by a depository bank appointed by the project. New as well as old shares of the project can be made available to an appointed depository's local custodian bank, which then instructs the depository bank, to issue depository receipts. These depository receipts are thereafter issued either publicly or through private placement. Depository receipts may trade freely, just like any other security. Depository receipts in American form are known as American depository receipts (ADRs) and in Global form they are known as global depository receipts (GDRs).

During 1997, out of the US $5642 million was raised through private placement of GDRs, the maximum amount was raised by India 19.3%, followed by Taiwan 17.0%, Hungary 10.4%, Poland 7.2% and Brazil and Korea 7.1% each.

Lease finance

A lease is a contractual arrangement between a lessee and a lessor. The agreement establishes that the lessee has the right to use an asset and in return must make periodic payment to the lessor, the owner of the asset. The lessor could be either the manufacturer of the asset or an independent leasing company. If the lessor is an independent leasing company, then it must buy the asset from a manufacturer and make it available to the lessee for the lease to be effective.

Leases are principally of two types:

Operating lease. Assets that are required occasionally can be hired under this form of lease. Its important features are the following:

Operating leases are usually not fully amortised. This means that the payments required under the terms of the lease are not sufficient to recover the full cost of the asset for the lessor. This is because the term or life of the operating lease is usually less than the economic life of the asset. The lessor therefore must either renew the lease or sell the asset for its residual value.
Operating leases usually require that the lessor maintain and insure the leased assets.
There is generally a cancellation option in this type of lease that gives the lessee the right to cancel the lease contract before the expiration date.

Financial lease. In this case the potential user identifies an asset in which they wish to invest, negotiate price, and delivery and then seek a supplier of finance to buy it. The asset is bought by the lessor and made available to the lessee for use. Its important characteristics are the following:

financial leases do not provide for maintenance or service by the lessor;
financial leases are fully amortised;

the lessee usually has the right to renew the lease on expiration;
generally financial leases cannot be cancelled.

The two other types of financial leases are sale and lease back and leveraged
lease.

In a sale-and-lease-back arrangement, the owner of an asset sells the asset to a
lease company and immediately leases it back. This helps the lessee to generate
cash from the sale of the asset and at the same time keep using the asset by
making periodic lease payments.

Leveraged lease is a three-sided arrangement among the lessee, the lessor and
the lender. The lessor buys the asset and provides it for use by the lessee. However, the lessor puts in no more than 40%–50% of the purchase price and the
remaining financing is obtained from a lender. The lender typically provides a
non-recourse loan and receives interest from the lessor. The lender has the first
lien on the asset, and in case of default by the lessor, the lease payment is made
directly to the lender.

Venture capital

Venture capital consists of funds invested in small, newly established enterprises
normally in high-growth markets that promise exceptional future profit levels.
The investment is normally made in the form of equity capital. However, some
venture capital funds invest in a mixture of debt and equity capital. Although
venture capital is thought to be providing capital only for new ventures, its role
is much wider. The types of investment which are being financed under venture capital can be put under five categories – start-up, early-stage development
capital, management buy-out (MBO), expansion and others.

Aid refers to a direct gift of money from one government or a multilateral supra
government agency such as the World Bank to another government. Grant aid
is usually intended to help less developed countries meet social or community
welfare objectives. It is seldom free from obligations on the recipient government.
It can have direct strings attached to the project or may influence the way a project
is managed. Other forms of aid are in the form of subsidised credit export or aid
plus credit package. They are best considered as debt financing. Aid is basically
of two forms.

(1) Project aid is specific, highly structured and formalised in accordance with
    the donor institutions project appraisal procedures.
(2) Programme aid intends to finance imports in return for sectoral policy
    reforms. Programme aid may involve individual projects.

Aid finance from one government to another government is known as bilateral
aid. It is usually, but not always, tied to supplies from the donor country. When
it is from a multilateral institution, it is known as multilateral aid. Use of funds
in the case of multilateral aid is not tied to any particular supplier.

6.4 Short-term financing instruments

Projects require short-term debt of two kinds. First, for working capital requirement, these funds are required once the project has been commissioned and there is a time lag between payments to be made for the purchase of raw materials, components and equipment for the operation of the project and the receipt from the sale of the product. Second, as bridging finance to meet temporary deficits in cash balance when there is a known source of funds which can be fully relied upon to liquidate the bridging loan. This may be due to various reasons such as the project may have negotiated a long-term loan but the formalities of documentation and fulfilling of various covenant conditions attached to the loan is taking longer than anticipated and the project is in need of immediate resources. Or the project may have raised equity capital through public offer but registration formalities with the local stock exchange authorities is taking longer, or the project is waiting to launch the issue when the stock market is in the right condition. In such situations the project raises bridging loans to fund the immediate requirement. The bridging loan is liquidated from the proceeds of the main loan or the equity capital raised.

Short-term (working capital) financing instruments

The short-term financing options available to a project for working capital requirements during the operation phase, other than the internal resources, can be broadly classified in three categories:

(1) Unsecured bank borrowing;
(2) Secured borrowing;
(3) Other sources.

Unsecured bank borrowing

The most common method to finance a temporary cash deficit is to arrange a short-term unsecured bank loan. This is either in the form of a non-committed or a committed line of credit. A non-committed line of credit is an informal arrangement that allows the project to borrow up to a previously specified limit without going through the normal loan documentation requirements. The interest rate on the line of credit is usually at the bank's prime lending rate plus an additional charge. Generally a compensating balance (about 2%–5%) is required to be kept at the bank.

Committed line of credit has formal legal arrangement and documentation requirement. It also involves a commitment fee. The interest rate is often linked to LIBOR or to the bank's cost of funds, rather than the prime rate.

Secured loans

Banks and financial institutions often require security for a loan. The security for a short-term loan usually consists of accounts receivable or inventory. In accounts receivable, the lender has a lien on the receivable of the project and also a recourse to borrowers assets.

Other sources

There are a variety of other mechanisms to secure short-term financing. But the most important are commercial paper and bankers' acceptance:

Commercial paper (CP). CP consists of short-term notes issued by large and highly rated projects. Typically these notes are of short maturity, ranging up to 270 days. The issue is normally backed by a special line of credit. Therefore the interest rate on CPs is below the prime lending rate.

Bankers' acceptance. This is an agreement by a bank to pay a sum of money. These arrangements typically arise when a seller, such as a supplier of raw materials or components, sends a bill or draft to a customer. The customer's bank accepts this bill and notes the acceptance on it, which makes it an obligation of the bank. In this way a project that is buying something from a supplier arranges for the bank to pay the outstanding bill. The bank charges a fee for this facility.

Finance can be sourced from a number of organisations or individuals.

Commercial banks.
Niche banks.
Insurance companies.
Stock market.
Individual shareholders equity.
Pension funds.
Bond market.
Vendors.
Suppliers.
Contractors.

It should be noted that the source of finance will determine the type of financial instrument used. For example, banks will lend on the basis of debt, stock markets on the basis of equity and bond markets on the basis of bonds. The art of determining which sources are best suited to provide project finance is known as financial architecture.

6.5 Project finance in bundled or portfolios of projects

Bundling is the grouping of projects or services within one managed project structure in a manner that enables the group to be financed as a single entity. The key benefits are that this allows small projects to be financed by increasing the overall debt within the bundle to an economic level and allows various projects to cross-collateralise each other. Key issues are that cash flows from the single project are robust (a single cash flow is often preferred) and the liabilities of each party, particularly those of the public sector partners, are adequately addressed in the event of, for example, partial or full termination.

Bundling projects can provide cash flows sufficient to provide a reasonable return after operating and debt service costs are addressed. It can also spread risk for funders between different projects and locations.

The financial analysis of these bundled projects can be considered as a portfolio of projects. Each individual project will have different cost and revenue implications and be subjected to different risk scenarios. When projects are considered individually, some may be commercially viable as stand-alone projects and others not so. However, when the projects are bundled together, the overall portfolio of projects may meet a promoter's minimum acceptable rate of return (MARR) and be deemed commercially viable. When projects are undertaken as commercially viable projects under the private finance initiative (PFI), they should piggyback non-commercially viable projects to ensure such projects are privately financed. These non-commercially viable projects can, however, be financed by cross-collateralisation of funds to make them viable as part of a portfolio of projects.

As with most project financings, refinancing is common after the construction of a facility is completed. Refinancing can be considered in two ways, one when a project is successful and the other when a project does not meet forecasted returns. The former can be through the provision of a new loan with a lower interest rate charge, the latter through debt for equity swaps.

6.6 Risks

In conventional financing methods, the lenders look not only at the prospect of the project becoming successful but also the general creditworthiness of the project sponsors. The risks associated with a particular project are not very critical because the lenders have access to the general assets of the project sponsors. In project financing, the borrower is usually an SPV.

An SPV will normally not have any history as it is created just to implement and operate a specific project. Further, the lenders and investors have no or limited recourse to the general assets of the sponsor company.

It is important to note that the key to project finance development, investing, and lending is the diligence in understanding the risks associated with a project and careful attention to how they are allocated among project participants. It is much easier to put money into an ill-conceived project than to pull it out. Careful reflection about how a proposed project is intended to work in good times and bad is just as appropriate for a power plant (power contract revenues) in the developing world as it is for an airport terminal serving a leading US city (rental income from airlines and concessions). The evaluation of the risks of diminution and interruption of the future revenue stream is the central question around which project finance revolves.

Merna and Smith (1996) classify the risks associated with a project financed under project finance techniques broadly under the following categories.

Elemental risks

These are the risks that are within the control of one or the other of the project participants and appropriate steps can be taken to mitigate them. These can be commercial risk of the project such as the project risk, political risk, those associated with the debtor (SPV), those associated with the sponsors and those associated with the local government or sovereign risk.

Project risk

Completion risks – risk of delay in completion of the project due to contractors delays.
Risk that project completion will involve cost overrun.
Risk that the project fails to meet the performance specification on completion.

Operation and maintenance risks

Risk that the project is unable to run at the desired efficiency because of deficiencies in equipment, personnel and maintenance, and risk that the cost of operation and maintenance of the project turns out to be more expensive than projected.
Risk of project operation being delayed because of legal issues such as environmental liabilities.
Risk of fire and other casualties.

Risk of inputs and outputs

Risk of inadequate, sub-quality and inconsistent supply of raw material and other utilities.
Risk of breach of contract by suppliers of raw material and purchasers of output.
Risk of increase in price of inputs as compared to estimated price.
Risk of inadequate demand for the output of the project.

Risk related to financing of the project

Risk related to increase in the servicing cost of money raised for the project.
Risks of exchange rate fluctuation.
Risk of inadequate funds in the event of cost escalation.

Both borrowers and lenders need to adopt a risk management programme. Risk management should not be approached in an ad hoc manner but in a structured manner. The five major steps of such a process are, according to Merna and Khu (2003):

identify the financial objectives of the project;
identify the source of the risk exposure;

quantify the exposure;
assess the impact of the exposure on business and financial strategy;
respond to the exposure.

6.7 Financial engineering

Just as engineers use special tools and instruments to achieve engineering perfection, the financial engineers use specialised financial instruments and tools to improve financial performance. Galitz (1995) describes the term financial engineering as

> the development and creative application of financial technology to solve financial problems and exploit financial opportunities. The author defines the concept of financial engineering as the use of financial instruments to restructure an existing financial profile into one having more desirable properties.

Financial engineering techniques are being put to wide application such as modelling and forecasting financial markets, development of derivative instruments and securities, hedging and financial risk management, asset allocation and investment management and asset/liability management.

The tools used by financial engineers comprise the new financial instruments created during the last two to three decades: forwards, futures, swaps and options. Further, these basic tools are being combined in different ways by the financial engineers to build more complex systems to meet specific requirements of their clients.

Caps, floors and collars

These are an important group of option instruments primarily to hedge interest rate risk. Caps, floors and collars among these are quite extensively used project financing.

An interest rate cap provides protection against an increase in interest rate, but at the same time allows the benefits of interest rate fall to be enjoyed. For example, a borrower has taken a 5-year loan at LIBOR + 50 BP and has also bought an 8%, 5-year interest rate cap. At each interest rate reset date, if LIBOR + 50 BP falls below 8%, the borrower simply pays the prevailing market rate and takes advantage of the lower rates. On the other hand, if the interest rates on any reset date are higher than the cap rate, the cap will provide a payoff to offset the consequences of the higher rate, effectively limiting the borrowing rate to the cap level.

An interest rate floor is used to limit the benefits from a fall in the interest rate once the floor level is reached. In practice, many users of interest rate caps seek to lower the cost of protection by selling a floor at a lower strike price. If interest rate falls through the floor level on any reset date, the floor is exercised against the seller, who must pay the difference between the prevailing rate and floor rate.

A collar is a combination of selling a floor at a lower strike rate, and buying a cap at a higher strike rate. It provides protection against a rise in rates and some benefits from a falling rate. A collar can be tailored to meet a compromise between interest rate protection and cost. By adjusting with the cap and floor rates, it is possible to create a zero-cost collar, for which no premium is to be paid.

6.8 Refinancing and restructuring

Common practices associated with project finance are the refinancing and restructuring of the original finance and terms and conditions of the concession and financing arrangement. As projects progress risks are either realised or diminished, generating either a loss or gain scenario set against a project base case. Actions either initiated by the principal or promoter may be taken to capitalise on the changing circumstances and risk profile of the project. For example, it is common practice to reduce the senior lending rates at the construction completion and once operational performance of the asset has been validated, often referred to as automated refinancing mechanisms.

While such action may be planned within the original terms of the concession and financing arrangement, unplanned events may act upon the project, which go beyond the capabilities of the original agreements. This often results in the restructuring of the project. For example, the Hungarian M1 and M15 motorway, which was scheduled to be refinanced in 2008, was restructured in 2003 to address the principal's desire to transfer the concession from a market-led tolled motorway to a contract-led payment mechanism structure. Initially the promoter sought to refinance the project prior to restructuring, securing shareholder value prior to negotiation.

6.9 Summary

Project finance is now used to finance many projects traditionally financed by government based on taxation. The characteristics of project finance are SPV, non-recourse and limited recourse finance, off-balance-sheet transaction and risk sharing.

More and more major projects involve project financing. Contractors, engineers and design builders are finding that their work, compensation and risks are shaped by this method of financing.

The sources of finance and the type of financial instruments used vary from project to project. The tailoring of financial instruments based on the perceived risk in a project is known as financial engineering. The financial engineering elements of a project are just as important as the construction and operation of an asset.

Project finance is dependent on the revenues of a project (the project's cash flow) to repay loans without affecting the balance sheet of the organisation. In the case of default in project finance, lenders only have recourse to the project's facilities and to the main organisation.

Project finance is a key factor in the successful development of infrastructure projects. Project finance provides an alternative source of finance to cash-starved governments to provide infrastructure to sustain economic growth.

References

Esty, B. C. (2004) *Modern Project Finance*, Wiley.

Galitz, L. (1995) *Financial Engineering: Tools and Techniques to Manage Financial Risks*, McGraw Hill.

Merna, A. and Dubey, R. (1998) *Financial Engineering in the Procurement of Projects*, Asia Law and Practice.

Merna, A. and Khu, F. L. S. (2003) The allocation of financial instruments to project activity risks, *The Journal of Project Finance*, **8**.

Merna, A. and Owen, G. (1998) *Understanding the Private Finance Initiative*, Asia Law and Practice.

Merna, A. and Smith, N. J. (1996) *Guide to the Preparation and Evaluation of Build-Own-Operate-Transfer (BOOT) Project Tenders*, second edition, Asia Law and Practice.

Nevitt. P. K. (1983) *Project Finance*, fourth edition, Bank of America, Financial Services Division.

Office of Government Commerce (2002) *Private Finance Unit. OGC Guide On Certain Financing Issues In PFI Contracts*, OGC.

Further reading

Frank, M. and Merna, T. (2003) Portfolio Analysis for a Bundle of Projects, *The Journal of Structured and Project Finance*, **9**.

Marti, S. and Keith, L. (2000) Cash flow volatility as opportunity: Added sophisticated insurance capital to the project mix, *Journal of Project Finance*, **6**.

Merna, A. and Lamb, D. (2004) *The Guide to Value and Risk Management in PPP Projects*, Euromoney Books.

Merna, T. and Njiru, C. (2002) *Financing Infrastructure Projects*, Thomas Telford.

Merna, A. and Smith, N. J. (1999) Privately financed infrastructure for the 21st century, Proceedings of the Institution of Civil Engineers, Civil Engineering, November, Thomas Telford.

Chapter 7
Cost Estimating in Contracts and Projects

The purpose of this chapter is to introduce the theory, techniques and practical implications of cost estimating throughout the stages of the project cycle.

7.1 Cost estimating

The record of cost management in the engineering industry is not good. Many projects show massive cost and time overruns. These are frequently caused by underestimates rather than failures of project management.

Estimates of cost and time are prepared and revised at many stages throughout the development of a project (Figure 7.1). They are all predictions, the best approximation that can be made, and it would be unrealistic to expect them to be accurate in the accounting sense. The degree of realism and confidence achieved will depend on the level of definition of the work and the extent of risk and uncertainty.

The objective of any estimate, no matter at what stage it is produced, is to predict the most likely cost of the project. It is essential to recognise that there is a range of possible costs within which the most likely cost lies. If the limits for the possible range of project costs over its time scale is plotted, an envelope similar to that in Figure 7.2 is obtained.

The key points to note are.

- The range of possible distributions. Generally these show narrowing range and increasing certainty as the project progresses, but certainty is only achieved following settlement of the final accounts and auditing of all project-related costs.
- The range of possible cost is much greater during the early stages. This arises because there is little information available to the estimator both in terms of scope, organisation and time and cost data.
- Many risks are latent in a project at its earliest stages. These often go unrecognised because the project team is intent on looking ahead towards the project's completion rather than reviewing the whole of its evolution.
- Risk decreases over the life of a project. This may not be continuous: from time to time there may be increasing risks or new risks that arise during

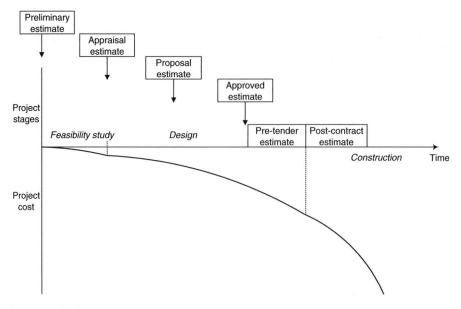

Figure 7.1 Estimates at each project stage.

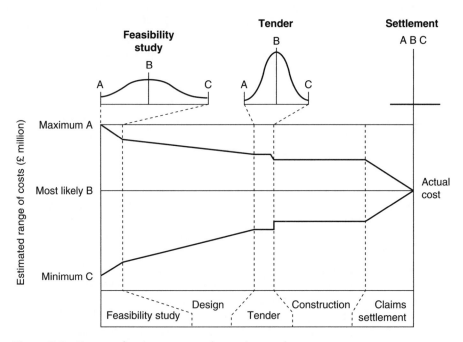

Figure 7.2 Range of estimates over the project cycle.

the project's development. Some of these are the result of failure to recognise their importance or presence earlier.

- The upper and lower bounds of the envelope are not absolute and in reality are not clearly defined.
- The estimated base cost plus contingencies that are likely to be spent must be close to the 'most likely' cost from the earliest stage if the common experience of underestimation is to be avoided. It follows therefore that the 'most likely' value is closer to the 'maximum' than the 'minimum' to allow for cost growth.

Ideally, any estimate should be presented as a most probable value, with a tolerance: a range of less likely values to emphasise that it is an estimate. Particular areas of risk and uncertainty should be noted and, if necessary, a specific contingency allowance or tolerance should be included in the estimate.

The requirements of an estimate are:

- to predict the most probable cost of the works;
- to predict the most probable programme for the works;
- to identify and quantify potential problems and risks;
- to state all assumptions and note all exclusions;
- to base a forecast of expenditure – the cash flow based on the project programme.

The most important estimates prepared are probably for a project, at sanction, and for a contract, at tender; for it is at these points that the promoter and contractor become committed.

7.2 Cost and price

Cost refers to the cost directly attributable to an element of work, including direct overheads, for example, supervision.

Price is the cost of an element of work, plus allowance for general overheads, insurances, taxes, finance (and interest charges) and profit (sometimes known as 'the contribution').

First, the most probable cost of the works must be ascertained. The traditional view in construction projects exemplified by the Schedule of Rates or Bill of Quantities (BoQ) approach is that costs are proportional to quantities. This is true only to a very limited extent.

In reality, the only truly quantity-proportional costs are those direct costs of materials in the permanent works, but care is required. For example, the cost of concrete may vary, depending upon whether it is site batched or not.

The cost of operating a hoist or crane is a time-related cost. It requires an operator, maintenance, fuel and lubricants when it is working, regardless of the volume of concrete it produces. The cost of setting up the hoist or crane and dismantling it are lump-sum start-and-finish, or on/off, costs.

Often time-related costs are dominant, but there are others that must be considered.

- Finance charges on money borrowed for the purchase of plant and equipment.
- Depreciation in the value of a piece of plant as it ages.

These are highest for new pieces of equipment and decrease steadily until the debt is repaid, or as in the second case, until the piece of equipment has only a residual or scrap value.

Time depends upon the output or productivity which can be obtained from the resources. Outputs and productivities are expressed in terms of a quantity of work in unit time, or time for a given quantity of work. Hence, the quantity of work to be done, say in the BoQ, is related to the level of resource that must be provided to carry out the work in the given time. If the time taken to carry out the work is longer, the cost will increase unless the provision of resource is reduced. It is important to realise that in reality major plant and equipment numbers and labour force sizes cannot be varied on a day-to-day basis; so the actual cost incurred is dependent on the provision of the resource, regardless of the quantity of work it performs.

The cost, therefore, of an element of engineering work comprises quantity-proportional material costs, time-related labour and plant and equipment costs and lump-sum costs. The latter may be for on/off costs for a particular item, or payment of a specialist subcontractor. These costs refer to the permanent works. Any temporary works must also be covered by the prices charged by the contractor (or subcontractor). Finally, the direct overhead costs of management, establishment and consumables that can be charged against a project can be assessed. These must then be spread over the other cost centres to arrive at the total cost to perform the work. The price of the work is derived from the cost. It is sometimes described as cost plus mark-up. This is a gross oversimplification.

There are general overheads incurred by an organisation that cannot be charged directly to a particular project, for example, administrative staff, senior management, cost of maintaining the head office buildings and insurances. The price must also cover payment of taxes, interest charges on monies borrowed and allowances or contingencies for risks and uncertainties. Finally, the contractor must make a profit after tax.

The amount of profit that can be made varies. Factors that affect it are the:

- type of work;
- size of project or contract;
- extent of competition;
- desire of the organisation to secure the work.

These may be grouped together very loosely as 'the state of the market'. It is possibly the factor influencing the price that is most difficult to assess, and experience, good judgement, knowledge of the market and not a little luck are required to make a realistic assessment.

Other factors that influence the overall cost and hence price of engineering works must not be overlooked, and these may include the:

- location;
- degree of innovation;
- type of contract;
- method of measurement;
- payment conditions;
- risk surrounding the project.

7.3 Importance of the early estimates

All estimates should be prepared with care. The first estimate that is published for review and approval has a particularly crucial role to play because:

- it is the basis for releasing funds for further studies and estimates;
- it becomes the marker against which subsequent estimates are compared.

There is further reason for the importance of early estimates, that is, the need to know, for the purposes of economic appraisal, the capital cost of the project. The capital cost may not be the most sensitive variable in such an appraisal, but it can be decisive in the key decision to proceed with a project or to invest the funds elsewhere.

The decision to sanction the project by its promoters should never be based on the very first estimate. Nevertheless, before significant monies are spent on developing designs and more detailed estimates, there is a need to have a realistic indication of the project's likely cost. By definition this must be done before a great deal of detailed information is available.

The earliest estimates are primarily quantifications of risks: there is little reliable data available to the estimator. To estimate effectively therefore requires the estimator not only have access to comprehensive historical data, be capable of choosing and applying the most appropriate technique, but also have, in conjunction with other members of the project team, the experience to make sound judgements regarding the levels of largely unquantifiable risk. For many projects, this will not be an onerous task, but for projects that are in any way unusual, it is essential that thorough exercises in risk identification, risk assessment and selection of the most appropriate responses are performed. Insofar as cost estimates are concerned, the latter is the quantification of allowances for uncertain items, specific contingencies and general estimating tolerances. In addition, the project's programme must be carefully reviewed and the costs of delay, the use of float or acceleration must be assessed and the appropriate contingencies included in the estimate.

It can be seen therefore that this can be a time-, resource- and cost-intensive exercise. It is not, however, required for all projects. The first step towards assessing the risk is to identify potentially high-risk projects. This can be achieved by considering a number of factors, regardless of the size, complexity, novelty or value of the particular project.

- The sensitivity of the promoter's business or economy to the outcome of the project in terms of the quality of its product, capital cost and timely completion.
- The need for new technology or the development of existing technology
- Any major physical or logistical restraints such as extreme ground conditions or access problems.
- A novel method of implementation.
- Large and/or extremely complex project.
- Any extreme time constraints
- Location.
- Parties involved – promoter, consultants and contractors – may be inexperienced.
- Sensitivity to regulatory changes.
- Developed or developing country.

Testing new projects against such criteria is the first step towards improving the realism of initial estimates.

If the result of the test outlined here is positive, that is, the project is inherently risky, further and detailed assessment must be performed of the potential sources of risk and their likely impact on the project.

There are two components to any estimate:

- the base estimate;
- contingency allowances that are required to cover the uncertainty in the base estimate.

Second, a list of the main sources of risk should be prepared: this may have as few as 5 items but should not exceed 15.

- Promoter.
- Host governments.
- Funding.
- Definition of project.
- Concept and design.
- Local conditions.
- Permanent installed plant (mechanical and electrical).
- Implementation.
- Logistics.
- Estimating data (time and cost).
- Inflation.

- Exchange rates.
- Force majeur.

It should be noted that these are both quantitative and qualitative. Following this, each source of risk should be developed into comprehensive lists of more specific, manageable and, where possible quantifiable, risk factors. It is useful at this stage to classify both the potential impact, major or minor, and probability of occurrence, high or low, as a guide to the most important factors, because the estimator's attention should be mainly directed at these high impact high probability risks.

7.4 Estimating techniques

Having distinguished between cost and price, it is necessary to consider in greater depth the estimating function. The five basic estimating techniques available to meet the project needs outlined earlier are summarised, together with the data required for their application, in Table 7.1.

Three of the techniques, global, factorial and unit rates, rely on historical cost and price data of various kinds. Comments on this aspect of each technique are given under the respective headings, but the associated risks are so important that it is worth making the following general warning points about the use of historical data in estimating:

(1) To obtain realistic estimates, the data must be from a sufficiently large sample of similar work in a similar location and constructed in similar circumstances.

(2) Cost data needs to be related to a specific base date. In the case of construction work carried out over a period of time, a 'mean' date has to be chosen, for example, two-thirds through the period. In the case of manufactured plant, the easiest date to determine will probably be the delivery date, whereas a 'mean' date during manufacture may be more relevant.

(3) Having selected the relevant base date, there remains the problem of updating the cost price data to the base date for the estimate. The only practical method is to use an inflation index, but there may not be a sufficiently specific index for the work in question. If there is not, recourse to general indices is usually made. In any event there is a limited length of time over which such updating has any credibility, particularly in times of high inflation and/or rapidly changing technology.

(4) Overlying the general effect of inflation is the influence of the 'market'. This will vary with the type of project being undertaken and, for international projects, with the host country and also with the supplying countries. The state of the market at the price date base will require careful consideration before historical data can be credibly applied to later or future dates.

Table 7.1 Data for the basic estimating techniques.

	Estimating technique				
	Global	Factorial	Man-hours	Unit rate	Operational
Project data required	Size/capacity Location Completion date	List of main installed plant items Location Completion date	Quantities Location Key dates Simple method statement Completion date	Bill of quantities (at least main items) Location Completion date	Materials quantities Method statement Programme Key dates Completion date
List of potential problems, risks, uncertainties and peculiarities of the project					
Basic estimate data required	Achieved overall costs of similar projects (adequately defined) Inflation indices Market trends General inflation forecasts	Established factorial estimating system Recent quotes for main plant items Inflation indices (for historical prices) Market trends General inflation forecasts	Hourly rates Productivities Overheads Materials costs Hourly rate forecasts Materials costs forecasts Plant data	Historical unit rates for similar work items Preliminaries Inflation indices Market trends General inflation forecasts	Labour rates and productivities Plant costs and productivities Materials costs Overhead costs Labour rate forecasts Materials costs forecasts Plant capital and operating costs forecasts

Source: Smith (1995).

Global

This term describes the 'broadest brush' category of technique, which relies on libraries of achieved costs of similar projects related to the overall size or capacity of the asset provided. This technique may also be known as 'order of magnitude', 'rule of thumb' or 'ballpark' estimating. Examples are:

- cost per square metre of building floor area or per cubic metre of building volume;
- cost per megawatt capacity of power stations;
- cost per metre/kilometre of roads/motorways;
- cost per tonne of output for process plants.

The technique relies entirely on historical data and therefore must be used in conjunction with inflation and a judgement of the trends in levels of prices, that is, market influence, to allow for the envisaged timing of the project.

The use of this type of 'rolled-up' historical data is beset with dangers, especially inflation, as outlined generally earlier; but more specifically, they are:

(1) Different definitions of what costs are included, for example,
   - Engineering fees and expenses by consultants/contractors/promoter, including design, construction supervision, procurement and commissioning
   - Final accounts of all contracts including settlements of claims and any 'ex gratia' payments
   - Land
   - Directly supplied plant and fittings
   - Transport costs
   - Financing costs
   - Taxes and duties
(2) Different definitions of measurement of the unit of capacity, for example,
   - Square metre of building floor area: measured inside or outside the external walls
   - Cubic metre of building volume: height measured from top of ground floor or top of foundations
   - A metre/kilometre of motorway an overall average including pro rata costs of interchanges or should these be estimated separately
   - The units included in associated infrastructure (e.g. transmission links/roads for power stations)
(3) Not comparing like with like, for example,
   - Differing standards of quality such as different runway pavement thicknesses for different levels of duty
   - Process plants on greenfield sites and on established sites
   - Different terrain and ground conditions – such as roads across flat plains compared with mountainous regions
   - Different logistics depending on site location

- Scope of work differences such as power stations with or without workshops and stores, housing compounds, and so on
- Item prices taken out of total contract prices (especially turnkey) may be distorted by front end loading to improve the contractor's cash flow especially for hard currency items

(4) Inflation
- Different cost base dates – as noted earlier, it is essential to record the 'mean' base date for the achieved cost and use appropriate indices to adjust to the forecast date required

(5) Market factors
- Competition for resources during periods of high activity
- Developing technology may influence unit costs

Many of these items are obvious, but for projects that do not have a continuous gestation period and possibly involve several different parties from time to time, it is essential that they are thoroughly checked. A scrutiny of all these dangers, especially the effects of inflation, must be made before any reliance can be placed on a collection of data of this type. It follows that the most reliable data banks are those maintained for a specific organisation where there is confidence in the management of the data. The wider the source of the data, the greater is the risk of differences in definition.

However, as long as care is taken in the choice of data, the global technique is probably as reliable as an over-hasty estimate assembled from more detailed unit rates drawn from separate, unrelated sources and applied to 'guesstimates' of quantities.

Factorial

These techniques are widely used for process plants, power stations and also where the core of the project consists of major items of plant that can specified relatively easily and current prices obtained for them from potential suppliers. The techniques provide factors for a comprehensive list of peripheral costs such as pipe work, electric, instruments, structure or foundations. The estimate for each peripheral will be the product of its factor and the estimate for the main plant items.

The technique does not require a detailed programme, but nevertheless one should be prepared to identify problems of implementation, lead times for equipment deliveries or planning approval, which will go undetected if the technique is applied in a purely arithmetical way.

The success of the technique depends to a large extent on four factors:

(1) The reliability of the budget prices for the main plant items. The estimator is still required to make a judgement on the value to include in his or her estimate depending on the state of the market and the firmness of the specification.

(2) The reliability of the factors, which should preferably be the result of long experience of similar projects by the estimator's organisation. During periods of significant design development, certain factors can change rapidly (e.g. instrumentation and controls systems).
(3) The experience of the estimator in the use of the technique and his or her ability to make relevant judgements.
(4) The adoption of the technique as a whole so that deficiencies in some areas will compensate for excesses in others.

The technique has the considerable advantage of being predominantly based on current costs, thereby taking account of market conditions and needing little, if any, reliance on inflation indices.

Factorial techniques are not normally reliable for site works, including most civil and building and mechanical and electrical installation work, except in a series of projects where the site circumstances are closely similar. In most overseas locations, the site works would need to be estimated separately using a more fundamental technique such as Operational.

Man-hours

Probably the original estimating technique is most suitable for labour-intensive implementation, construction and operations, such as fabrication and erection of piping, mechanical equipment, electrical installations and instrumentation – work where reliable records of productivity of different trades per man-hour (e.g. process plant construction and fabrication of offshore installations) exist. The total man-hours estimated for a given operation are then costed at the current labour rates and added to the costs of materials and equipment. The advantages of working in current costs are obtained.

The technique is similar to the operational technique. In practice, it is often used without a detailed programme on the assumption that the methods of construction will not vary from project to project. Experience has shown, however, that where they do vary (e.g. the capacity of heavy lifting equipment available in fabrication yards), labour productivities and, consequently, the total cost can be affected significantly. It is recommended that a detailed programme be prepared when using this technique to identify constraints peculiar to the project. The prediction of cash flow requires such a programme.

Unit rates

This technique is based on the traditional BoQ approach to pricing construction work. In its most detailed form, a bill will be available containing the quantities of work to be constructed, measured in accordance with an appropriate method of measurement. The estimator selects historical rates or prices for each item in the bill using either information from recent similar contracts, published information (e.g. price books for building or civil engineering), or 'built-up' rates from his or her analysis of the operations, plant and materials required for the measured item. As the technique relies on historical data, it is subject to the general dangers outlined earlier.

When a detailed bill is not available, quantities will be required for the main items of work, and these will be priced using 'rolled-up' rates, which take account of the associated minor items. Taken to an extreme, the cruder unit-rate estimates come into the area of global estimates as described earlier (e.g. unit rate per metre of motorway).

The technique is most appropriate for building and repetitive work where the allocation of costs to specific operations is reasonably well defined and operational risks are more manageable. Sophisticated methods of measurement for building have been developed in the UK, and internationally many contractors are able to tender by pricing bills of quantities using rates based directly on continuing experience. Nevertheless, it is essential that the rates are selected from an adequate sample of similar work with reasonably consistent levels of productivities and limited distortions arising from construction risks and uncertainties, for example, access problems.

It is less appropriate for civil engineering where the method of construction is more variable and where the uncertainties of ground conditions are more significant. It is also likely to be less successful for both civil engineering and building projects in locations where few similar projects have been completed in the past. Its success in this area depends much more on the experience of the estimator and his or her access to a well-understood data bank of relevant 'rolled-up' rates.

Unit rates quoted by contractors in their tenders are not necessarily directly related to the items of work they are pricing. It is common practice for a tenderer to distribute the monies included in the tender across the items in the bill to meet objectives such as cash flow and anticipated changes in volume of work, as well as to take some account of the bill item descriptions. It is unlikely that similar 'weighting' is easily carried out by all tenderers in an enquiry, and therefore it is not easily detected. It follows that tendered bill unit rates are not necessarily reliable guides to prices for the work described.

From the promoter's point of view, the technique does not demand an examination of the programme and method of construction, and the estimate is compiled by the direct application of historical 'prices'. It therefore does not encourage an analysis of the real costs of the work of the kind that would need to be undertaken by a tendering contractor for any but the simplest of jobs. Nor does it encourage consideration of the particular peculiarities, requirements, constraints and risks affecting the project.

There is a real danger that the precision and detail of the individual rates can generate a misplaced level of confidence in the figures. It must not be assumed that the previous work was of the same nature, carried out in identical conditions and with the same duration. The duration of the work will have a significant effect on the cost. Many construction costs are time related, as are the fees of supervisory staff, and all are affected by inflation.

It is therefore recommended that a programme embracing mobilisation and construction be prepared. This should be used to produce a check estimate in simplified operational form where there is any doubt about the realism of the unit rates available.

Despite its shortcomings, unit-rate estimating is probably the most frequently used technique. It can result in reliable estimates when practised by experienced estimators with good, intuitive judgement, access to a reliable, well-managed data bank of estimating data and the ability to assess the realistic programme and circumstances of the work.

Operational cost (resource cost)

This is the fundamental estimating technique as the total cost of the work is compiled from consideration of the constituent operations or activities revealed by the method statement and programme and from the accumulated demand for resources. Labour, plant and materials are costed at current rates. The advantage of working in current costs is obtained.

The most difficult data to obtain are the productivities of labour and plant and equipment in the geographical location of the project and especially the circumstances of the specific activity under consideration. Claimed outputs of plan are obtainable from suppliers, but these need to be reviewed in the light of actual experience. Labour productivities will vary from site to site depending on management, organisation, industrial relations, site conditions and other factors, and also from country to country. Collections of productivity information tend to be personal to the collector, and indeed this type of knowledge is a significant part of the 'know-how' of a contractor and will naturally be jealously guarded.

The operational technique is particularly valuable where there are significant uncertainties and risks. Because the technique exposes the basic sources of costs, the sensitivities of the estimate to alternative assumptions or methods can be investigated easily and the reasons for variations in cost appreciated. It also provides a detailed current cost and time basis for the application of inflation forecasts and hence the compilation of a project cash flow.

In particular, the operational technique for estimating holds the best chance of identifying risks of delay as it involves the preparation of a method of implementation and sequential programme including an appreciation of productivities. Sensitivity analyses can be carried out to determine the most vulnerable operations and appropriate allowances included. Action to reduce the effects of risks should be taken where possible.

It is the most reliable estimating technique for engineering work. Its execution is relatively time consuming and resource intensive compared with other techniques. However, estimating organisations geared up to this technique accumulate data in an operational form, which enables them to prepare even preliminary estimates with some appreciation of the more obvious risks, uncertainties and special circumstances of the project.

7.5 Suitability of estimating techniques to project stages

The objective should be to evolve a cost history of the project from inception to completion, with an estimated total cash cost at each stage close to the eventual outturn cost. This can be achieved if the rising level of definition is balanced by reducing tolerances and contingency allowances that represent uncertainty. Ideally each estimate should be directly comparable with its predecessor in a form suitable for cost monitoring during implementation and for a usable record in a cost data bank. This may, in practice, be difficult to achieve.

There is some correlation between the five estimating techniques that have been described and the estimating stages that have been defined, which is related to the level of detail available for estimating:

Preliminary. An initial estimate at the earliest possible stages where there is likely to be no design data available and only a crude indication of the project size or capacity and the estimate is likely to be of use in provisional planning of capital expenditure programmes.

At this stage, the global estimating technique can be used, which is a crude system that relies upon the existence of data for similar projects assessed purely on a single characteristic such as size, capacity or output. Widely used on process plants is the Factorial method, where the key components can be easily identified and priced and all other works are calculated as factors of these components.

Feasibility. Sometimes known as an appraisal estimate, these are directly comparable estimates of the alternative schemes under consideration. It should include all costs that will be charged against the project to provide the best estimate of anticipated total cost, and if it is to be used to update the initial figure in the forward budget, it must be escalated to a cash estimate.

A price can be described by the following equation:

$$\text{Price} = \text{Cost estimate} + \text{Risk} + \text{Overheads} + \text{Profit} + \text{Mark-up}$$

The size of the profit margin and the commercial decision making behind the selection of the percentage mark-up or mark-down are not discussed in detail here.

The basic cost estimate is the largest of all these elements, often accounting for more than 90% of the total price. Usually the basic cost estimate is derived from the unit rate or operational assessment of the labour, plant, materials and subcontract work required. Quotations are required for materials and subcontractors. Typically materials account for between 30% and 60% of a project's value, and sub-contractors can typically account for between 20% and 40%.

The cost of the company's own labour is usually calculated either per hour, per shift or per week. The cost to the company of employing their own labour is greater than that paid directly to the employee. The elements in the calculation could include such factors as plus rates for additional duties, tool money, travel monies, national insurance, training levies, employers and public liability insurance and allowances for supervision.

Erection or implementation. The plant may be obtained for a contract either internally or externally. Quotations for the plant required are therefore obtained from external hirers or from the company's own plant department. It is rare for the UK contractors in the domestic market to calculate the all-in plant rate from first principles. This calculation is usually undertaken by the plant hire company.

Overheads (or on-costs) for the project could include allowances for site management and supervision, clerical staff and general employees, accommodation, general items and sundry requirements.

Design. Estimate for the selected scheme. Usually evolves from a conceptual design until immediate pre-tender definitive design is completed.

A man-hours method is most suitable for labour-intensive operations, like design, maintenance or mechanical erection, and work is estimated in total man-hours and costed in conjunction with plant and material costs.

The decisions made in design on the size, quality and complexity of a project have the greatest influence on the final capital costs of a project. As a design develops, more and more capital cost is committed on behalf of the promoter until, at tender stage, with the design complete, or virtually complete, the promoter is committed to a high percentage of the tender value. Unless a redesign is undertaken, with the consequent loss of fees and time, the ability to save money while maintaining the original design concept is very limited. The design budget estimate should confirm the appraisal estimate and set the cost limit for the capital cost of the project.

Implementation or construction. A further refinement to reflect the prices in the contract awarded. This would require some redistribution of the money, for example, in the BoQ in a unit-rate contract and assist more efficient management of the contract. The Unit Rate method is a technique based on the traditional BoQ approach where the quantities of work are defined and measured in accordance with a standard method of measurement.

7.6 Estimating for process plants

The most significant difference between estimating for a process plant project and estimating for other types of engineering project is that the base cost of the process estimate is derived from material and equipment suppliers, plant vendors, specialist contractors and subcontractors. These components commonly account for about 80% of the total cost of the project. Consequently, the project estimator must ensure that firm quotations, together with guaranteed delivery dates and installation schedules, are confirmed with all suppliers and subcontractors.

The main process plant contractor, the engineering contractor, usually carries out the design, procurement and management functions, which accounts for most of the remaining cost, about 20%. The first task is for the engineering contractor to estimate his own base costs as accurately as possible for the work to be undertaken at the detailed design, procurement, project management and site supervision stages of the project. Typically the erection of process plants is labour intensive and hence this assessment is made using the Man-Hours estimating technique.

The engineer contractor's control philosophy must be aimed so as to minimise change at the detailed engineering phase. The initial global estimate can be made by comparative means with similar types of plants based on throughput. The estimate will have considered the principal quantities of the work, the items to be subcontracted, the materials and plant for which quotes are required, critical dates for actions by subcontractors and suppliers and whether any design alternatives should be pursued.

At this stage the effects of layout and location are considered by the preparation of the base or net cost estimate. Costs will be established for project and engineering services, which is usually done in-house, and any remaining quotes are substantiated and confirmed. It is important to have a good definition of the scheme, including general arrangement, piping and instrument diagrams, equipment lists and material specifications.

In the production of a net cost, allowances and contingencies need to be considered. Allowances can be defined as costs added to individual estimated costs to compensate for a known shortfall in data or to provide for anticipated developments. It is often the case that, when firm quotes are replaced with contracts and orders after the successful tender bid, more detail is available than was the case earlier. Contingency is an adjustment to the net cost and has to be considered no matter how much detailed work has been completed during the preparation of the estimate as there will undoubtedly be uncertainties that will affect the final cost. Contingency adjustments allow for the unknown element and also for any factors outside the control of the engineer contractor, which are perceived as likely to affect the final cost.

The cost of preparing a fixed-price estimate can be as high as 1–3% of total cost, compared with 0.5–1% for building and civil engineering work, and can take up to 6 months to prepare. Typically, a 3-month tender period for most major process plant projects is allowed by most promoters, which is really too short to enable all the requirements of the estimate to be completed.

7.7 Information technology in estimating

The estimating techniques described were developed mainly as manual methods. The development of computer hardware, dedicated computer software packages and increased computer literacy amongst the professionals engaged in project management and estimating has resulted in the application of information technology (IT) tools to facilitate and assist in the estimating process.

As in manual estimating, the role of computer estimating will depend on the user's requirements. Those of a promoter at the design stage will be very different from those of a contractor at tender. The data available for the production of an estimate will also be different. This section does not contain details of specific hardware or software products, which are well marketed and are almost continually updated and improved, rather it focuses on some of the main functions that IT can adopt in this process.

IT is concerned with the storage, transfer and retrieval of data, manipulation and calculation using data, and presentation of output. The degree of complexity will vary as will the operation of the user interface, ranging from data libraries to fully automated expert systems. Most software estimating packages incorporate a data library for storing data, a range of methods for manipulating direct costs to produce a price, the ability to update or alter any of the input data and appropriate reporting formats.

The software packages attempt to be user-friendly by permitting data input or price build-up by the estimator to closely resemble the normal way of working. The estimator has to spend less time performing routine calculations and arithmetical checks while the basic method remains unchanged, but has more time to use judgement and experience where appropriate. It is important to note that a computer-based estimate is only as accurate as the input data and that the use of a computer in itself will not necessarily increase the accuracy of estimates. Indeed if the data library is not kept up to date or if it is applied without careful thought, then the estimate produced might suffer from a decrease in accuracy.

The application of IT cannot replace the role of the estimator in the project, nor is it intended to do so. Software packages are readily available to assist with most types of estimating technique, but there are also other advantages for the project manager. One of the recognised sources of errors and misunderstanding in project management has been identified as the linking interface between feasibility study, project estimate and detailed planning. Often these stages were tackled by different people using different estimating methods with associated differing assumptions and where the previous estimate was usually abandoned once that phase of the project had been completed. Current estimating software packages can be related to a single database and also be linked with other computer programs. This allows all the work done at feasibility stage to be refined, modified and developed as the project progresses, hence retaining all the information within the system. Typical interfaces relate to measurement and valuation, to BoQs, to the valuation of variations and to delays and disruptions. An estimating package can be linked with a planning package, and data is automatically transferred, facilitating the understanding of the relationship between time and money for the project.

Expert systems are attracting considerable interest as potential aids to decision making. In an advisory role they could provide the necessary assistance to produce the cheapest cost estimate by allowing automated decision making to select resources to match different work loads. The 'expert' approach purports to provide tools for the effective representation and use of knowledge developed from experience enabling the optimisation of resource selection. A computer simulation then begins with a preliminary questions module in which the user responds to a set of questions posed by the system. Estimating systems are evolving within an integrated environment as a decision making or decision support.

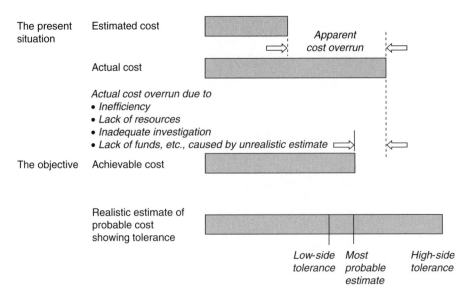

Figure 7.3   Estimating performance and objectives.

7.8 Realism of estimates

The use of the word realism in this context, rather than on accuracy, is important. As noted earlier, estimates are not accurate in the accounting sense and the makeup of the total must be expected to change. The realism of estimates will depend greatly on the nature and location of the work, the level of definition of the project or contract, and particularly on the extent of the residual risk and uncertainty at the time, as discussed earlier. Studies have determined that a standard deviation of 7% was common for contractors' bid estimates in the UK process plant industry, but performance of particular companies varied from 4% to 15%. The ranges of accuracy for high-risk projects, and in particular development projects, may be much greater.

The variation on individual items within the estimate is much greater, and consequently any system of cost control based on comparison of actual and estimated cost must be of dubious value. It is preferable to use performance measurement systems based on the integration of cost and programme. These systems are based on the concept of earned value.

Their simulation studies suggested that improvement in estimating accuracy will produce a corresponding improvement in company performance.

Many estimating problems can be addressed by adopting:

- a structured approach;
- choice of the appropriate technique;
- use of the most reliable data;
- consideration of the risks.

The improvement in establishing performance that should be obtained is illustrated in Figure 7.3.

Reference

Smith, N. J. (ed.) (1995) *Project Cost Estimating*, Engineering Management Guide Series, Thomas Telford Services Ltd.

Further reading

Bower, D., Thompson, P. A., McGowan, P. and Horner, R. M. W. (1993) Integrating project time with cost and price data, Developments in Civil and Construction Engineering Computing, The 5th International Conference on Civil and Structural Engineering Computing.
Thompson, P. A. and Perry, J. G. (eds) (1992) *Engineering Construction Risks – A Guide to Project Risk Analysis and Risk Management*, Thomas Telford.

Chapter 8
Programme Management

Programme management is a value-adding business function that interfaces strategic management and project management to provide sustained organisation-wide capabilities and benefits over time. It provides a structured framework or 'gestalt' supporting organisational learning and the core business by grouping, ranking, allocating resources to and coordinating projects as vehicles for change.

This chapter provides an overview of strategic management and the link with programme management. Subsequently, by exploring alternative definitions of programme management the chapter locates its position within the overall organisational structure between strategic and project management. Interfaces between strategy, programmes and projects are explored with a particular emphasis on programme management. The factors involved with developing a programme are discussed with special prominence being accorded to planning and control, benefits, ranking and value. The chapter concludes with a consolidated overview of programme management and its relationship to procurement strategies.

8.1 Strategic management, managing change and programme management

The intention here is only to provide a very brief overview of the concept of strategic management, concentrating on its interface with programme management.

The strategic management process answers questions about what the organisation ought to be doing and why and where should it be going and why. Strategic management also involves making choices, that is, decisions and managing change. An organisation's strategic decisions are likely to be long term in nature and with distant horizons. Distant horizons equate to uncertainty, and therefore there is the need for flexible or adaptable solutions. Other characteristics of strategic decisions include the need for integration between conflicting objectives and divisions within the organisation and often the need for radical and possibly unpopular actions. Change, forming part of the strategic management process, is inherent in business organisations.

Change can be classified into two major and three contingent types.

Recurrent change is incremental and routine, and, requires no major realignment of the organisation with its external environment. Operational change, which occurs at the lower levels of an organisation through its day-to-day activities, is a form of recurrent change.

Transformational change creates a fundamental shift between the organisation and its external environment. Transformational change can comprise strategic or competitive change. Strategic change is immediate, fundamental, radical and discontinuous. Most managers in an organisation are unlikely to see it coming. It affects the organisation from top to bottom. Competitive change creates a fundamental shift between the organisation and its environment in the medium to longer term. This will normally be felt at a sustained, deep-seated and continuous pressure on the firm to readjust its activities.

The role of an organisation's executive, for example, the board of directors, trustees or governors, is to manage change and place the organisation at the optimum position within its environment. That will first require deciding where that position is. 'Placing' implies a forward position, action and change. Establishing where or what that forward position is requires a strategic decision. A strategy is a plan – a way of doing things and as such strategy is pervasive. A plan could quite feasibly be to maintain a status quo, or it could be a vision of change. To be realistic and capable of achievement, the strategy must be matched to available resources and so the plan or scope of activity is constrained within a boundary. Strategic management can therefore be described as 'defining a future position and matching resources to that vision'.

The process of creating a strategy is generally as follows:

(1) Investigate the situation – define the decision to be made.
(2) Develop alternative decisions – to ensure the right problem is being addressed.
(3) Evaluate alternative decisions – options appraisal.
(4) Select.
(5) Implementation and follow up.

This strategy creation process can be simply modelled as shown in Figure 8.1.

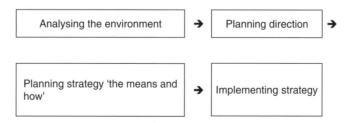

Figure 8.1   A simple model of the strategy creation process.

Items 1 and 2 in the preceding list are normally termed 'strategy formulation'; items 3 and 4 are termed 'strategic choice', while item 5 is termed 'strategic implementation'. It has been recognised for many years that implementation is frequently the necropolis of strategy. Strategy formulation tends to dominate over implementation.

Projects are an outcome of the strategic management process; they are experienced through strategic implementation. The increased use of projects has also brought the need to marshal project activity in some coherent and beneficial way. Programme management offers a structured, integrated and central approach to project selection and resource allocation so that the aims and objectives of the organisation as a whole can be balanced. Programme management is, therefore, a business function, providing information, not data, and is designed to support the core business of an organisation. As a business function it must have a demonstrable cost and time benefits. Programme management can also be viewed as providing the link between strategic and project management. It has developed in response to the widespread use of projects as a means of realising strategic change. As a link, its role is two directional. First, it assists the strategic decision-making process and second, it delivers the changes necessary. Its central position within the organisation enables it to ensure that the strategic delivery of projects is consistent with one another, and their own strategies. This means that the right projects must be selected, appropriate scarce resources allocated as and when necessary and that appropriate monitoring and control procedures are established.

8.2 Basics of programme management

Programme management provides an umbrella under which several projects can be coordinated. It does not replace project management but rather it is a supplementary framework. Managed in this way, projects are more likely to be driven by business needs rather than personal agendas, the chance of duplication is reduced and interdependencies become more explicit and recognisable.

Part of the role of programme management is to contribute to that strategic decision-making process. By merging the boundaries between 'strategy' and 'programme' the relevance and importance of this interface, as the first step towards effecting change, becomes apparent and should not be undervalued. The need for this interface, as a communication conduit, is essential given the volatile, responsive nature of some strategic decisions (e.g. the reaction to an OPEC decision or major corporate bankruptcies) on the one hand and the relatively long lead-in times of major projects and the complexity of their interdependencies on the other. Strategic decisions are by definition broad-brush, based on often incomplete or ambiguous information. The potential for changes to strategic requirements during the project development phase could be quite considerable. A permanent channel for communicating changed strategic requirements through to the implementation of projects is therefore essential.

It is argued here that it is not enough to just develop a strategy without considering its implementation. The interface between strategic management and programme management facilitates an opportunity for senior programme managers to advise on the feasibility of implementation. Equally, programme management may also influence the creation of strategy. By grouping together several small construction projects, for example, and adding in their facilities management requirements, a strategic decision to adopt a private finance initiative (PFI)-type procurement route could be made. In such a case it could be said that Programme Management had influenced, along this interface, a change in financial strategy. That example serves also to show another important face of programme management, which is its ability to add value or benefit to the organisation as a whole.

Although strategic management can be considered superordinate to programme management in that it provides a framework for the latter, it is not superior in the sense of being more important per se. It is rather that they have different roles to play in the same organisation and they both complement one another. Having introduced programme management through its interface with strategic management, the next section examines in more detail the definitions, nature and scope of programme management.

8.3 Defining programme management, its nature and scope

Programme management is not a new concept. It was probably first consciously used to deliver a defence initiative following the Second World War that became known as the Polaris Missile Programme. While network analysis and, largely, international peace have also been attributed to the Polaris Programme, sadly the benefits of programme management have been largely overlooked since that time. Reasons might include that the early developments in programme management were in the defence industries, areas that are often of a sensitive or secret nature and their functions are essentially non-economic. Nevertheless, the need for an effective interface between strategy and projects has continued to exist, albeit largely unrecognised. However, with the current advent of several public sector initiatives and major infrastructure investments, the need for a distinct 'discipline' that can accommodate flexible management structures has been highlighted, capable of responding efficiently to uncertainty and multiple goals has once more been recognised.

Project management is now a relatively mature discipline, having developed many excellent tools for the delivery of predefined objectives. It is concerned with the delivery of a unique piece of work. Programme management is extensively interactive with strategic management and individual projects and is instrumental in adding value to the organisation. It is the role of the organisation's strategic management team to identify and articulate the need for change. Projects are the instrument of change, a conduit through which the organisation's desires to improve, expand, adjust, etc., are delivered or converted into reality. The need for programme management becomes apparent when the scope of change is so extensive as to require delivery through several projects over protracted periods. The position of programme management in the organisation is between strategy formulation and delivery. Its ethos is a hybrid of strategic management and project management. It is a permanent, pivotal function in any changing or developing organisation providing full-circle vision across the boundaries of strategy development and project delivery.

Unfortunately, there is no convenient dictionary definition of the word 'programme' for use within the context of business management. However, several of the limited number of writers on the subject of programme management provide insights into this emerging discipline. They have attempted to remedy that situation by contributing their own definition or description, which include:

- the process of co-ordinating the management, support and setting of priorities on individual projects, to deliver additional benefits to meet changing business needs ... a portfolio of projects which are managed in a co-ordinated way to deliver benefits which would not be possible were the projects managed independently' (Turner and Speiser, 1992);
- a grouping of projects, either for purposes of co-ordinated management or simply as a hypothetical construct to facilitate aggregate reporting at the strategic level' (Gray, 1997);
- the co-ordinated management of a portfolio of projects that change organisations to achieve benefits that are of strategic importance' (OGC, 1999);
- a collection of projects related to some extent to a common objective' (APM, 2000);
- a framework for grouping projects and for focusing all the activities required to achieve a set of major benefits' (Pellegrinelli, 1997).

While there has not yet emerged an unambiguous and universally accepted definition of programme management, this should not be seen in a negative light. In fact the opposite is more appropriate – unencumbered by definition, programme management is free to grow, innovate and develop to find is own level and identity.

It is tempting to consider large or multi-projects as programmes because there is an existing and extensive body of knowledge and experience covering projects and project management. Conversely, there is an absence of a coherent and widely recognised body of knowledge surrounding programmes and programme management. The difference between a 'programme' and a 'multi-project' can be illustrated by comparing the Polaris Missile 'Programme' and an Olympic Games stadium 'project' by looking at their respective time and strategic objectives/benefits aspects. Polaris was to be completed 'urgently', there was no defined date. An Olympic stadium has to be completed by a fixed time. Polaris was designed to benefit the whole of the free world over a long period; a stadium's benefit can be said to be more parochial. A stadium could serve a limited population and alternative arrangements are conceivable, albeit inconvenient.

However, there was no option of transfer for a failed Polaris programme. A stadium's total funding arrangements are put in place before work starts. Polaris was funded by annual budget allocation.

The danger of attempting to employ project management techniques to the management of a programme is in the level of detail. This will destroy the inherent flexibility that programmes offer and more importantly ignore their involvement with the organisation's strategic management. Thus, programme management supports both strategy and projects. There are three parts to programme management:

- selecting projects;
- assigning priorities to projects;
- coordinating those projects by managing their interfaces.

The first two parts are clearly the more important and require a knowledge of and interaction with strategic management. A programme will provide a capability, that is, a benefit, whereas a project (regardless of size or complexity) is only a process and enables a benefit to be obtained. A project has (or should have) a clearly defined objective that enables a benefit to flow. A programme will deliver a benefit. Programme management also provides a repeat business/project framework, or a way of thinking, that will absorb and retain the benefits of organisational learning. Among other things, that framework brings together related projects and maintains a strategic view over them, aligns and coordinates them with a programme of business change.

Having identified the required projects, it will then be necessary to implement them. That is of course the role of project management, but under the overall control and direction of programme management.

8.4 Project ranking within a programme

Selection of the projects to be included in that programme is an exercise of major importance and the process of benefits appraisal, ranking and prioritising achieves that aim, the subject of subsequent sections.

However identified, any range of beneficial and feasible projects will need to be scheduled into a rank order. This process should not be confused with prioritising, which is concerned with the timing of activities within a programme or the allocation of resources. Ranking involves placing the projects into a hierarchy, reflecting the effect they will have on or contribute to delivery of the organisation's objectives, creating a new value activity or enhancing an existing value activity. If, for example, a car manufacturer wanted to introduce a new model and it had been decided that the optimum way to do so was via a purpose built factory, that project would probably be placed higher in the hierarchy than re-roofing the existing factory. Clearly, without a clear strategic direction, it is impossible to establish a rank order. Without such direction, the usual result is to keep the options open or do nothing. The rank order within the programme will be subject to change through circumstances beyond the programme manager's control. In such a circumstance it will be necessary to review, and, possibly realign priorities. Human resource opposition, discontent and demotivation will invariably accompany this action at least, which the programme manager must anticipate and attempt to cater for.

A project's rank may be considered its key to success. A highly ranked project will probably attract more resources, thus making its success more probable, provided it is feasible. It is also important to know how a ranking, be it high or low, is assigned and often organisational politics will creep into this process. Unfortunately, many projects are considered to have failed because they do not achieve what was anticipated of them. Reasons for this could include:

- lack of clearly defined objectives;
- unrealistic or over-optimistic objectives;
- driven by technology rather than business need.

There are four ways by which a project may secure preference.

Differentiation – establishing unchallengeable claims on valuable resources by distinguishing an organisation's own products from those of competitors, that is, convince others that only their project will satisfy the need.

Co-optation – attempting to absorb new elements into the decision-making structure as a means of averting threats to organisational stability or existence. This can involve either the claiming of power or simply sharing the burden of power.

Moderation – attempting to build long-term support by sacrificing short-term goals. The key to this strategy is an ability to estimate and compare short- and long-term gains and losses.

Managerial innovation – an attempt to achieve autonomy in the direction of a complex and risky project through the introduction of management techniques that appear to indicate unique management competence. Often seen as high-level intervention, giving the impression of the project being important. However, this can often stifle innovation from lower down in the organisational hierarchy.

There is no formula that will uniquely identify those projects deserving a high ranking, but some formal approach will form the integrating factor in the array of choice. The evaluation framework must itself be a part of the strategic plan and be flexible enough to allow pre-selection of techniques and methodologies as appropriate. Also, programme management efficiency can be improved by improved procedures for project evaluation, that is, a requirement for a rigorous and competitive analysis of the technical problems and benefits likely to be produced and their cost implications. The knowledge that there exists competition to secure a superior ranking will force competing project sponsors to examine their case and arguments more carefully. Finally, there is a likelihood that there may be many proposals to select from and a screening process is required to eliminate early on any projects that are clearly deficient against preordained criteria, for example, a minimum rate of internal rate of return.

Generally speaking, the selection criteria should include the questions that focus on whether or not the project:

- takes advantage of an organisational strength;
- avoids dependence on a known organisational weakness;
- offers an opportunity to gain competitive advantage;
- contributes to internal consistency;
- presents an acceptable level of risk.

Finally, it should be noted that even if an individual project delivers according to its plan, it does not necessarily mean that the programme has added value.

The project-ranking process provides an excellent opportunity to implement the techniques of value management to give a programme or an individual project the best chance to add value. Value management is a structured, systematic, challenging, team-based process that keeps function and purpose to the fore and attempts to make explicit the package of whole life benefits an organisation is seeking from a service, project, programme or product. It is not the intention of the chapter to discuss value management in detail; it has also been mentioned in other chapters.

The process of strategic management and its link with projects and programme management is demonstrated in Figure 8.2. Programme management has been defined in this chapter as an integrating business function to support the core business of an organisation. As a business function, it must have a demonstrable cost,

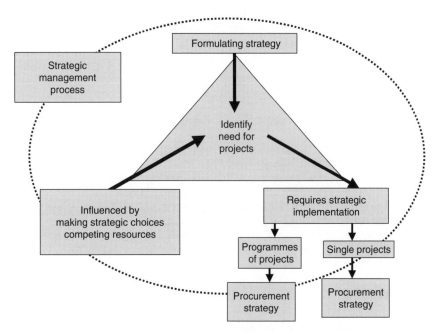

Figure 8.2 The link between strategy and programme management.

quality and time benefits, and developed in response to the widespread use of projects as a means of realising strategic change. To be effective, programme management requires the right projects to be managed, selected, coordinated and appropriate scarce resources allocated as and when necessary. It also requires appropriate monitoring and control procedures. It has already been stated that programme management's position in the organisation is between strategy formulation and delivery. As such, it has interfaces, distinct links, or communication conduits between strategic management in one direction and project management in the opposite direction.

8.5 Summary

This chapter has demonstrated that, while programme management is not a new concept, it is an extremely useful value-adding addition to any organisation contemplating change. The absence of an accepted definition of this management function is considered a powerful strength, allowing its future, natural development. Programme management contributes to strategic and project management. Both of these managerial functions have been reviewed, but more importantly, the interfaces between them have been explored. It is by the effective management of those interfaces that considerable efficiency gains can be made. There is also scope for improvement provided by the umbrella approach of programme management. It provides an effective 'gestalt' for managing projects, as resources for implementing change, in a coordinated way to deliver business benefits. Programme management is a value-adding business function, responsible in part for the selection of projects, potentially using the application of value management tools to assist with this task.

References

APM (2000) *Association for Project Management Body of Knowledge*, fourth edition, APM.

Gray, R. J. (1997) Alternative approaches to programme management, *International Journal of Project Management*, **15**.

OGC (Office of Government Commerce) (1999) *Managing Successful Programmes*, HMSO.

Pellegrinelli, S. (1997) Programme management: organising project-based change, *International Journal of Project Management*, **15**.

Turner, J. R. and Speiser, A. (1992) Programme management and its information systems requirements, *International Journal of Project Management*, **10**.

Further reading

APM (2006) *Association for Project Management Body of Knowledge*, fifth edition, APM.

Clark, P. (1990) Social technology and structure. In Hillebrandt, P. M. and Cannon J. (eds), *The Management of the Modern Construction Firm: Aspects of Theory*, Palgrave Macmillan.

Cleland, D. I. and King, W. R. (1988) *Project Management Handbook*, second edition, Van Nostrand Reinhold.

Ferns (1991) Developments in programme management, *International Journal of Project Management*, 9.

Grundy, T. (1998) Strategy implementation and project management, *International Journal of Project Management*, 16, 43–50.

Johnson, G. and Scholes, K. (2006) *Exploring Corporate Strategy*, seventh edition, Longman.

Langford, D. and Male, S. P. (2001) *Strategic Management in Construction*, Blackwell Science.

Lansley, P., Quince, T. and Lea, E. (1979) *Flexibility and Efficiency in Construction Management*, Final Report, Building Industry Group.

Payne, J. H. and Turner, J. R. (1999) Company-wide project management: the planning and control of programmes of projects of different types, *International Journal of Project Management*, 17.

Platje, A. and Seidel, H. (1993) Breakthrough in multiproject management: how to escape the vicious circle of planning and control, *International Journal of Project Management*, 11.

Turner, J. R. (1993) *The Handbook of Project Management*, McGraw Hill.

Chapter 9
Planning

This chapter reviews programming and planning techniques for design, procurement, construction and commissioning of projects. The role of information technology is examined together with suggestions on the identification and selection of appropriate software packages.

9.1 Planning

The successful realisation of a project will depend on careful and continuous planning. The activities of designers, manufacturers, suppliers, contractors and all their resources must be organised and integrated to meet the objectives set by the promoter and/or the contractor.

The purposes of planning are to persuade people to perform tasks before they delay the operations of other groups of people, and in such a sequence that the best use is made of available resources, and to provide a framework for decision making in the event of change. Assumptions are invariably made as a plan is developed: these should be clearly stated so that everyone using the plan is aware of the limitations on its validity. Programmes are essentially two-dimensional graphs and in many cases are used as the initial, and sometimes the only, planning technique.

Packages of work, usually referred to as 'activities' or 'tasks', are determined by consideration of the type of work, the location of the work or by any restraints on the continuity of the activity. Activities consume 'resources' which are the productive aspects of the project and usually include the organisation and utilisation of people (labour), equipment (plant) and raw material. Sequences of activities will be linked on a time scale to ensure that priorities are identified and that efficient use is made of expensive and/or scarce resources, within the physical constraints affecting the job.

A degree of change and uncertainty is inherent in engineering and it should be expected that a plan will change. It must therefore be capable of being updated quickly and regularly if it is to remain a guide to the most efficient way of completing the project. The plan should be simple, so that updating is straightforward and does not demand the feedback of large amounts of data, and flexible, so that all alternative courses of action can be considered. This may be achieved either by allocating additional resources or by introducing a greater element of float and extending the contract duration, as necessary. In either case, the estimated cost will increase; hence, it is essential to link the programme with the cost forecast.

It is difficult to enforce a plan that is conceived in isolation, and it is therefore essential to involve the people responsible for the constituent operations in the development of the plan. The plan must not impose excessive restraint on the other members of the organisation: it should provide a flexible framework within which they can exercise their own initiative. The plan must precipitate action and must therefore be available in advance of the task.

9.2 Programming

Programmes are required at various stages in the contract; when considering feasibility or sanction, at the pre-contract stage and during the contract. They may be used for initial budget control or for day-to-day implementation work. They may pertain to one contract or to a number of contracts in one large project.

The planner must therefore decide on the appropriate level of detail for the programme and the choice of programming technique. Important factors in the choice include the purpose of the programme, the relevant level of management and the level of detailed required. Simplicity and flexibility are the keys; a programme of 100 activities is easy to comprehend, whereas a programme of 1000 activities is not. Often it is good practice to ensure that the number of individual activities should relate closely to the basic packages of work required or to the cost centres defined in the estimate and should all have durations of a similar order of magnitude.

The period of time necessary to execute the work of an activity, the 'duration', depends on the level of resources allocated to the activity, the output of those resources and the quantity of work. The duration may also depend on other outside restraints, such as the specified completion date for the whole or some part of the work, the delivery date for specific material or restrictions on access to parts of the works.

A number of common forms of programme used in engineering project management are reviewed as follows:

Bar charts. The most common form of plan is the bar chart, also known as the Gantt chart; an example is shown in Figure 9.1. Each activity shown in its scheduled position to give efficient use of the resources; the logic and float are shown by dotted lines and important constraints or key dates are clearly marked. The space within the bars can be used for figures of output or plant costs, and there is room beneath to mark actual progress. Frequently it is useful to plot a period-by-period histogram of the demand for key resources directly under the bars at the bottom edge of the programme.

Line of balance. This simple technique was developed for house building and is useful for any repetitive type of work. The axes are the number of completed units and time: the work of each gang appears as an inclined line, the inclination being related to the output of the gang, as shown in Figure 9.2.

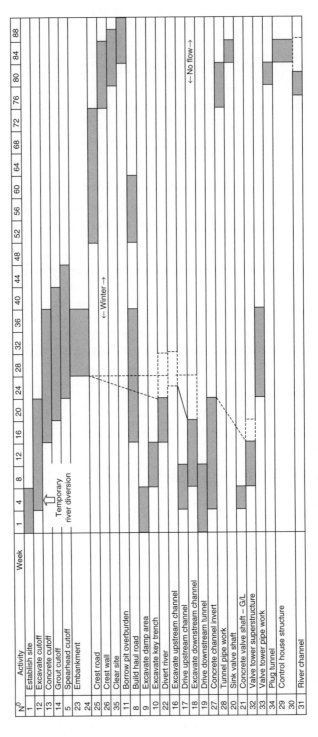

Figure 9.1 Construction programme in bar chart form.

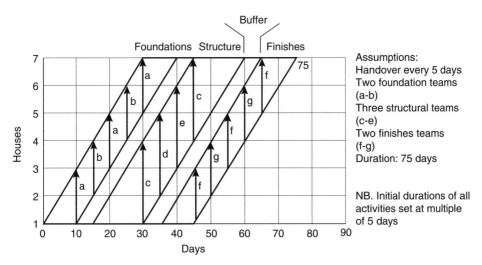

Figure 9.2 Line of balance.

Location–time diagram. In cross-country jobs, such as major road works, the erection of transmission lines, or pipe laying, the performance of individual activities will be greatly affected by their location and the various physical conditions encountered. Restricted access to the works, the relative positions of cuttings and embankments, sources of materials from quarries and temporary borrow pits, the need to provide temporary or permanent crossings for watercourses, roads and railways, and the nature of the ground will all influence the continuity of construction work and the output achieved by similar resources of men and machines working in different locations (Figure 9.3).

9.3 Network analysis

There are two basic forms of network analysis techniques: precedence diagrams, sometimes called activity-on-node networks; and arrow diagrams, sometimes called activity-on-line networks.

Although both methods will achieve the same answer, the precedence diagram is recommended by the authors. The advantages of the precedence system over arrow diagrams may be listed as follows:

- flexibility, logic is defined in two stages;
- dummy activities are eliminated;
- revision and introduction of new activities are simple;
- overlapping and delaying of activities is easily defined;
- use of preprinted node sheets possible.

The network diagram, Figure 9.4, resembles a flowchart: activities being represented by egg-shaped nodes and the inter-relationships between activities by

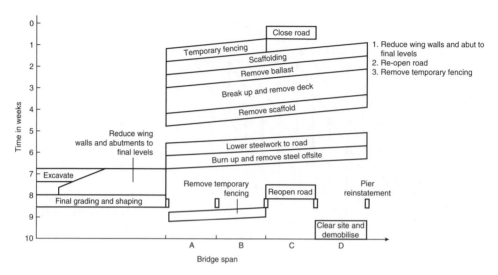

Figure 9.3   Time–location diagram.

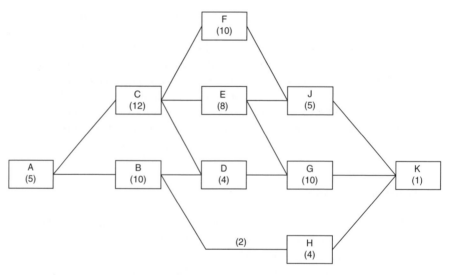

Figure 9.4   Precedence diagram: worked example.

lines known as dependencies. The dependencies are developed by moving to each activity in turn, asking 'What can start once this activity has started?' and drawing the relevant lines. The convention is that dependencies run from nose to tail of succeeding activities that is from the finish of one activity to the start of the succeeding activity. Figure 9.4 relates to the following worked example.

Apart from the 'critical path' a degree of choice exists in the timing of the other activities – a characteristic called 'float'. It is the float that will be later utilised to adjust the timing of activities to obtain the best possible use of resources. The 'total float' associated with an activity is the difference between its earliest and latest starts or finishes.

'Free float' is the minimum difference between the earliest finish time of that activity and the earliest start time of a succeeding activity. Total float is a measure of the maximum adjustment that may be made to the timing of an activity without extending the overall duration of the project: free float is that part of the total float that can be used without affecting subsequent activities.

To illustrate the techniques needed to perform the precedence diagram critical path method, a worked example based on a simple 10-activity network is described in the following section.

Worked example

Table 9.1 shows the durations and interdependence of 10 activities required to carry out a project:

(1) Construct a precedence diagram assuming no resource restrictions and calculate the minimum duration of the project.
(2) Schedule the earliest, latest start and finish for each activity and show the critical path.
(3) If activity F is extended to a duration of 15 days, what is the effect on the critical path?

Solution

(1) Start with the first activity A, at the left-hand edge in the centre of the preprinted sheet. As a start activity it has only successor activities. Using the logic in Table 9.1, the start of activities B and C is dependent upon the completion of activity A; so these activities are plotted immediately adjacent to A. This process is repeated until activity K is plotted. The network should

Table 9.1   Network example.

Activity	Duration (days)	Logic
Depends only on the completion of:		
A	5	–
B	10	A
C	12	A
D	4	B and C
E	8	C
F	10	C
G	10	E and D
H	4	B (2 days)
J	5	E and F
K	1	H and G and J

then be checked to ensure that there are no loops of activities and that there are no unconnected strings of activities.

The next step is to complete the forward pass. The project will commence at the end of time period zero. Hence the early start for activity A will be 0. The early finish is calculated by adding the activity duration to the early start, that is, $0 + 5 = 5$. The early finish of A is then the early start of both B and C, day 5, and the process is repeated. Activity D has two predecessor early finishes from activity B and activity C; as D cannot start until both B and C are completed, the larger value for the early finish from C, day 17, becomes the early start for D.

The link between activity H and activity B is subject to an overlap of 2 days, that is, the start of H can overlap the finish of B by 2 days. The early finish for B is day 15 and so the early start for H is day 13. The forward pass continues until activity K is reached and an early finish for the project is calculated as 36 days.

(2) The process is then reversed to calculate the backward pass. Using the same logic but starting at activity K, a late finish is determined. In this example, as in the majority of cases, the late finish is taken to be the same as the early finish, day 36. It should be noted that it is possible to insert a higher number as the late finish, but the backward pass would then not end at time period zero. By the same logic, the late start is calculated by subtracting the activity duration from the late finish. This process is repeated, taking the lower value on the backward pass and adding an overlap until a late start for activity A is calculated. Unless this value is also zero an error has been made.

Each activity will now have an earliest and latest start and an earliest and latest finish. A number of activities, including the start activity and the finish activity will have identical early starts and late starts. The difference between the late start and the early start is known as float. The critical path or paths consists of the activities with zero float logically linked between the start activity and the finish activity. This gives a critical path of A–C–E–G–K.

(3) Activity F is not critical – it has a float of 3 days. By extending the duration by 5 days, the early finish of activity F is extended by 5 days also. The forward pass is continued to show that activity K now finishes at day 38, an extension of 2 days. When the backward pass is completed, it is shown that there is a new critical path and changed float durations for non-critical activities. The new critical path is A–C–F–J–K.

9.4 Updating the network

Updating should be done frequently to ensure that the network is relevant; the procedure is as follows:

(1) Identify and mark all activities on which work is currently proceeding as 'live' activities. This is important as it focuses attention on the future. What has

happened to the present date must be accepted and concentration given to re-planning and scheduling future activities.

(2) Introduce a new activity at the origin of the programme, having duration equal to the time interval between the start of the programme and the date of updating.

(3) Change the durations of all completed activities to zero.

(4) Calculate revised durations from the date of update for all live activities and all future activities, taking into account any changes in requirements or actual performance. Note that a completely new estimate of the amount of work remaining to be done should be made for each 'live or future' activity at each update. The revised activity duration is derived from this figure and a reassessment value of the probable output of the relevant resources.

(5) Evaluate the programme in the normal way.

9.5 Resource scheduling

In the initial stages of planning a project, a precedence network has been constructed. The data for this has usually been derived using most efficient sizes of gangs, the normal number of machines with the assumption that the resource demands for each activity can be met. Bar charts have then been drawn using earliest start dates and showing maximum float.

The next step is to consider the total demand for key resources. When considering the project or contract as a whole, there will be competition between activities for resources, and the demand may either exceed the planned availability or produce a fluctuating pattern of their use. This is known as 'resource aggregation'.

The next stage, usually known as 'resource levelling' or 'smoothing', utilises the project float. The float can be used to adjust the timing of activities so that the imposed resource limits and the earliest completion date are not exceeded. If the float available within the programme is not sufficient to adjust the activities, the planner could consider the resources given to each activity and assess whether the usage can be changed, so altering the individual activity durations. It is clear that the levelling of one resource will have an effect on the usage of others. In consequence resource levelling is usually only applied to a few key resources.

In some cases it will still not be possible to satisfy both the restraints on resource availability and the previously calculated earliest completion date, and the duration of the project is then extended. The lower the limits placed on resources, the greater is the extension of the completion period of the project when it is 'resource scheduled'. Once the key resources have been adjusted, a new completion date results. If this is not acceptable, the resource limits must be adjusted and the process repeated. When resource scheduling has produced a satisfactory solution, the start and finish dates for each activity are said to be their 'scheduled' values. It is probable that few scheduled activities will offer float.

9.6 Planning with uncertainty

The planner is often faced with a degree of uncertainty in the data used for planning. Most planning techniques inevitably use single-point, deterministic values for duration and cost, though practically it is realised that there may be a range of values for these parameters.

There may be times when it is more appropriate to consider the uncertainties of the project. There are a number of ways in which this can be achieved, ranging from the simple techniques of using pessimistic, optimistic and most-likely three-point values, through sensitivity analysis, to the more sophisticated Monte Carlo random sampling method. In all cases it causes the planner to adopt a more statistical approach to the data, and it focuses attention on the uncertainties in the project.

When using such techniques, the planner is probably forced into using a computer. The mechanics and logistics of multi-value probabilistic models eliminate manual calculation on all but the simplest of construction programmes.

One planning technique not necessarily requiring such powerful computation tools is the decision tree approach. Used as a means to judge between decisions, this technique can be of assistance in the early stages of planning. Full details of the technique can be found in many standard texts.

9.7 Information and communication technologies for project planning

The complexity of modern projects and the competitive markets in which projects must be delivered has compelled businesses to explore all possible options for improving the delivery of their products and services. Advancements in information and communication technologies (ICT) have facilitated the planning phase. ICT allows the planner the possibility to interrogate and verify information at any stage. Furthermore, the speed with which information is processed and printed out either in numerical or graphical form allows the project manager to explore different possibilities before the project commences and allows the flexibility to amend project strategies as the project progresses.

A number of project planning and control packages have been produced and marketed. The software ranges in cost from hundreds to thousands of pounds. These programs may vary in terms of their modelling flexibility and simulation options, but all are designed to provide project managers with the powers to plan and to aid decision making. All have the ability to analyse networks and produce the basic standard outputs.

The main difference between the various programs available seems to be the additional facilities available and the degree of sophistication of the output. Many of the programs can be linked to 'add-on' programs to integrate other aspects of project management. It is impossible to describe the many intricacies of all the available systems or indeed comprehensively compare the different systems. If such detailed comparisons are required, it is advisable to consult some of the specialist magazines or periodicals of the computer industry that often carry out such studies. When the need for such detailed comparison is not necessary, selection of the appropriate package may be made by clearly identifying the intended use of the planning software. Other factors such as costs, user friendliness, computing power, the platform of the operating system of the PC, output sophistication or range of 'add-ons' may then be considered against the primary intended use of the package.

In considering the costs, one should look not just at the upfront costs but also the service and maintenance costs, the training costs, the need for hardware, the cost of upgrading, etc. At times an annual licensing agreement may offer better value for money and flexibility than owning the software package.

With the proliferation of PCs and expansion of ICT, a number of integration options are available to link the use and output of planning software to other programs such as word processors, presentation tools and spreadsheets. The use of e-mail, Internet and intranets allows information to be easily shared with many stakeholders and, in the case of projects, run from different geographical locations, and such instantaneous communication can enhance project success. Two-way information may be readily relayed between the project manager and project teams. The use of handheld microcomputers and wireless Internet connectivity can enable sharing of information between remote sites and main offices. In addition, information needed for project reports can be sent back and forth. This, however, does require all information to be carefully checked before dissemination as multi-user access can be a source of errors often complicated by the false belief that computers are infallible.

9.8 The planner and project teams

In the days of bar charts, planning was carried out by engineers or production staff using simple techniques to record their ideas on paper and transmitting them to other members of the team. Nowadays, however, the specialist planner or scheduler has come to the fore in the planning process. Planning does not exist in its own right. It is always associated with another activity or operation, that is design planning, construction planning, production planning, etc. It is logical, therefore, that a design planner should be or should have been a designer, a construction planner should be familiar with construction methods and techniques, and a production planner should be knowledgeable in the process and manufacturing operations of production. As long as the specialist planner has graduated from one of the engineering disciplines and is familiar with the problems of a particular project, a realistic network will probably be produced.

A real problem, however, arises when the planner does not have the right background or experience of the processes that he or she is planning. A network that is not based on sound technical knowledge is not realistic, and an unrealistic network is dangerous and costly, as decisions may well be made for the wrong reasons. In planning engineering projects, the effects of a programme decision based on wrong assumptions and inadequate understanding of the project details may not be felt for months, and so it may be very difficult to ascertain the cause of the subsequent problem or failure.

The advent of personal computers significantly changed the whole field of computer processing. It is claimed that modern programs eliminate the need to draw the network manually. Unfortunately, the use of PCs has enabled inexperienced planners to produce impressive outputs that are not very useful and often misleading. This has been partly because the computing industry has created an aura of awe and admiration around itself: anyone who familiarises himself with the right jargon can give an impression of considerable knowledge – for a time at least!

There is a great danger in shifting the emphasis from the creation of a genuine project plan to the analysis by the PC, so that many people believe that to carry out an analysis of a network one must have a computer. The heart of project planning is the drafting, checking, refining and redrafting of the plan itself, an operation that must be carried out by a team of experienced participants of the job being planned. The most significant drawback of using PCs is the removal of the project team from the network planning process. The manual approach of sitting around a table has often provided an opportunity for the team to explore alternative strategies for project delivery and to discuss potential risks. A better approach would be to use PCs only after the draft plan has been produced. The power of the PC would then be beneficially applied to check whether the overall cost and time targets are achievable. If unsatisfactory, the project team can then be given an opportunity to brainstorm alternative options.

Further reading

Lester, A. (2006) *Project Management, Planning and Control*, fifth edition, Elsevier Butterworth-Heinemann.
Pinto, J. K. (2006) *Project Management: Achieving Competitive Advantage*, Prentice Hall.
Meredith, J., Samuel, J., Mantel, J. R. (2003) *Project Management: A Managerial Approach*, fifth edition, Wiley.

Exercises

1. *New housing estate*

The new housing estate comprises four blocks of low-rise flats, a block of five-storey flats and a shops and maisonettes complex. New water and sewerage systems are required and an existing drain requires diverting. Attractive landscaping, fencing and screening will complete the site.

Produce a programme for the project and resource level the programme for (a) bricklayers and (b) total labour

- Restraints

  Target completion: 82 weeks
  Flats A and B to be completed as early as possible.
  Shops, maisonettes not required until all other work on site is completed.

- Resource demand

Flats, Blocks A, B, C and D, per block:
Excavation, foundation   40 man-weeks
Brickwork   120 man-weeks
Carpenters   30 man-weeks
Finishing trades   70 man-weeks

- Shops and maisonettes

Excavation, foundations   120 man-weeks
Concrete frame   150 man-weeks
Brickwork, block work   150 man-weeks
Finishing trades   250 man-weeks

- Five-storey block

Excavation, foundations   190 man-weeks
Concrete frame   300 man-weeks
Brickwork   350 man-weeks
Finishing trades   500 man-weeks
Site establishment   4 men × 3 weeks
Drain diversion   30 man-weeks
Completion of main drain   24 man-weeks 300-mm-diameter drain
Drain connections   40 man-weeks connections to main
Roads – first phase   60 man-weeks
Road surfacing   50 man-weeks
Paths   75 man-weeks
Fences and screens   20 man-weeks
Water services   80 man-weeks 150-mm-diameter ring main
Landscaping   100 man-weeks

---

2. *Pipeline*

The contract comprises the laying of two lengths of large-diameter pipeline with flexible joints 16 km from pumping station A to reservoir B and 8 km from B to an industrial consumer C. There is a continuous outcrop of rock from chainage 10 to 13 km, and special river, rail or road crossings are to be constructed at chainages 2, 7, 13, 18 and 22 km.

Isolating valves are to be installed at ends and at 3 km intervals along the entire length of the main, each providing a suitable flange and anchor for test purposes. Water for testing will be supplied free by the promoter following completion of the pumping station during week 33. The reservoir will be commissioned during week 35. The industrial plant will be completed by week 40, but cannot become operational until water is available. Pipes are available at a maximum rate of 1000 man/week (commencing week 1). The contractor is responsible for off-loading from suppliers' lorries, storing and stringing out. Each stringing gang can handle 1000 man/week.

Produce a time–location programme for the contract on the assumption that one stringing gang and four separate pipe-laying gangs are to be employed.

The average rate of pipe laying per gang may be taken as 300 and 75 man/week/gang in normal ground and rock, respectively. Testing and cleaning each of length 3 km will take 2 weeks. A river, rail or road crossing is estimated to occupy a bridging gang for 4 weeks. Because of access problems, pipe-laying and stringing gangs should not be operating concurrently in the same 1 km length. Attention should be paid to manpower resources and continuity of work.

3. Industrial project

(a) Construct a precedence diagram as the master programme for an industrial project using the logic and estimated duration given in the following table. Determine the earliest completion date and mark the critical path.

Activity	Duration (month)	Precedence activity
1. Promoter's brief	3	–
2. Feasibility study	18	1
3. Promoter considers report	12	2
4. Land purchase	12	3
5. Site investigation	4	3
6. Design stage I	6	3
7. Design stage II	4	5, 6
8. Civil tender documents	3	5, 6
9. Specify mechanical plant	3	6
10. Mechanical plant tender	4	9
11. Civil tender	2	7, 8
12. Specify electrical plant	3	9
13. Electrical plant tender	5	12
14. Manufacture mechanical plant	18	10

Continued

Continued

Activity	Duration (month)	Precedence activity
15. Design stage III	4	7, 10
16. Design stage IV	4	13, 15
17. Manufacture electrical plant	20	13
18. Construction stage I	6	4, 11, 15
19. Construction stage II	12	16, 18
20. Install plant	6	14, 16, 17, 18
21. Test and commission	3	19, 20

Estimate the float associated with activity 16.

(b) The tenders for the electrical and mechanical plant have been received and are as follows:

Plant		Period of manufacture	Cost (months)
Electrical A	*Activity 17*	20	444,000
Electrical B		16	600,000
Mechanical C	*Activity 14*	18	600,000
Mechanical D		16	700,000

If the promoters profit is estimated to be £50,000/month from the date of the completion which two tenders would you recommend?

4. Bridge

Using the precedence diagram method you are to produce networks for the construction of the bridge.

Activity number	Activity description	Duration (weeks)	Resource demand
1	Set up site	1	
2	Excavate left abutment	6	Excavation
3	Excavate left pier	4	Excavation
4	Excavate right pier	4	Excavation

Continued

Continued

Activity number	Activity description	Duration (weeks)	Resource demand
5	Excavate right abutment	6	Excavation
6	Pile driving to right pier	7	
7	Foundations left abutment	8	Concrete
8	Foundations left pier	6	Concrete
9	Foundations right pier	6	Concrete
10	Foundations right abutment	8	Concrete
11	Concrete left abutment	9	Concrete
12	Concrete left pier	7	Concrete
13	Concrete right pier	7	Concrete
14	Concrete right abutment	9	Concrete
15	Place beams left span	4	Crane
16	Place beams centre span	4	Crane
17	Place Beams right span	4	Crane
18	Clear site	1	Crane

The exercise is in three parts.

(a) Produce a network assuming unlimited resources are available. Evaluate the network, showing the critical path and earliest completion date.
(b) By consideration of the network produced in (a), evaluate the effects of reducing the duration of pile driving to 5 weeks.
(c) As (a) but assume the resources are heavily constrained.

The resources available in total are one excavation team, one concrete team for foundations, one concrete team for abutments and piers and one crane team.

Suggested answers to the exercises

1. New housing estate

The project consists of a small development of a new housing estate, familiar to many urban or suburban sites. The project includes four blocks of low-rise flats, a block of five-storey flats and a shops and maisonettes complex, with support services.

The key issue is to identify where most of the work is required. In this case, the five-storey block absorbs the majority of the site man-hours, and unless careful assumptions are made regarding the gang sizes and the overlapping of the different stages of construction, it would not be possible to finish within the time allowed. Therefore, despite having blocks A and B as a priority, as soon as the drain has been diverted, work must start on the five-storey block.

There is no single correct solution as the precise answer depends upon the assumptions made. The model solution is shown in Figure S1. Assumptions made include having two excavation teams of eight men, the first team commences on block A and then goes to the five-storey block; the second team goes to block B and then also to the five-storey block. The concrete frame for the five-storey block is assumed to have 16 men, the brickwork 12 men and the finishers 20 men. It was further assumed that brickwork could overlap the frame by 14 weeks and that finishing could overlap brickwork by 15 weeks.

This results in a base demand for 12 bricklayers on site, rising to a plateau of 24 bricklayers for 20 weeks before falling back to 12 again. The logic is shown by dotted lines and important constraints or key dates are clearly marked. The space within the bars has been used for figures of output. A histogram of the demand for bricklayers and for total labour has been plotted directly under the bars at the bottom edge of the programme.

This solution completes in 78 weeks, allowing 4 weeks float to compensate for an optimistic programme and has a maximum number of people on site of 65. Keeping the total on site low is important in practice as accommodation and equipment has to be provided for all workers. There is a conflict between smoothing the bricklayers and a smooth demand for total labour, and computer-based methods are more efficient in resolving complex resources conflicts. With this solution flats A and B are ready by week 34.

It is important to note that, for A and B to be habitable, assumptions have to be made about the proportions of work for the drain connections, paths, fences and screens, water services and landscaping, which will have to be completed by week 34 also.

2. Pipeline

The time–location diagram is particularly suitable for cross-country jobs such as pipe laying, where the performance of individual activities will be greatly affected by their location and the various physical conditions encountered. In this exercise the assumption was given that one stringing gang and four separate pipe-laying gangs were to be employed. The remaining problem is to decide if a single bridging gang A can cope with all restrictions placed on the programme.

The programme has to adopt a trial-and-error approach to producing the diagram, to balance the needs for bridging with the productivity of the gangs. The time–location diagram, as in Figure S2, shows a possible solution for one bridging gang, one stringing gang and four pipe-laying gangs. Pipe-laying gang one starts at 0 km, gang two at 9 km, gang three at 12 km and gang four at 16 km. It can be seen from the slopes of the progress lines, falling from 300 man/week/gang in normal ground to 75 man/week/gang in rock, that there is a need for two gangs to tackle the rock between 10 km and 13 km. The bridging gang move to keep ahead of each of the pipe-laying gangs approaching an obstruction and hence move from the culvert at 2 km, to the culvert at 18 km, to the thrust bore at 13 km, to the river crossing at 7 km and finally to the railway crossing at 22 km.

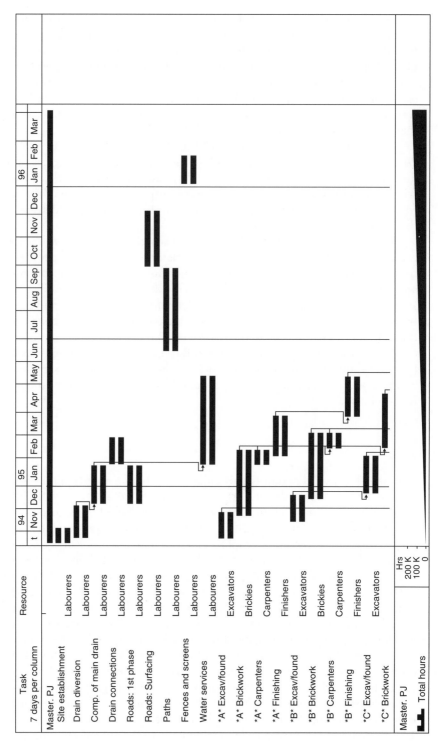

Figure S1 New housing estate: suggested solution.

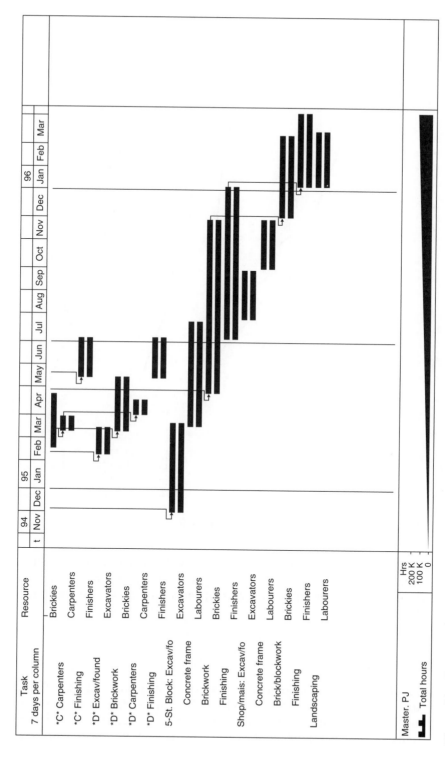

Figure S1 Continued.

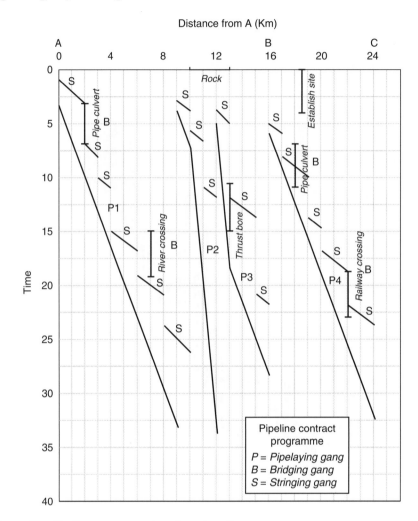

Figure S2   Pipeline contract: time–location diagram.

Pipes are available at a maximum rate of 1000 man/week, remembering the constraint to keep at least 1 week ahead of the pipe layers a sequence for off-loading from suppliers' lorries, storing and stringing out is shown on a weekly basis.

3. *Industrial project*

(a) The precedence diagram for this exercise is shown in Figure S3. The activities are represented by egg-shaped nodes and the inter-relationships between activities by lines known as dependencies. The diagram is constructed and the forward and backward passes undertaken in exactly the same way as the worked example

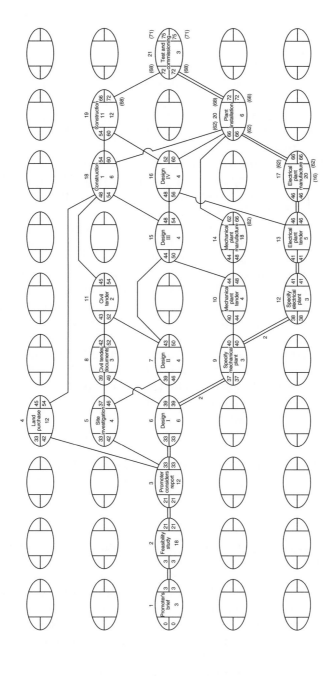

Figure S3 Industrial project: outline project network (precedence diagram). All durations are expressed in months.

in the chapter. Minimum duration is 75 months and the Critical Path goes from activities 1–2–3–6–9–12–13–16–20–21.

Note: It is unlikely that you will get the most useful diagram at the first attempt. When redrawing a critical path diagram, try to minimise the crossing of logical dependencies, although often some crossing is unavoidable, but more importantly think about the role of the plan in communicating to the project team. In Figure S3 each level or row of the diagram represents a particular responsibility: top level, land purchase and legal department; second level, site investigation and civil engineering; main level, project activities and design; fourth level, mechanical department; and fifth level, electrical department. This is a way of presenting the information, but there are many other ways that might be particularly useful in a given situation.

The float associated with activity 16; TOTAL FLOAT, the difference between its earliest and latest starts or finishes, 8 months and FREE FLOAT, the minimum difference between the earliest finish time of that activity and the earliest start time of a succeeding activity, 2 months.

(b) The option to spend more money to save time is frequently encountered in practical project management. In this simple example the combinations of A and C represent the existing situation and A and D, B and C, and B and D require investigation. However, activity 14 is not on the critical path, and therefore A and D would not be beneficial and can be ignored.

Take activity 17 first and substitute B for A. Recalculate the remainder of the forward pass to show an early finish of 71 months. Calculating the backward pass provides two critical paths: 1–2–3–6–9–12–13–16–20–21 and 1–2–3–6–9–10–14–20–21. The extra cost of substituting B for A is £156,000, but the profit earned would be $(75 - 71)$ 4 × £50,000 = £200,000, a gain of £44,000. Therefore, option B should be adopted for activity 17.

Activity 14 is now on one of the critical paths. However, as there are two paths, a separate reduction in the duration of activity 14 would not reduce the project duration and hence would not be cost-effective and option C should be retained.

Therefore, options B and C are the recommended choice.

4. Bridge

(a) Taking a step closer to the real world, no logic dependencies are indicated and the planner has to use judgement to construct the network. There are a number of possible networks, but it is recommended that the diagram be kept as free and unconstrained as possible. It is suggested that 'set up site' would be the start activity and this would then link to all the excavation activities. The pile driving for the right pier should be inserted after 'excavation' but before 'foundations'. All other 'excavation' activities are followed by 'foundations'. 'Foundations' are followed by the next stage of the 'concrete' process. Next, beams can be placed, but remember the logic that for a beam to be placed it must have the supports at both sides completed. Activity 18 'clear site' is suggested as the finish activity.

No figure is included as there are a number of viable solutions. One of the optimal solutions gives completion in week 30 with two critical paths, both going through activity 6 – the right pier pile driving.

(b) The effects of reducing the duration of pile driving to 5 weeks will vary depending on the network drawn for part (a). Nevertheless, the overall effect on most networks is to reduce the completion time to 29 weeks with two different critical paths, one going through the north and the other the south, abutment activities.

(c) The resources are heavily constrained: one excavation team, one concrete team for foundations, one concrete team for abutments and piers and one crane team severely limit the activities that can operate in parallel.

The question is slightly unfair – as this type of resource scheduling is difficult to undertake without the assistance of a computer. Many people find the use of resourced bar charts of assistance, but there is no easy way. In this case the obvious action is to employ resources away from the right pier such that the extra work required for the pile driving does not delay or disrupt the use of any resource.

If this is achieved, a completion time for a 'resource-constrained' project of 46 weeks can be achieved. However, any solution under 50 weeks represents a reasonable attempt at the exercise.

Chapter 10
Project Control Using Earned Value Techniques

The purpose of this chapter is to first describe project-control theory in general and then to explain and discuss the system of earned value that is now used by a number of organisations. The section on earned value covers all aspects of the approach, from the definition to applications in a variety of situations.

10.1 Project control

The purpose of project control is to ensure that project status is reported in a consistent, cost-effective and timely manner to the project manager so that any necessary actions can be taken and the status of the project reported to senior management. To do this the project manager will need to hold regular meetings with the project team and have regular, meaningful reports provided in an efficient and timely manner. In addition to this a process must be put in place to control change, including schedule controls, change requests, budget controls and re-plan. The model here has to be plan–do–review. The 'do' could take the following forms:

- do nothing – if the work is progressing as planned;
- refer on – if the change lies outside the scope of the control of the project team;
- change the plan – this has the inherent problem that future measurements will be against a new baseline. This may be the preferred option if it can be clearly demonstrated that it will benefit the project;
- change the work – either scope or method of execution. For example, change quality, functionality or seek an alternative.

The role of the project manager is to:

- establish the system;
- allocate responsibilities;
- ensure costs are properly allocated;
- ensure payments are authorised as appropriate.

This can be monitored by:

- obtaining regular, verbal reports;
- progress reports;
- internal measurement procedures;
- external audits.

This has to be balanced by the needs of the project team: although they do need feedback on how they are performing, they have to be clear as to the need for the information that they are providing (that it is not just more bureaucracy) so that they do not feel constrained by the reporting system. If the focus is on reporting by exception, when there is a clear deviation from expected, then 'information overload' can be avoided and the reports should be timely.

In 1967 the US Department of Defence published a set of cost/schedule control system criteria, known as the C-Spec. These criteria define minimum earned value management control system requirements. Previous management control systems assumed a direct relationship between lapsed time, work performed and incurred costs. This chapter describes how the earned value system analyses each of these components independently, comparing actual data to a baseline plan, set at the beginning of the project. First, however, there will be a brief description of the *cost-control cube* (Figure 10.1), which forms the basis of the methodology for controlling project costs.

All costs can be assigned to a cell of the cube, and through the cube all costs have a position in each of the three breakdown structures. A project aggregate can then by prepared by summing along any of the three directions. Some of the cells will have no costs.

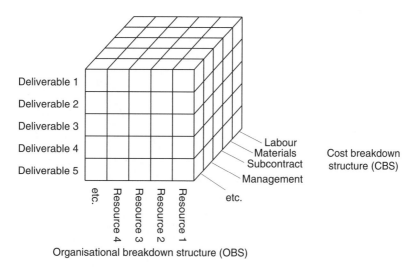

Figure 10.1   Cost-control cube.

10.2 Earned value definitions

Earned value analysis (EVA) compares the value of work done with the value of work that should have been done.

Budgeted cost of work scheduled (BCWS) is the value of work that should have been done at a given point in time. This takes the work planned to have been done and the budget for each task, indicating the portion of the budget planned to have been used.

Budgeted cost of work performed (BCWP) is the value of the work done at a point in time. This takes the work that has been done and the budget for each task, indicating what portion of the budget ought to have been used to achieve it.

Actual cost of work performed (ACWP) is the actual cost of the work done.

Figure 10.2 illustrates a typical S-curve plot comparing budget and actual costs.

Productivity factor is the ratio of the estimated man-hours to the actual man-hours.

Schedule variance (SV) is the value of the work done minus the value of the work that should have been done (BCWP − BCWS). A negative value implies that work is behind schedule.

Cost variance (CV) is the budgeted cost of the work done to date minus the actual cost of the work done to date (BCWP − ACWP). A negative value implies a current budget overrun.

Variance at completion is the budget (baseline) at completion (BAC) minus estimate at completion. A negative value implies that the project is over the budget.

Schedule performance index (SPI) is (BCWP/BCWS) × 100. Values under 100 indicate that the project is over the budget or behind schedule.

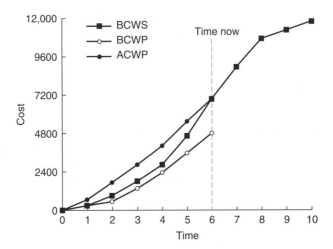

Figure 10.2 Typical S-curve.

Cost performance index (CPI) is (BCWP/ACWP) × 100. Values under 100 indicate that the project is over the budget.

10.3 The theory and development of EVA

S-curves examine the progress of the project and forecast expenditure in terms of man-hours or money. This is compared with the actual expenditure as the project progresses, or the value of work done. If percentages are used, the development of useful data from historic records of past projects is simplified, since size, and hence total time and cost, is not significant. All projects, whatever their size, are plotted against the same parameters (namely, time and cost) and characteristic curves can be more readily seen.

The form of the S-curve is determined by the start date, the end date and the manner in which the value of work done is assessed. Once a consistent approach has been established and the historical data analysed, there are three significant variables that need analysing: time, money and the shape of the S (known as the route). As the expectation is that the route is fixed, only two variables remain. The route is as much a target as the final cost. If the movement month to month is compared, the trend can be derived.

If the axes are expressed as a percentage, then the percentage of what must be carefully defined. If review estimates are made at regular intervals, then the value of work done will always be a percentage of the latest estimate. Revision of the estimate automatically revises the route of the S-curve.

The assessment and precise recording of the value of work done is crucial to project cost control. This is described by accountants as 'work in progress'. For example, design man-hours are usually measured weekly. The hours that have been booked can be evaluated at an average rate per man-hour, that rate being the actual costs in the project. Materials are delivered against a firm order so that normally an order value is available. The establishment of realistic targets is very important if the analysis is to be meaningful. A low-cost estimate often leads to a low estimate of the time required to carry out the work. The immediate target is not the final target.

A series of standard S-curves have been developed by companies in the oil and gas industries so that the performance of existing projects can be monitored. These curves have been derived empirically and, when projects within certain categories do not follow the norm, investigations ensue to identify the source of the discrepancy. These curves have been put to a number of uses, including monitoring, reporting and payment. They are not always used in the pure form, that is, further curves, such as productivity factors, can be developed so that certain aspects of the project can form the focus of attention.

Developments have been made from using the S-curve simply as a method for controlling the progress of the project to using it proactively in the evaluation of indirect costs associated with project changes introduced by the promoter. This method is known as 'impact or influence'.

Influence may be applied whenever there is a variation and is applied to the estimate of additional man-hours taking into account the indirect costs for the whole of the variation, including the additional indirect costs associated with parallel activities. The revised estimated cost of the variation may then be issued to the promoter.

The influence is composed of:

- time lost in stopping and starting current activities in order to make the change;
- special handling to meet a previously scheduled activity;
- revisions to project reports and documents;
- unusual circumstances which could not be foreseen;
- recycle (lost effort on work already produced);
- other costs that may not appear to be related to a particular change.

Influence is incorporated by multiplying the direct cost for the variation by the influence factor and adding this to the direct cost to give the total cost for the change.

$$T = V(1 + \text{IF})$$

where V is the direct cost of variation; T is the total cost of variation; IF is the influence factor; and Cost is the cost to the promoter.

Any organisation wishing to adopt this approach would need to derive standard curves (Figure 10.3).

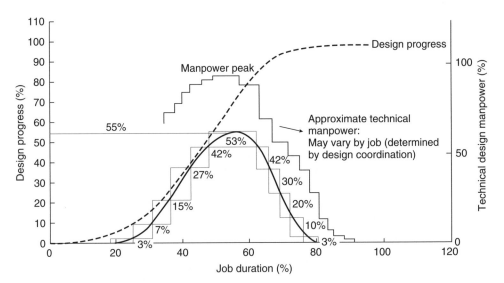

Figure 10.3   Influence curve.

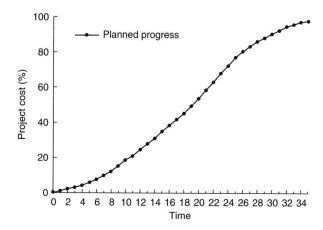

Figure 10.4   S-curve showing the value of work done.

10.4 Relationship of project functions and earned value

Monitoring and cost control can be described as identifying what is happening and responding to it. Cost planning involves forecasting how money will be spent on a project to determine whether the project should be sanctioned and has sufficient money available when required.

Typically there are three major areas of control: commitments, value of work done and expenditure. These are all controlled in relation to their progress over time and may be illustrated diagrammatically as shown in Figure 10.4, which presents the typical S-curve for the value of work done. Similar S-curves may be developed for both commitments and expenditure.

Planning is primarily responsible for establishing the time target, both overall and in detail. The primary, although not the only, task of project cost control is to establish the exact position of the project from time period to time period in terms of value of work done, and to compare this with the targets for each time period. Finance will maintain expenditure records, working closely with project cost control. Finance will be responsible for maintaining a record of the commitments, as they have to ensure that payments are within the approved limits.

Project control can comment on the validity of planning work, by comparing planned and actual progress via the value of work done. The commitment record, if properly maintained, provides a ceiling at any point in time during the project life.

10.5 Value of work done control

Value of work done is not expenditure, although it eventually equates to expenditure. It is often considered as the work in progress but is, most of the time,

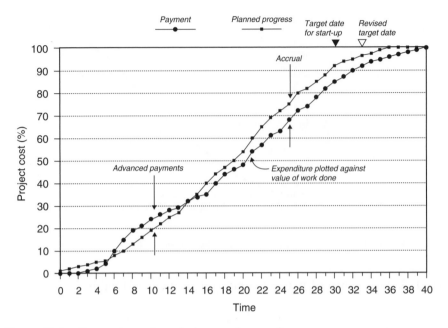

Figure 10.5 Comparison of work done and expenditure.

greater than the expenditure. Figure 10.5 illustrates the relationship between value of work done and expenditure as recorded over the life of a project.

There is normally a considerable time lag between having a project ready for start-up and the final payment of invoices and retention when the project is completed in financial terms. The value of work done can be summarised as design and head office costs plus the value of material delivered to site and the work done at site.

The techniques for an approximation of value of work done from month to month can be related to three major areas:

Head office. This deals with design, procurement and project supervision activities. The work done can be measured in man-hours and is usually recorded on a weekly basis. Hours booked are valued at an average man-hour rate and the value of work done is then assessed. The simple way to do this is on a cumulative basis. If the latest cumulative booked cost comes perhaps 3 weeks later, then that is compared with the cumulative including the last month's estimate, a correction can be made resulting in the approximation being only 1 month behind the actual cost at that time.

Materials delivery to site. Materials are normally delivered to site against firm orders so an order value is available. As materials arrive on site the material receipt note can be valued, using information on the order, and a progressive total maintained. As the value shown on the order may not be the absolute invoice value, for reasons such as discounts, freight charges, insurance or escalation, this approach gives a nominal and close approximation of the value.

In the case of bulk materials it can be very time consuming to value all items on a receipt note from the itemised prices on an order, and it is often permissible to use a weighbridge weight multiplied by an average price per kilogram calculated from the orders.

Erection. The approach to erection contracts is very similar, but this time the value is erection man-hours. Erection man-hours, exclusive of site super-vision, for all contracts on site, should be recorded from week to week. Progress on site by various erection contractors is measured by teams of sched-ulers or measurement engineers in accordance with certain standard methods of measurement to enable progress payments to be made. Such evaluations often run late, so erection man-hours worked are often valued on the basis of standards multiplied by an average rate per hour, applicable to the erection activity.

10.6 EVA techniques

EVA is often presented in the form of progress, productivity or S-curve diagrams. Many of the current reporting procedures were developed by the US Corps of Engineers. EVA has been adopted by the oil, manufacturing, gas and process industries, where man-hours and material deliveries to site are used to monitor projects from inception to completion.

Actual/estimated man-hours are made available to determine the progress and productivity factors at any stage of the project. The productivity factors are used by both promoter and contractor organisations to monitor the progress of a project and forecast the outcome.

Productivity is the ratio between output and input and provides a measure of efficiency. Ideally, productivity is always unity with both the output and the input measured in the same units. In the oil and gas industries man-hours are used as the common unit to determine productivity. Productivity factors are used to monitor variance and trends for individual activities.

To establish a trend actual progress must be measured and compared with forecasted progress, that being the baseline S-curve. As the forecasted progress depends upon the end target, revised targets will influence the progress required to be made each month. Once it becomes evident that a work package is going to cost more, or less, than the original (or earlier) estimate, targets should be revised and the potential influence on monthly progress evaluated and a new target computed.

Table 10.1 illustrates typical 'progress of value of work done' expressed both in man-hours and completed items, common to manufacturing industries. The revised number of man-hours is forecast against the actual work completed and is based on progress and productivity. Often this information is displayed on a screen to manufacturing operatives as a guide to production rates and the time to complete. If a bottleneck occurs at any point in the production line, the effect on the costs, time and productivity of the completed item can be identified. In the case of production being carried out at a number of locations with final assembly

Table 10.1 Progress of value of work done.

Estimated man-hours/units to date	Actual man-hours/units to date
360,000	295,000
4,750	4,900
Estimated total man-hours/units	**Revised total man-hours/units**
500,000	440,000
6,000	6,000

at one specific point, it is imperative that any deviation from the activity S-curve is reported immediately and acted upon.

In many manufacturing industries, especially automobile assembly lines man-hours are allocated for reworking items to meet quality and standards. For example, Mercedes Benz reportedly spent up to 30% of the total man-hours allotted to reworking to meet defined quality and standards. This 'man-hour float' is often the determinant for price setting, to the customer, for each model. The hours expended from the 'float' of man-hours is then used to forecast the final costs, times and margins. In many cases reworking is analogous to commissioning and as such may not be allotted targeted man-hours in the project estimate.

10.7 Application of EVA

The information required from analysis of the curves will vary depending on its end use. At project level the aim will be to identify any areas where the project is underachieving so that action can be taken to improve the performance of the problematic resource. This can be done by examining the BCWP and ACWP curves. If the curves are showing cause for concern, then productivity curves can be plotted and examined in greater detail. Work packages or sets of work packages can be examined and the productivity derived.

A project manager can adopt this system through analysis of completed time sheets. It is also important to note the percentage complete status of a task at the recording date, as described earlier. This is where a proactive role is required by the project management team in gleaning information from the various disciplines. The team leader should progress the work of his section with the planners taking a recording and reporting function. If progress is to be expedited, then the project team must actively pursue information, check its accuracy and take action when low productivity becomes apparent.

Optimum workforce requirements will become apparent as data for specific projects is collated. Productivity increases or decreases will permit management decisions based on actual measurement rather than optimistic or pessimistic forecasts based on 'gut feeling'. The team leaders of various disciplines must be aware of the significance of the information they are reporting and why it is being recorded; if they do not realise that it will be used for control and not just for monitoring, then they may not give it a high priority, resulting in erroneous data.

Cost codes both for direct and indirect costs will need to be considered on the basis of compatible codes allocated to different work packages. In a number of major process organisations in the oil and gas industries for example, the number of multidisciplinary functions and development stages of a project have led to confusion of the cost codes, resulting in inaccurate data and delays in transmitting data for analysis. Ideal cost coding systems are those that allow the system to be used at all stages in the development of a project.

One very important aspect of coding and recording man-hours is the comparison of man-hours expended against a code or codes and the remaining man-hours expended as overheads. If for example a design engineer is allocated 40 hours per week on say three projects and only 32 hours are recorded, then the remaining 8 hours must be reconciled somewhere within the organisation. Unfortunately, the remaining man-hours are often distributed between the project codes. In cost plus contracts it is standard procedure to book additional man-hours against the contract code.

To achieve the introduction of EVA a basic requirement is the collection and processing of data related to existing and past projects. The importance of developing standard curves cannot be overemphasised. If there are no standard curves then it is difficult for the project manager to set realistic targets.

This data forms the basis for estimating and allotting man-hours to each activity or work package. The effort required at project level to undertake EVA is such that it is not recommended for 'small' projects. Below a certain size the effort required to gather and process data on the progress of small value items may be greater than the value of the package of work. The proportion of the inaccurate part of the assessment of the work left will also be greater, invalidating the approach.

If the decision is made to adopt EVA for a given project, then the management team must be dedicated to fully implementing the system. Taking action on inaccurate data can be worse than taking no action at all. A standard reporting technique should be developed for all projects so that a given productivity rate means something.

10.8 Examples of EVA

Table 10.2 illustrates the historical data of a typical six package project. Estimated man-hours are often based on historical data from completed projects against which the actual man-hours expended to complete each work package and the overall project are plotted.

Figure 10.6 illustrates the S-curve for the prototype work package shown in Table 10.2. The S-curve is prepared on the basis of estimated man-hours. The actual man-hours expended are then plotted on a regular basis to determine the variance in man-hours, productivity factors and forecast the work package trend.

Table 10.2 Work packages, man-hours and productivity factors.

Work package	Actual man-hours	Estimated man-hours	Productivity factor
Feasibility study	550	450	0.82
Design	3000	2500	0.83
Prototype	1500	1700	1.13
Manufacture	2300	2200	0.96
Erect	900	600	0.66
Commission	300	300	1.00

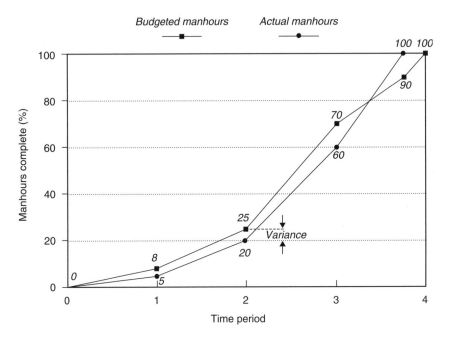

Figure 10.6 S-curve for prototype work package.

On a number of civil engineering projects EVA has been used primarily to monitor the construction of individual units and as a reporting tool to both construction management and the promoter. In its simplest form bar charts are prepared from which histograms of man-hour estimates for each construction activity are generated. The cumulative man-hours are plotted in the form of an S-curve and progress is monitored as man-hours are expended. Plant usage is often added to man-hour schedules based on weekly costs, usually plant hire rates, for specific types of plant/activity usage. The type of plant and estimated usage often forms the basis of plant allocation for a number of projects. Clearly if the productivity of one particular work package falls below the estimate, illustrated by the S-curve, then additional plant time and usage will be required to reverse the trend.

The diagrammatic presentation facilitates the comparison of actual man-hours expended in construction with the man-hours estimated. For example, if in a fixed-price construction project the foundation works of a particular unit has a programme duration of 4 weeks and 480 h is completed in 3 weeks utilising only 320 h, then the saving in man-hours can be utilised on another construction unit. The saving in man-hours and plant usage and the reduction in duration can then be plotted as actuals against the estimated S-curve. The 'productivity factor' for this completed activity would be 1.5.

If, however, the man-hours expended are 480 h after 4 weeks with only 50% of the activity completed, then completion would require another 4 weeks duration and an additional 480 h based on the previous productivity factor. This often results in manpower being reallocated to an activity falling below the estimated productivity factor. Also, the requirement of plant to complete the work package will often result in delays to other work packages as it is normally very difficult to acquire plant at short notice.

A major problem area is the accuracy of reports of percentages of work completed. As a result of this, activities or work packages are often reported as being completed when, in fact, they are not. This is often referred to as the 'persistent 99% complete syndrome', which results in the saying that 99% of tasks in 99% of projects are 99% complete for 99% of the time. It should also be noted that the method of appraising work completed should be timely and consistent for each section of work and always kept up-to-date. In many civil engineering contracts, managers prefer to rely on the 'remaining duration' as a good measure of progress for a work package as this does not assume anything about the accuracy of the estimate or when the work package started. This approach also ensures the manager is aware of the remaining duration to meet any contractual milestones, and acts accordingly.

EVA is also used in the manufacturing and production industries to monitor production rates. In these industries, work is normally performed in controlled environments, where it is a simple task to measure the physical work done and materials used especially where the delivery of an item is vital to the success of a project. By monitoring, often on a daily basis, both the number of man-hours estimated and expended on a number of activities delivery dates can be identified at an early stage of the project. Cumulative man-hours expended for each activity are compared with the estimated target of man-hours and the completed items, which permits adjustments to the overall target over the manufacturing period. Changes in the base rates per hour for each activity can easily be amended to identify the cost to completion. In production line industries man-hours expended can be collated on a daily basis through time sheets and completion verified, and hence are accurately reported on the basis of a completed item.

Many of the US companies involved in the oil and gas industries in the Gulf of Mexico introduced EVA to major projects carried out in the Middle East. In most cases these projects were undertaken on a cost-plus basis. Data, in the form of man-hours relating to specific tasks were analysed and then utilised to

Table 10.3 Estimated/actual man-hours and productivity factors.

Month	Estimated man-hours	Actual man-hours	Productivity factor
1	10,000	18,000	0.55
2	40,000	50,000	0.80
3	60,000	70,000	0.86
4	50,000	48,000	1.04
5	40,000	30,000	1.33
6	20,000	15,000	1.33
7	10,000	8,000	1.25
8	9,000	10,000	0.90
9	1,000	2,000	0.50
	240,000	251,000	0.96

provide 'productivity factors' relating to work carried out on projects in the US. To compensate for the required time to learn specific tasks, productivity factors were factored to take into account the additional times required.

For example, the work package of cold insulation of pipe work may have a US base rate of 1 man-hour to insulate 1 m of 200-mm-diameter pipe. To take account of the possible learning period for this activity, often performed by Third Country Nationals (TCN), a factor of say 1.2 is used. The S-curve is then factored by 1.2 – this factor is called the US Gulf Factor (USGF). If the total number of estimated man-hours is say 200,000 h, then the factored total will be 240,000 h.

The man-hours can then be scheduled for this work package against time as illustrated in Table 10.3.

Clearly the overall 'productivity factor' for this work package is 0.96; however, productivity over the 9-month duration has varied from 1.33 to 0.5. This method of reporting and planning at timely intervals provides management with a clear view of progress and productivity and provides the basis for management decisions. In this particular example of a cost plus project, the additional man-hours expended to meet the contract programme are paid for by the promoter.

It is very important that work packages utilised in EVA are defined correctly within the disciplines to be adopted, the allocation of man-hours and the interdependency on other work packages. The overall project is determined by the worst work package. In a project consisting of say 17 work packages, productivity of 16 of the packages may be above that estimated. In a number of projects contractors have superimposed individual S-curves to provide a total contract S-curve, normally as the basis for reporting to the promoter. This often results in a false picture of the project's progress as the total contract S-curve implies that the project is ahead of schedule.

In one example, schedulers on one particular work package reported that productivity was as estimated; however, the man-hours recorded against the actual work completed were optimistic. This resulted in the project being reported as being ahead of time and below the budget. When an audit of the work package was performed, it was found that, unless this critical work package was brought back on programme, the remaining 16 work packages could not be completed.

An additional 30,000 man-hours had to be allotted to the work package, which resulted in an overall productivity factor of 0.33. Clearly, the accuracy of reporting actuals and progress is a prime function of EVA. Should effective reporting and updating have been carried out, then at some time during the activity, negative progress would have alerted management of the apparent problems.

A further problem on this particular project was monitoring the receipt of materials delivered to site. The numerous cost codes, often with more than one cost code being applied to pre-assembled units, resulted in delays in progressing invoices. In some cases a 3-month lag occurred between receipt of materials and the processing of invoices.

To ensure that the same mistakes did not occur again, the promoter instructed the contractor to change the scales on each work package S-curve. Clearly, when a project has an estimated man-hour expenditure of 11,000,000 h a slight deviation on a graph does not fully represent the impact on the project as a whole. Over the remaining project duration each work package was expressed in total histogram form with 'productivity factors', time to complete and trend analysis illustrated for presentation to the promoter.

Project reporting

The control reports should facilitate the following:

- agreement of customer requirements;
- optimisation of data requirements;
- standard formats;
- minimal distribution lists;
- common understanding of terms;
- paperless systems, where possible.

The project manager should agree reporting requirements as early as possible in the project's life. These requirements will be dependent upon size, complexity and customer requirements and need to cover the following:

- level of detail;
- recipients;
- schedule;
- level of reporting.

Wherever possible, data should be collected from existing systems. It is sensible to format reports so that detailed reports can be summarised as necessary for upward transmission.

10.9 Summary

Standard procedures need to be prepared as the EVA system is tested and the requirement of each discipline addressed. As with many reporting and scheduling systems, it is essential that the organisation use and develop the system to suit their needs, and not to create mountains of irrelevant information.

EVA is primarily a system of approximation, the accuracy of which depends on the time and costs prepared in the estimate compared with the actual time and costs as work progresses. The accuracy of the estimated data and actual data and the time intervals for auditing are of paramount importance to its successful application.

Further reading

Harrison, F. L. (2004) *Advanced Project Management: A Structured Approach*, fourth edition, Gower.

Kerzner, H. (2003) *Project Management: A Systems Approach to Planning, Scheduling and Controlling*, John Wiley & Sons Inc.

Maylor, H. (2005) *Project Management*, third revised edition, Financial Times Prentice Hall.

Chapter 11
Contract Strategy and the Contractor Selection Process

Contracts are used to procure people, plant equipment, materials and services and consequently are fundamental to project management. This chapter outlines the main components in the selection process for both contract and contractor. The processes that are influenced by the nature of the parties include the project objectives and the equitable allocation of responsibilities and risk against other factors. This chapter is not meant to be a substitute for the expertise and thinking needed in the drafting and the administration of the individual contract. It is an introduction to the relationship between legal means and engineering ends.

11.1 Context

Every project involves starting from one point and getting to a different point. In other words, the project is always about achieving a result. The problem for the promoter of the project is that he or she cannot or does not wish to provide all the resources necessary to get from the start point to the final result from within internal sources. Therefore, there is a need to obtain resources, work, materials, equipment or services from other organisations to achieve that result. The method by which the promoter will obtain those resources is that of the contract. A proper understanding of how contracts work and how contracts must be managed is therefore fundamental to the management of virtually all projects.

In addition it has to be recognised that different industries use different kinds of contract to achieve different results. Indeed there is almost a complete cultural divide between some industries. The standard type of activity/work-based contract used within the building and construction industries is totally different to the result-based contracts used within the equipment process or oil industries and within service industries, such as the telecommunications or software industry.

In addition most complex projects can be treated in several different ways. They can be carried out under a single 'turnkey' contract. They can be broken down into separate contracts on a time/stage basis, with preliminary design being carried out under one contract, site preparation under a second contract, equipment manufacture under a third contract, and so on. They can be broken down into separate contracts on the basis of contractor skills, with one specialist contractor being responsible for all civil engineering and site preparation work, another being responsible for the supply of the main plant and equipment, a third being responsible for the supply of ancillary equipment and a fourth being responsible for installation/erection. They can be broken down into a large number of small contracts or a small number of large contracts. Each will give the promoter different advantages and disadvantages. The promoter must decide.

Finally, different contracts can operate in different ways. The way in which a contract operates will generally depend to a large extent, in practice, upon the way in which it provides for payment by the promoter to the contractor. In very broad general terms, contracts provide for payment in three different ways. The first is that of the price-based contract, under which the contractor will provide the bulk of the work materials/equipment or services for a stated price or fee. At the opposite end of the scale is the reimbursable contract, under which the contractor is reimbursed the costs of carrying out the work plus profit. Finally, somewhere in between the two is the quantities- or rates-based contract, such as the typical civil engineering/building contract, in which payment is based upon the prices stated in a Bill of Quantities or a Schedule of Rates. The different types of contract tend to create different relationships between promoter and contractor, and therefore produce totally different results.

Therefore, the type(s)/size(s) of contract should be selected by the promoter only after due consideration of the nature of the parties to the project, the contract management resources available, the project objectives and the skills required to achieve them, the time available to carry out the project and the appropriate allocation of duties, responsibilities and risk.

11.2 Factors affecting strategy

In deciding the choice of contract strategy a number of factors should be considered. Clear definitions of the promoter's objectives are required so that the significance of these factors can be established.

The responsibilities of the parties need to be determined. Responsibilities may include design, quality assurance and control, operating decisions, safety studies, approvals, scheduling, procurement, construction, equipment installation, inspection, testing and commissioning. Then the risks must be allocated between the parties: in other words who is to bear the risks of defining the project, specifying performance, design risks, selecting subcontractors, site productivity, mistakes and many more. Then the basis for payment for design, equipment, construction and services must be decided. These are all major influences on the choice of contract strategy.

It is also important to consider the provision of an adequate incentive for efficient performance from the contractor. This must be reflected by an incentive for the promoter to provide appropriate information and support in a timely manner.

The contract may need to be flexible. The prime aim is to provide the promoter with sufficient flexibility to introduce change that can be anticipated but is not defined at the tender stage. An important requirement is that the contract should provide for systematic and equitable methods for introducing managing and pricing changes.

All these considerations are closely interrelated.

It is apparent that, generally, the interests of the promoter and the contractor tend to be opposed to each other. For example, a lump sum contract minimises the cost/price risk carried by the promoter, but it imposes maximum cost/price risk on the contractor. The converse is true at the other extreme of a cost-reimbursable plus percentage fee contract.

Many or almost all of these factors could be important on a project but it is likely that certain factors will dominate. All factors are significant but they may also conflict. If working on an offshore oil energy project timescale might be the dominant objective to meet a weather window or to avoid disruption of the flow of crude petroleum; this would have obvious implications for the cost of the project. If working on a project for a plant to manufacture pharmaceuticals the quality of plant and its ability to manufacture the correct product might be dominant.

There will often be a number of possible strategies that could satisfy these objectives and it is the task of the project manager to advise the promoter on which strategy to adopt. This selection is possibly one of the most important decisions in any project.

There can be other criteria to consider and every project has to be assessed individually. For example, changes and innovations in contract arrangements have followed the privatisation of what were formerly public services in the UK to try to meet commercial rather than political accountability.

11.3 Contractual considerations

The project manager should always remember the three Rs of contract management (Figure 11.1).

- *Relationships*. Relationships form part of all commercial dealings. Some relationships are permanent and others are temporary. Some relationships make for collaboration between the parties. Some relationships make for arm's length dealings. Some are highly structured. Some are unstructured and will develop according to the way that circumstances dictate. Different types of relationships need to be managed in different ways. A collaborative relationship requires a very different management style to an adversarial relationship.
- *Risks*. Risks are present in all businesses and transactions. Risk does not go away. Risk is always there in one form or another. Whatever contract strategy the promoter chooses, it will bring with it its own particular set of risks for the promoter and contractor. It is always important for the promoter to select a strategy that produces a risk set that he or she is able to manage competently. (It is also important to select a contractor who is also competent to manage that risk set. A contractor who can carry out work competently, but who cannot manage risk properly, is potentially disastrous.) Every risk has to be

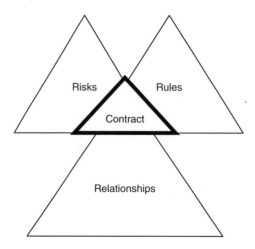

Figure 11.1   The three Rs.

managed by one or other of the parties involved. If risk is not managed, and something goes wrong, then the problem will be much more serious.

- *Rules*. Rules are created in order to regularise the way in which organisations and individuals work in relation to each other. Comply with the rules and you have a reasonable chance of being successful. Fail to comply with the rules and, however brilliant you are, you will probably fail to achieve success, unless the other party is prepared to be forgiving.

Contracts are one area where the three Rs overlap each other. Contracts formalise a set of risks rules and relationships into one set of words that will govern all dealings between the parties while carrying out that contract.

Second, the promoter needs to remember the constraints imposed by the law. This is obviously not the place to discuss the law of contract in any detail. However, it is very important to understand that the law, whatever the country, takes a very artificial and simplistic view of the contract relationship. This view is best explained by using two similes.

- 'Making a contract is like jumping off a cliff'. It is entirely up to the parties to decide whether or not they wish to enter into the contract. However, once they have made the decision and entered into the contract, there is no going back. They have to live with the deal that they have made, as set out in the contract, and carry out the terms of that contract, unless both parties agree not to do so. If one party gets itself into the 'wrong' contract, there is no easy option.
- 'The contract is an egg'. Once an egg has been laid, there can be no direct contact between what is outside the eggshell and what is inside the eggshell until the egg hatches. The egg is its own little world with no relationship to anything else during that time. The law sees a contract in exactly the same way. The written contract is seen as a complete and precise statement of all

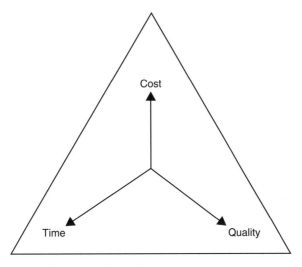

Figure 11.2   Project objectives (derived from N. M. L. Barnes, 1991).

the provisions agreed between the parties relating to that contract. Nothing else can apply, and what goes on inside the contract is entirely divorced from any other relationships that the parties may have between them.

11.4 Contractor choice

'Normal' contract objectives are usually said to be cost, time and quality. They are often pictured as a triangle (Figure 11.2).

The use of the triangle is intended to illustrate the conflict between these three objectives. If the promoter wants the lowest possible cost, then he or she will not get the shortest timescale nor can he or she obtain the highest quality. If the promoter wants the best quality, the timescale will not be short and the cost will be high. If the promoter wants the shortest possible timescale, then equally quality may suffer and cost will also have to be sacrificed. These statements are really so obvious as to be virtual truisms. However, they do illustrate the basic problem of contract strategy. Everything has to be a compromise between conflicting objectives.

In the planning of a contract the promoter needs to consider carefully the reason for employing a contractor. A promoter generally selects a contractor for one or more of the following.

- To use the particular management, technical and organisational skills and expertise of the contractor for the duration of the project.
- To use the skills of the contractor after the project has been completed.
- To have the benefit of the contractor's special resources, such as licensed processes, unique design or manufacturing capability, plant, materials in stock, and so on.

- To get work started quicker than would be possible by recruiting and training direct employees.
- To get the contractor to take some of the cost risks of a project, usually the risks of planning the economical use of people, plant, materials and subcontractors.
- To use the contractor to provide the resources, physical and financial, needed for the project.
- To be free to use his own (limited) resources for other purposes.
- To encourage the development of potential contractors for the future.
- To deal with a contractor who is already known to the promoter.

Whatever the reason, the promoter should always make a positive decision for a clear reason. That clear reason should then govern the promoter's decisions on the number, scope, type and terms of contracts.

11.5 Project objectives

The most important decision that the promoter must always make in relation to any project is to define precisely what the objectives of that project are to be. This means decides the answers to a number of questions. The questions are very simple. The answers may be very complex indeed.

What does the organisation want to have?
What does the organisation need to have?
What can the organisation afford to spend on the project?
How will the money spent on the project be recovered?
What can the organisation provide from within its own internal resources?
What does the organisation wish to provide from within its own resources?
What is the timescale within which the project must be complete?
What are the management resources that the organisation has available to manage the project?
What types of contract are those resources competent to place and manage?
What types of contract does the organisation usually use?
Can the organisation define accurately what it wants/needs to have, or is producing that definition part of the project?

The answers to these questions will tend to define the project and its objectives, and to define them in a way that enables a choice of contract strategy and contract type to be made.

The basic project objective will usually be to produce a usable asset, of one form or another. Other objectives may also be important, or even dominant in some circumstances, such as:

- the management and control of risk;
- ensuring high standards of quality assurance or plant/process validation;
- ensuring high standards of site safety;
- ensuring high standards of environmental protection;

- assistance by the contractor in operating or maintaining equipment;
- the performance of plant;
- the long-term operability of plant;
- guaranteed supplies of spare parts for plant;
- the ability to modify or expand the plant;
- training of the promoter's staff;
- the acquisition of know-how;
- influencing or controlling design by the contractor;
- influencing or controlling the identity of any subcontractors;
- retention of construction plant by the promoter;
- promoter involvement in project and construction management;
- cooperation in construction and implementation;
- establishment of an operating/maintenance force.

Project management should bear all the objectives in mind, rank them in importance and seek a strategy to optimise their achievement.

It is always essential that the project manager ensures that the promoter clearly defines the project objectives. Furthermore, once those objectives have been defined, it is equally important that those objectives are properly understood by management within the promoter. The likelihood of a successful project is greatly improved when all the managers of the design, construction, user and supporting groups within the organisation are fully informed of the objectives of the project, and are committed to those objectives.

11.6 Contract selection

Figure 11.3 seeks to encapsulate the considerations that must underlie the choice as to contract type that the promoter has to make.

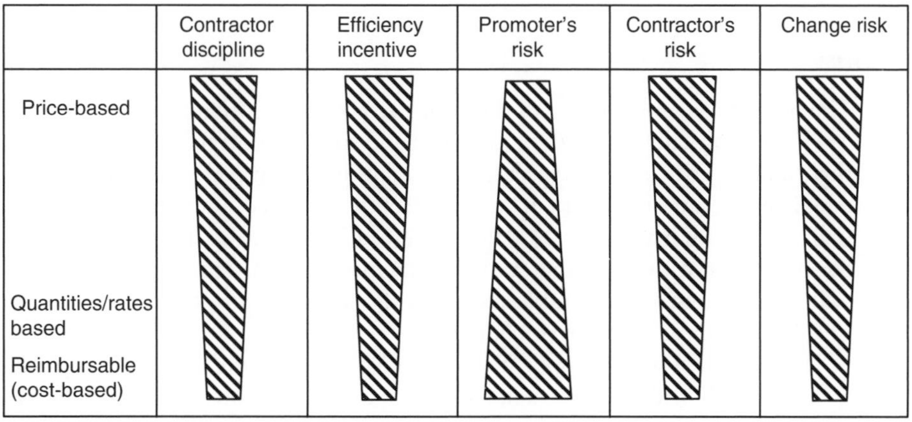

Figure 11.3   Contract types.

Note that at one end of the scale there are the 'price-based' contracts, while at the other 'quantities/rates-based' contracts and reimbursable contracts operate in a broadly similar fashion. A reimbursable contract may operate on the basis of 'cost + profit' or 'agreed rate' payments. The principal considerations are described in the following sections.

Discipline

The price-based contract imposes 'discipline' upon the contractor. It requires a specification to be met, a price and (usually) a fixed timescale within which to comply. Any failure to comply is automatically penalised by the terms of the contract. Therefore, the contract focuses the mind of the contractor on the need to be efficient in carrying out the work, and the need to provide a result that meets the specification included in the contract. Anything less and the contractor risks not making a profit. Therefore, in theory at least, the task of the project manager is straightforward, simply to observe the progress made by the contractor and administer the terms of the contract.

The promoter must choose

The quantities/rates-based and reimbursable contracts impose far less discipline on the contractor. Even if there is a specification to meet and perhaps also a timescale within which to do so, there are no price constraints. Indeed the more chargeable units of work or man-hours spent in carrying out the contract, the higher is the profit made. Therefore, if there is to be 'discipline' on the contractor, it will not come from the basic contract structure. It will need to be written into the contract or imposed by the promoter, or project manager, in such forms as a target cost arrangement, detailed contract reporting procedures or arrangements to control and monitor the contractor's work. (The admeasurement procedure, common within civil engineering contracts, is a typical example of this.)

Incentive

The price-based contract gives the contractor two incentives.

The first, already referred to, is to plan and run the contract efficiently, because this reduces cost and increases profit. Planning, however, creates a potential problem for the promoter, in that a carefully planned and organised contract will never result in the shortest delivery period. (Minimising costs means avoiding the problems of overlapping activities for instance.) A tightly planned contract will also always be vulnerable to severe disruption if work is affected by such problems as force majeure or change.

The second is only to supply the minimum amount of work or equipment necessary to comply with the contract, to 'design, or supply, down to the specification', because this also saves cost. In one sense the inherent risk of the price-based contract is under-design/supply (just as the inherent risk of the quantities/rates-based and reimbursable contracts is over-design/supply). Therefore, the fixed/firm-price contract must always be based upon an adequate specification. An adequate specification is one that describes the result required from the contractor sufficiently accurately to minimise any opportunity for significant under-design/supply.

The quantities/rates-based and reimbursable contracts, on the other hand, give the contractor no incentive to under-design/supply: it gives the incentive to over-design/supply, as the more work done, the greater the profit. Therefore, what the quantities/rates-based or reimbursable contract should do is to give the promoter proper project management powers to ensure that the contractor performs efficiently.

Risk

The price-based contract automatically imposes a considerable degree of price/money risk on the contractor and will usually impose a high level of other risk as well. Conversely, the promoter carries a low level of financial risk. The promoter knows exactly what the contractor will charge and when payment is needed – and it is not normal to have to pay the full price until after the contractor has met the contract requirements.

Under the quantities/rates-based and reimbursable contract, on the other hand, the contractor usually bears a rather lower degree of price/money risk and a much higher level of this risk is borne by the promoter.

Change

The great weakness of the price-based contract, from the promoter's point of view, is that, as it sets up a rigid contract structure, it may not cope very well with anything more than the minimum degree of change during the life of the contract. This makes the tasks of the project manager and contractor much more demanding if a substantial degree of change does happen. The quantities/rates-based and reimbursable contracts, by their very nature, cope much better, and are therefore inherently more suitable where significant change can be foreseen. However, that change will still need to be managed.

The promoter must also remember the implications of these different types of contract. Some of the more obvious implications that can be identified are described in the following sections.

Timescale

By its very nature the price-based contract requires a longer timescale to put into place, and to carry out than the other types. To place a price-based contract requires the promoter to put together a detailed inquiry document, including in particular a detailed and comprehensive specification. This takes time to prepare. Then the contractor will need time to examine that specification, finalise designs, bring in prices from potential subcontractors, and so on, and then put together the tender in response. Finally, the negotiations of the final contract document will also, then, almost certainly take more time. As has been shown already, once that contract is in place, it will probably also require more than the absolute minimum time in to carry out. The net result is that price-based contracts do take a significant time to carry out. If time (or quality) is the main consideration, the price-based contract is not always the best choice. If cost is the main consideration, however, the time-based contract will always be the best choice – provided that the contract can incorporate an adequate specification *and* the level of voluntary change is low.

Relationship

In the fixed/firm-price contract, even when highly cooperative, the relationship must always be 'arm's length'. In the reimbursable contract, the relationship should be much more collaborative. This is a much more sophisticated relationship and requires a very different style of project management on both sides. In particular it requires more, and higher-quality, project management time from the promoter.

Change

It has already been shown that the price-based contract copes badly with change. From the point of view of the contractor implementing change part way through the contract within the context of a tightly planned contract has considerable time, cost and organisational implications. This means that the arm's length relationship between the two parties will always come under a certain amount of stress if the managers on both sides find themselves in a series of negotiations about the cost and time implications of changes imposed on the contract by the promoter. It is all too easy for the arm's length relationship to become both adversarial and abrasive, unless carefully managed.

Change is much easier to implement within the context of a reimbursable contract, as the payment structure of the contract means that the cost/profit recovery problem for the contractor is minimised. Change is therefore much less adversarial. Indeed, the problem is that change can become too easy and therefore too tempting for the promoter, so that too many changes are made and the cost of the project becomes too high.

A particular problem has arisen with change in quantities/rates-based contracts within the UK construction industry in recent years. The UK civil engineering and building industries experienced the growth of a 'claims culture'. Many contracts have become the subject of adversarial claims for extra payments, which consume considerable resources on both sides to settle. This claims culture is directly contrary to theory. (Change within the quantities/rates-based contract should be no different to change within the reimbursable contract.) It was to some extent instrumental in the production of the engineering and construction contract (ECC) conditions (see Section 11.10).

11.7 Project organisation

The number and sequence of contracts for equipment and services can vary greatly from project to project, dependent upon the capabilities or preference of the promoter, Figure 11.4. Often a consultant or design contractor may be employed to advise on the feasibility stage of a proposed project, or a project management consultant to advise on risks and contract strategy. For implementation, it may be appropriate for a promoter to employ a single contractor for the whole project. Or a project might be so large or diverse that no one contractor is appropriate to share the risks, so that the project might be let to a consortium of companies or be shared out among two or more separate contractors. For the

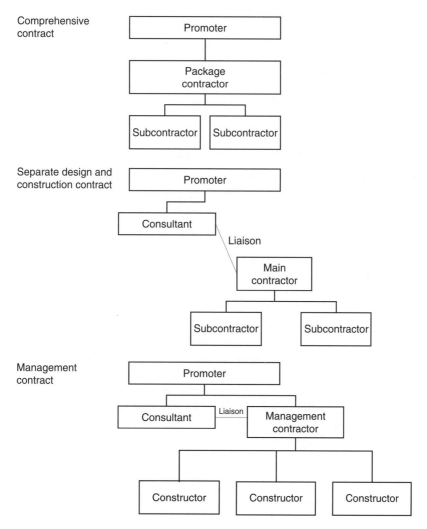

Figure 11.4   Organisational structure.

equipment required for a new factory, one contractor might be employed to install equipment supplied by others.

A series of contractors can be employed in turn for construction work, for instance one for demolition work, another for new foundations, the super-structure and building work and others for designing and supplying equipment, installing it, completing systems and services, and testing and commissioning for replacing part of a factory, each under different terms of contract. For a building project, different contractors could be employed separately for the structural, finishing and services work, instead of one main contractor and these specialist tasks being sublet, or a joint venture or consortium approach could be adopted.

The choice of appropriate organisation for any project should consider the:

- ability to meet project objectives;
- resources and services offered;
- resources and expertise the promoter is able to commit to the project;
- balance between management, design and implementation.

The common types of project organisation are discussed in the following sections.

Package deal ('turnkey', design and supply)

The simplest arrangement contractually is that one single main contractor is responsible for the carrying out of everything necessary for the project from the start to the handover of the completed work to the promoter. Payment is gener-ally on a lump sum or reimbursable basis, although this is often broken down into elements or phases of work. Although the main contractor is responsible for engineering, procurement, construction and installation of equipment, testing and commissioning, parts of the work may be subcontracted to specialists. The main contractor may also be responsible for ancillary activities, such as financing, obtaining public approvals, purchasing process materials or other functions.

The main strengths of the package deal contract are:

- pricing/costing the project at an early stage may be possible provided the promoter's requirements are known;
- price/cost of the project will probably be reduced;
- time can be shortened from design/construction overlap;
- design integration is much easier;
- overall project organisation tends to be better;
- the promoter has to deal with only one organisation for both design and construction;
- fewer promoter's resources need to be involved;
- design should be tailored to give the most efficient construction;
- operating requirements are better dealt with;
- the causes of defects are less a matter of dispute.

There are some weaknesses, which primarily relate to the role of the promoter in the project, including:

- the contractor may not always do what the promoter expected and the promoter's ability to control the contractor will be fairly low;
- managing the contractor will require high-quality skills;
- the promoter must commit to the whole package at an early stage;
- the promoter will have little contact with subcontractors;
- the promoter is in a relatively weak position to negotiate change;
- the extent of competition is likely to be reduced;
- in-house practices of the promoter may constrain the contractor.

In the process/plant and service industries, where contractors are widely experienced in this type of contract, package deal contracts are very widely used. In civil engineering, etc., the package deal contract may be used when the building is of a standard or repetitive nature, for example, houses, warehouses and office blocks, and it can also be used where contractors offer specialist design/construction expertise for particular types of project. It can satisfy a need for an early start to a project. When the promoter and his or her advisors have insufficient specialist management resources for the project, or if the promoter wishes to place his or her work with a single organisation, this form of contract is effective.

Privately financed or concession contracts

The privately financed or build, own, operate and transfer (BOOT) concession concept can be defined as a major start-up business. The contractor, or consortium of contractors, undertakes to build, own and operate a facility that would normally be undertaken by government or a private promoter organisation for a fixed concession period. The ownership of the facility then returns to the government or private promoter organisation at the end of that concession period. These public and private sector contracts are described in further detail in Chapter 17.

Separation of design and implementation

Separation of design and implementation results in a series of two or more contracts for the successive stages of work for a project. A common example is the employment of an architect or consulting engineer for the design of a project, and then a main contractor for its construction/implementation. This is still the conventional or traditional contract arrangement for many building and civil engineering projects within the UK. Often, the design contractor will supervise the construction/implementation contract on the promoter's behalf. Construction/implementation is usually undertaken under a quantities/rates-based contract or occasionally a lump sum or reimbursable contract. Management responsibilities are divided.

Management contracting

Management contracting is an arrangement where the promoter appoints an external organisation to manage and coordinate the design and construction phases of a project. The management organisation does not normally himself execute any of the permanent works. They are packaged into a number of discrete contracts, but may provide specified common user and service facilities.

When management contracting is used, the promoter is creating a contractual and organisational system that is significantly different from the conventional approach. The management contractor becomes a member of the promoter's team, and the promoter's involvement in the project tends to increase.

Payment for the management contractor's staff is reimbursable plus a fee, but the engineering contracts, let by the management contractor, will usually be lump sum or quantities/rates based. The management contractor is appointed early in the project life and has considerable design input. Designers and contractors are employed in the normal way. Another benefit is that the work can be divided into two or more parallel simultaneous contracts, offering reduced duration and risk sharing. Separate contracts for independent parts of the construction of a project divide a large amount of work amongst several contractors.

Some contracts would be consecutive, but it is possible to award two or more parallel simultaneous contracts. Examples in industrial construction and maintenance are separate contracts to employ different contractors' experience on different types of work.

Management contracts have considerable advantages in large civil engineering and building contracts, where the package deal contract cannot normally be used.

- Time saving is one of the main advantages of this form of contract, by the extensive overlapping of design and engineering. By saving time, the impact of cost inflation can be reduced.
- Cost savings can also arise from better control of design change, improved buildability, improved planning of design and engineering into packages for phased tendering, and keener prices due to greater competition and better packaging of the work to suit the contractor's capabilities.
- Further, time saving arises from the management contractor's special experience, for example, overcoming shortages of critical materials and delays arising from a scarcity of particular trades.
- Flexibility for promoter's design changes during construction to improve the fitness of the project for their needs and the contractor's design changes to improve buildability.
- Discipline can be imposed on the promoter's decision-making procedures and on the management of design, in particular to ensure proper coordination of the availability of design information with the sequence of work, particularly on multidisciplinary projects.

There are also some potential drawbacks with management contracting.

- The promoter may be exposed to greater risk from the contractors than in a conventional arrangement.
- Risks may be present due to the absence of an overall tender price for the complete works at the start of work.
- There is a tendency to produce additional administration and some duplication of staff on supervision.

Management contracts can be particularly advantageous when there is a need for the following:

- an early start to the project for political reasons, budgetary or procurement policy;
- an early completion of the project but design cannot be completed prior to construction. This requires good planning and control of the design/construction overlap and careful packaging of construction contracts – the normal skill of the management contractor;
- innovative and high-technology projects when it is likely that design change will occur throughout;
- organisational complexity. Typically this may arise from the need to manage and coordinate a considerable number of contractors and contractual interfaces and possibly several design organisations and it is useful when the promoter and his or her advisors have insufficient specialist management resources for the project.

Offshore oil engineering

The offshore oil industry employs the same range of contract types used in other industries, with the exception of the BOOT contract, which is not used. However, the industry has always used terminology totally different from that in other industries when classifying its contracts. Contracts within the industry are usually defined in terms of acronyms based upon the work content to be carried out by the contractor. Thus, we have the EPC (engineer [i.e. design], procure, construct) contract, the PC (procure and construct – to a design supplied by the promoter) contract, the EPIC (engineer, procure, install, commission) contract and the PIC (procure, install, commission) contract. In broad terms, the EPIC contract is equivalent to what other industries would describe as a turnkey contract.

The only substantial difference between offshore oil contracts and others is that oil industry contracts, even EPIC contracts, always provide for a high level of promoter involvement at all stages of the work.

Project services and management services contracts have been developed to deal with very large and complex projects such as oilfield development. These operate in a similar way to management contracts in building and civil engineering but with some significant organisational differences, for example:

- very large numbers of staff are involved, and 'management of management' becomes crucial to the success of the project;
- the project services contractor is formally integrated into the promoter's management structure.

Direct labour

Instead of employing an external organisation, a promoter may have equipment made or installed or some projects constructed by an in-house maintenance or construction department, known in the UK as 'direct labour' or 'direct works'. If so, in all but the smallest organisations the design decisions and the consequent manufacturing, installation and construction work are usually the responsibility of different departments. To make their separate responsibilities clear, the order instructing work to be done to the design may be in effect the equivalent of a contract that specifies the scope, standards and price of the work as if the departments were separate companies.

This arrangement should have the advantages of making clear the responsibilities for project costs due to design and due to the consequent work. These internal 'contracts' may be very similar in outward form to commercial agreements between organisations, but of course any dispute between the two departments would be settled by managerial decision within the organisation rather than by external legal or other dispute resolution procedures. Contractual principles and these notes should therefore be applied to them.

As described in Chapter 17, if the contractor both promotes and carries out a project he should consider separating the two roles. The expertise and responsibilities involved in deciding whether to proceed with a project are totally different to those required when implementing that project. Separation of these responsibilities may also be required when other organisations are involved, for example, because they are participating in financing the project. For all such projects, except small ones, an internal contract may again be appropriate to define responsibilities and liabilities.

11.8 Risk allocation

A prime function of the contract is to allocate, as between the parties, both the responsibility and liability for risk. Risk may be defined as the possible adverse consequences of uncertainty. The identification of potential areas of risk/uncertainty and consideration of the appropriate way to manage them is a logical part of the development of organisational and contractual policies for any project. Some of these uncertainties will remain whatever type of contract is adopted and the tender must include a contingency sum for them.

Some risks can be due to a promoter's inability or unwillingness to specify what is wanted at the start of the project. If, for example, the contract is to be for the design and supply of equipment that is to be part of a system, and that system cannot be properly defined until after work on the contract has begun, or if the initial prediction of the purpose of a project may have to be varied during the work in order to meet changes in forecasts of the demand for the goods or services that it is to produce when completed, then a high level of uncertainty will be implicit in the project.

Promoters rarely invite tenders on the basis of an organisation being committed to complete the work regardless of risks. Contractors would have to cover themselves by prices far in excess of the most probable direct and indirect costs they might incur. Governments and other promoters in countries with less engineering expertise do, at times, ask bidders to carry very high levels of risk, but the trend in industrialised countries since early in the twentieth century has always been that a risk should be the responsibility of whichever party is best able to manage it to suit the objectives of a project.

All parties to a project carry risk to some extent whatever the contracts between them, for example, work may be frustrated by forces beyond their control. If so, the time lost and all or some of their consequent costs may not be recoverable. The basic division of risks between the parties is usually established in the conditions of contract specified by promoters when inviting bids.

11.9 Terms of payment

Price-based: 'Lump sum', 'fixed' or 'firm' prices are stated in the contract.

Quantities/Rates-based: The contract contains a Bill of Quantities or Schedule of Rates.

Cost-based: Cost-reimbursable, either on the basis of actual cost (plus a profit) or agreed unit rates, (sometimes combined with a target cost mechanism). The actual costs incurred by the contractor are reimbursed, together with a fee for overheads and profit.

Lump sum

A lump sum payment is based on the single tendered price or fee for the whole of the work.

Again, it is important to remember that different industries and organisations use the terminology in totally different ways. Like many other words used in contract management, the same terms can be used to mean different things. What matters in each contract are the terms of payment in that particular contract.

The words 'lump sum' are used in civil engineering to mean that the contractor is paid on completing a major stage of work or milestone, for example on handing over a section of a project. In the process industry; on the other hand, 'lump sum' is used to mean that the amount to be paid is fixed, but perhaps subject to change for escalation.

The words 'firm' and 'fixed' can again be given different meanings. Some industries use the word 'fixed' to mean that the contract price is to be the final price because it will not be subject to escalation, while a 'firm' price will be subject to escalation. Others use the words in the opposite way.

Payment to a contractor under a price-based contract will usually be in stages, in a series of lump sums each paid upon his achieving a 'milestone', meaning a defined stage of progress. Use of the word milestones usually means that payment is based on progress in completing what the promoter wants. Payment based on achieving defined percentages of a contractor's programme of activities is also known as a 'planned payment' scheme.

Although relatively inflexible, contracts will generally include a 'Variation Clause', which provides the promoter the right to make changes within the scope of the works, coupled with the negotiation of consequent price/time/specification changes.

The advantages to the promoter of using a price-based contract are:

- a well-understood and widely used type of contract in the equipment plant process and offshore industries;
- high degree of certainty about final price;
- easy contract administration provided there is no or little change;
- facilitates competitive bidding by contractors;
- the promoter's management resources are freed for other projects.

The disadvantages are:

- unsuitable when a high level of change is expected;
- there is always the possibility that a contractor who has accepted too low a price may be faced with a loss making situation, especially where considerable risk has been placed with him. This may lead to cost cutting, claims and, in the extreme, project collapse;
- the promoter and advisers will have minimal involvement/influence upon the implementation of the project by the contractor.

Lump sum contracts provide an incentive to the contractor to perform when design is complete at tender and little or no change or risk is envisaged. It can be adopted if the promoter wishes to minimise resources involved in contract administration. It should be noted this type of contract is rarely used for main civil engineering contracts. However, it is much more common in process plant contracts, and is virtually standard practice within the equipment industries.

Quantities/unit rate (admeasurement)

This is based on 'Bill of Quantities' or 'Schedules of Rates', typically used in civil engineering and building contracts. Items of work are specified with quantities. Contractors then tender rates against each item. Payment is usually monthly and is derived from measuring quantities of completed work and valuing those quantities at whatever rates are finally agreed in the contract.

Quantities/unit rate contracts provide a basis for paying the contractor in proportion to the amount of work completed and in proportion to the final amounts of the different types of work actually required by the promoter, if different from the amount predicted at the time of agreeing a contract.

Within the UK, the predicted amounts are usually stated in a Bill of Quantities, which lists each item of work to be done for the promoter under the contract, for example, the quantity of concrete required, to a quality defined in an accompanying specification. An equivalent in some industrial contracts is a 'schedule of measured work'. In these contracts the total amount to be paid to the contractor is therefore based on fixed rates but changes as the quantities change.

In some contracts, what is called the 'schedule of rates' is very similar to a Bill of Quantities in form and purpose, as contractors when bidding are asked to state rates per unit of items on the basis of indications of possible total quantities in a defined period or within a limit of say $\pm 15\%$ variation of these quantities. In other cases, the rates are to be the basis of payment for any quantity of an item that is ordered at any time.

The principal strengths of the quantities/rates-based contract are:

- a well-understood, widely used type of contract within the civil engineering and building industry;
- some flexibility for design change;
- some overlap of design with construction;
- good competition at tender;
- the tender total gives a good indication of final price where the likelihood of change, disruption and risk is low.

The principal weaknesses of the quantities/rates-based contract are:

- difficult claims resolution; quantity based and adversarial;
- limits to flexibility, new items of work difficult to price;
- limits to involvement of promoter in management;
- the final price may not be determined until long after the works are complete, especially when considerable change and disruption has occurred and major risks have materialised.

The admeasurement contract can be used with the separate organisational structure. It requires the design be complete but can accommodate changes in quantity. It is used on many public sector civil engineering projects like roads and bridges, where little or no change to programme is expected and the level of risk is low and quantifiable.

It should be noted that admeasurement contracts are sometimes used on high-risk contracts or where considerable change and disruption are expected but the promoter's procedures and regulations prevent the use of a cost-based contract. In these circumstances promoters should proceed with caution and note the deficiencies of the traditional Bill of Quantities approach in allowing for extensive change and particularly disruption to programme.

Cost reimbursable

This is based on payment of actual cost plus a specified fee for overheads and profit. The contractor's cost accounts are open to the promoter (open-book accounting). Payments may be monthly in advance, in arrears, or from an imprest account.

The simplest form is that of a contract under which the promoter pays ('reimburses') all the contractor's actual costs of his or her employees employed on the contract ('payroll plus payroll burden') and of materials, equipment and payments to subcontractors, plus usually a fixed sum or percentage for financing, overheads and profit. Some items, such as computer time or telecommunication costs will be paid for at unit rates. Often agreed-upon low-profit rates will be used as the basis for reimbursement of the contractor's employee costs as well. Sometimes only the costs of satisfactory or acceptable work are reimbursed, but none or only some of the costs of rejected work.

Contractors working under this low-risk arrangement should not expect to be rewarded at the same rate of profit as in price-based contracts because the risk burden is less.

In project management terms, the cost-reimbursable contract tends to produce perhaps the most sophisticated and collaborative relationship between promoter and contractor of these three types of contract. Within the process and offshore industries, it is widely used in large projects, where the detailed design/specification of the plant does not exist at the date of contract. It is also used for 'fast-track' contracts, and as the basis for 'partnering' contracts. Finally, it is also used in contracts for development, repair and maintenance work and a range of other work where the scope, timing or conditions of work are uncertain. Cost reimbursement is therefore the basis of payment in many 'term contracts' for providing construction or other work when ordered by a promoter at any time within an agreed-upon period (the 'term').

Under all such contracts, the promoter in effect employs a contractor as an extension of his own organisation. To control the cost of cost-reimbursable work, the promoter's project team must direct the contractor's use of resources. The contractor's risks are limited, but so is his or her prospective profit.

The advantages of cost-reimbursable contracts are as follows:

- extreme flexibility, plus a high degree of collaboration between the parties;
- fair payment for and good control of risk;
- allows and requires a high level of promoter involvement;
- facilitates joint planning;
- knowledge of actual costs is of benefit to the promoter in estimating and control and in evaluating proposed changes.

The disadvantages of cost-reimbursable contracts are as follows:

- little incentive for the contractor to perform efficiently;
- no estimate of final price at tender;

- administrative procedures may be unfamiliar to all parties. In particular, the promoter must provide cost accountants or cost engineers who must understand the nature of a contractor's business.

In one sense, cost-reimbursable contracts are weak contracts from the promoter's viewpoint. Unless both promoter and contractor are highly professional in their approach, they should be restricted to contracts containing major unquantifiable risk and to projects where no other form of contract is feasible.

There are a number of situations when the use of this type of contract can be justified – when the work is of an emergency nature or when the work is innovative and productivities are unknown, for example, involving research and development. Cost-reimbursable contracts can be used if the work is of exceptional organisational complexity, for example, when there are multi-contract interfaces, to the extent that definition of a target cost is impossible.

Target cost

Target cost contracts are a development of the reimbursable type of contract. The promoter and contractor enter into a reimbursable contract, but also agree to a probable (or target) cost for the probable scope of work, together with an incentive payment mechanism dealing with any difference between actual out-turn and target cost. The contractor's actual costs are monitored and reimbursed under the reimbursable contract. Any difference between the final actual cost and the final target cost is then shared by the promoter and the contractor in accordance with the incentive mechanism. Before reaching the stage of tender documents, the promoter is recommended to investigate the implications of several different incentive mechanisms. If the incentive is to be maintained, the target cost will subsequently be adjusted for major changes in the work and cost inflation. A fee, which is paid separately, covers the contractor's overheads and any other costs specified in the contract documents as not being allowable under actual cost and the profit.

The following simple example and Figure 11.5 shows the effects of a 50/50 share incentive mechanism and a fixed fee on the payment by the promoter and on the financial incentive to the contract. The target cost is £150 million, sharing of difference between actual cost and target cost is 50/50 promoter and contractor, and the fixed fee is £15 million.

In the case where actual cost equals £135 million, the total payment by the promoter is £157.5 million, which is less than the sum of £165 million the contractor would have paid if actual cost had equalled target cost. The contractor receives a lower total payment, but his or her margin, as a percentage of actual cost, is increased from 10% to 16.7%. Conversely, if actual cost exceeds target cost, the promoter is partly cushioned by the incentive mechanism. The cushioning effect also applies to the contractor, but as the contractor has to bear 50% of the excess costs, his or her margin is much less attractive (4.5%). This illustration clearly demonstrates two powerful features of a target cost contract: first, the

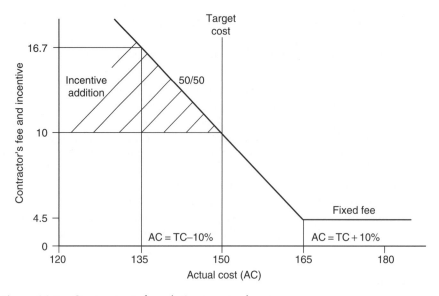

Figure 11.5   Contractor's fee relative to actual cost.

motivational effect of the incentive mechanism on the contractor and, second, the benefits to both parties of working together to keep actual cost under strict control and to bring it below target cost whenever possible. Thus, the target cost mechanism remedies the principal weakness of a pure cost-reimbursable contract by imposing an incentive on the contractor to work efficiently.

The advantages of this form of contract are:

- fair payment for and good control of risk;
- a high level of flexibility for design change;
- identity of interest: both parties have a common interest in minimising actual cost. Fewer claims result and settlement is easier;
- promoter involvement: the contract offers an active management role for the promoter or their agent. Joint planning aids integration of design and construction, efficient use of resources and satisfactory achievement of objectives;
- realism: the facility to require full supporting information with bids coupled with thorough assessment ensures that resources are adequate and the methods of construction are agreed;
- knowledge of actual costs is of benefit for estimating and control, and in evaluating proposed changes.

There are some potential problems in using this type of contract that might make it unsuitable for certain projects; important points to consider include:

- promoter involvement is essential and must be taken a different attitude from that adopted on the price-based contracts;

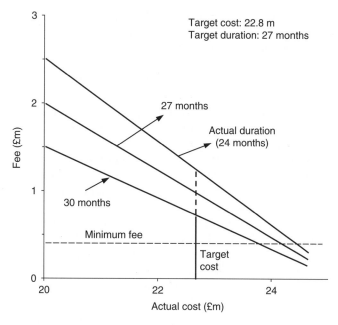

Figure 11.6 Combined cost/time target.

• unfamiliar administrative procedures and a probable small increase in administration costs.

The use of this form of contract should be considered when there is inadequate definition of the work at time of tender due to emphasis on early completion or an expectation of substantial variation in work content, when the work is technically or organisationally complex or when the work involves major unquantifiable risks. The role of the promoter is significant and a target contract can be used if the promoter wishes to be involved in the management of the project or wishes to use the contract for training of their own staff or for development of a local skilled construction labour force.

Targets can be based on other project parameters instead of, or as well as, cost. Time-target contracts are popular for certain types of work and Figure 11.6 shows a graphical representation of a combined time and cost-target mechanism.

11.10 Model or standard conditions of contract

The great majority of projects will use contracts that are governed by 'standard' or 'model' sets of contract conditions. By 'standard' contract conditions, we mean 'in-house' conditions. In-house conditions have the advantage, or disadvantage, that they will always be biased to a greater or lesser extent in favour of the company that writes them. By 'model' conditions, we mean industry-wide conditions, generally accepted as a fair or reasonable basis for contracts within a particular industry.

The advantages of using standard and model conditions are obvious. It saves enormously on time and resources if the contract can be based on pre-existing contract conditions that are already known and understood by both sides. If both sides understand the conditions from the start, it is easier and quicker for the contractor to tender and easier for both sides to manage the contract. Model conditions have the added advantage that contract negotiation will be easier and less adversarial and that fair conditions generally produce a better contract relationship, and probably a more equitable result. Also, model conditions generally have accepted and understood meanings. They will have been used in hundreds or thousands of contracts before. They have been debated by lawyers and managers in dispute after dispute. Their meaning is well established.

Promoters and contractors need to understand how such conditions come to be written and operate.

No set of standard or model conditions exists in isolation. Every set is written by a committee. Standard conditions will generally be written by a committee drawn from within the organisation. Model conditions are written by committees that represent the various interests within the industry. However, when the committee is made up, it is given the task of writing a set of conditions that is suitable for the *typical* contract. In the case of standard conditions, it may be for the typical contract within the particular company. In the case of model conditions, it will be for the typical contract within the industry as a whole.

The conditions will then be written to describe the work that the parties will usually carry out in that kind of contract and the various necessary stages of that type of contract, such as design, manufacture of equipment, site preparation and construction. The conditions will also allocate the various commercial technical and economic risks involved between the parties to the contract, and to provide appropriate solutions to the more predictable problems that may arise during it.

The consequence of this is that the different sets of conditions, especially model conditions, often deal with completely different issues and problems. Civil engineering contracts, for example, deal with contracts where almost all the work is carried out on the site. They deal with the problems of managing a large complex site operation and of dealing with such issues as 'bad ground' conditions. Contracts for the supply of electrical or mechanical equipment, such as a computer system, on the other hand, deal with completely different problems. The contractor will spend many months designing the system, manufacturing equipment, writing and testing software, and so on. Then he or she will bring everything to the site and half a dozen people will install the equipment in a few days. A building contract might cover the erection of a building; the building must be constructed to a good standard, and it will be inspected to ensure that this is so. A process industry contract might cover the design and supply of a plant to manufacture pharmaceutical or food products. The plant will be inspected to ensure that it is properly constructed, but the design/construction would then also need to be validated, and the plant will be subjected to a series of exhaustive operating and performance tests to prove that it can manufacture the correct products for the correct cost.

Different types of commercial/contract situation require different conditions of contract because the work is different, the problems are different and, even more important, the risks are different.

To understand any set of model conditions, and therefore to be able to manipulate and make the best use of them in practice, the promoter needs to understand the following:

- what the basic commercial/contract situation is which that set is principally designed to cover. (In other words, we have to read through the conditions and ask ourselves 'What kind of story are the conditions telling us?');
- how flexible the conditions are, and therefore how easily and how far they can be adapted to cover other commercial/contract situations;
- the commercial attitude (or, almost, the bias) of the conditions. Do they favour the promoter or contractor, and if so how? Do they try to hold some sort of balance between the two? Do they treat the parties as having an equal degree of skill, or do they treat one party as having a lower degree of skill than the other?
- what the more important sections of the conditions actually say; in other words how the conditions actually operate in practice.

There are a variety of model forms within the UK, and Table 11.1 shows some of the forms commonly used in the UK for engineering projects. Note the range of industries represented.

Promoters can, of course, if they wish, use in-house contracts or commission new contracts for a specific project. Sometimes this will be the only appropriate course to take, simply because the project is so unusual that no existing conditions are appropriate. However, there are drawbacks with this approach, primarily that the contract will not have been 'tested at law' and the precise meaning and interpretation of any clause in the new contract could, potentially, be the source of lengthy litigation. It is more likely that the promoter will select a set of conditions of contract that most closely matches his requirements and then modify clauses or add additional 'special' clauses as necessary. (Even then the meaning of the modified or additional clauses may be open to question.)

From time to time there are attempts to produce sets of model conditions designed to bridge the gap between different industries, so that they could be used for a wide range of projects in many different industries. The most recent within the UK has been the third edition of the *New Engineering Contract* (NEC3) originally published in 1993. This was sponsored by the Institution of Civil Engineers, but was intended for all types of engineering and construction projects. NEC3 consists of relatively brief core terms for all contracts and modules of optional terms for use with the core terms as needed, for example, to make a cost-based rather than price-based contract. With this flexibility it was hoped that it would replace the variety of models at present in use in the UK and other countries. It has not so far been successful.

Table 11.1   British model forms of contract.

Institution of Chemical Engineers
Conditions of Contract for Complete Process Plants for Lump Sum Contracts, third edition, 1995; known as the IChemE 'Red Book'.
Conditions of Contract for Complete Process Plants for Reimbursable Contracts, second edition 1992; known as the IChemE 'Green Book'.
Conditions of Contract for Minor Works, first edition, 1998; known as the IChemE 'Orange Book'.

The Institution of Mechanical Engineers, The Institution of Electrical Engineers and The Association of Consulting Engineers
General Conditions of Contract, model form MF/1, 1988.
General Conditions of Contract, model form MF/2, 1991.

Institution of Civil Engineers, The Association of Consulting Engineers and The Federation of Civil Engineering Contractors
Conditions of Contract for Civil Engineering Works, seventh edition, 1999; known as 'the ICE 7th edition'.

Institution of Civil Engineers
Conditions of Contract for Minor Works, 1988.
New Engineering and Construction Contract, third edition, 2005.

CRINE/Institute of Petroleum
CRINE conditions for the offshore oil and gas industry, various 1997 et al.
GC/Works/1 Revision 3 Conditions of Contract for Construction, prepared by the Property Services Agency and published by HM Stationery Office, 1988.

Joint Contracts Tribunal for the Standard Form of Building Contract (the JCT)
Standard Form of Building Contract 1980. Alternatives for private and local authority promoters, and with and without bills of quantities or approximate quantities; known as 'JCT 80'.
Intermediate Form of Building Contract, 1984 (for smaller building projects).
Management Contract, 1987.
Works Contract 3 1987 (for use with the JCT Management Contract).

Note: CRINE – cost reduction in the new era.

11.11 Subcontracts

The same contract strategy principles apply to decisions made by a main contractor to employ subcontractors. Hence, the section above can be interpreted at a number of levels with different parties filling the 'promoter' and 'contractor' roles. For example, in a subcontract the contractor fills the promoter role and the subcontractor fills the contractor role, but in a further supply contract the subcontractor could fill the promoter role and the vendor, the contractor role.

A common principle is that in a main contract a contractor is responsible to the promoter for the performance of his subcontractors. Practice varies in whether a main contractor is free to decide the terms of subcontracts, choose the subcontractors, accept their work and decide when to pay them. It also varies in whether and when a promoter may by-pass a main contractor and take over a subcontract.

In nearly all engineering and construction, the main contractors employ subcontractors and suppliers of equipment, materials and services in parallel.

Reference

Barnes, N. M. L. (1991) The New Engineering Contract, *International Construction Law Review*, 8, 247–255.

Further reading

Allwright, D. and Oliver, R. W. (1993) *Buying for Goods and Services*, Chartered Institute of Purchasing & Supply.

Merna, A. and Smith, N. J. (1990) Project managers and the use of turnkey contracts, *International Journal of Project Management*, 8, 183–189.

Morledge, R., Smith, A. and Kashiwagi, T. (2006) *Building Procurement*, Blackwell Publishing.

Thompson, P. A. and Perry, J. G. (eds), 1992, *Engineering Construction Risks – A Guide to Project Risk Analysis and Risk Management*, Thomas Telford Ltd.

Wright, D. (1994) *Guide to IChemE's Model Forms of Conditions of Contract*, Institution of Chemical Engineers.

Chapter 12
Contract Policy and Documents

Any organisation that wishes to employ others to undertake work on its behalf must have a coherent policy for their selection, together with appropriate procedures for placing and managing contracts with the contractors that it selects. The possible strategies for the identification and selection of the most appropriate contractors and the broad principles that apply to the selection process are considered in the following sections.

12.1 Tendering procedures

Several different procedures may be used for selecting vendors, suppliers, tenderers and contractors.

(1) Competitive – open or select (a restricted number of bidders).
(2) Two stage – a bidder is selected competitively early in the design process. The tender documents contain an outline specification or design and approximate quantities of the major value items. As design and planning proceeds, the final tender is developed from cost and price data supplied with the initial tender.
(3) Negotiated – usually with a single organisation but there may be up to three.
(4) Continuity – tendering competitively on the basis that bidders are informed that the successful party may be awarded continuation contracts for similar projects based on the original tender.
(5) Serial – the bidder undertakes to enter into a series of contracts, usually to a minimum total value. A form of standing offer.
(6) Term – the bidder undertakes a known type of work, but without knowing the amount of the work, for a fixed period of time.

In all cases a pre-qualification procedure may be adopted.

12.2 Contracting policy

It is important that every promoter involved in contracting has a formalised contracting policy established by management, broadly defining why, what, when and how work should be contracted out by the promoter.

The objective of going out to contract is to obtain specific works, equipment, services and goods required to support the promoter's general business. The promoter will, of course, provide some of these requirements from within their own resources. There may, however, be very good commercial reasons why the promoter should obtain them from others, even if the promoter is able to provide them internally. Perhaps a contractor can supply items at lower cost/risk than the promoter. Perhaps a contractor can provide staff, labour and expertise that cannot be made available from within the promoter's own resources, or cannot be made available in the quantity or time required. Perhaps the type, quality, enhancements and fluctuation of required works, services or goods are inherent to the contractor's special skills and not to the promoter's. Contracting out should encourage the optimum utilisation of the promoter's own executive staff and keep administrative overheads to a practical minimum.

As a general rule contractors should be independent, self-sufficient and 'at arms length', and the promoter's aim should be to manage the contract, not the contractor.

Matters that merit review for a consistent policy approach include:

(1) qualification of contractors by qualities such as reputation, experience and reliability;
(2) qualification of contractors by product, location, price/cost, previous performance, and so on;
(3) circumstances under which negotiated, rather than competitive, tenders may be appropriate and/or acceptable;
(4) types of contracts to be preferred in given circumstances;
(5) facilities and services that may be provided to contractors by the promoter;
(6) the promoter's project/contract management resources and skills;
(7) commercial aspects to be included in the contracts themselves.

Procedures

It is always important to remember the basic principles. Contract procedures are written for the purpose of outlining the objectives and scope of contract policy and the manner in which it is to be applied. They are designed to ensure that:

(1) The promoter's policies and procedures are clearly stated and applied.
(2) Individual authorities and responsibilities for the preparation, award and control of contracts are defined.
(3) Appropriate input from advisory and client functions is incorporated at each stage of the contracting process.
(4) Information is passed in an orderly manner to all parts of the organisation, as well as to and from appointed contractors, on a need to know basis.
(5) The best interests of the promoter and its staff are protected and safeguarded.

Business ethics

Management should establish a code of conduct which is brought regularly to the attention of all staff, and should cover:

- the general policy to be adopted in relationships with contractors and their staff who should be dealt with in an equitable and business-like manner;
- the obligation to declare any conflicts of interest, potential as well as actual, should they arise;
- the importance of confidentiality and security in all matters concerned with contracting activities.

Contractors should also be made aware of promoter policies with regard to conflicts of interest and the giving and receiving of gifts and so on, and promoters may consider it appropriate to incorporate a suitable clause on the subject in their general conditions of contract.

12.3 Contract planning

The promoter will usually differentiate between 'projects', and 'purchasing/procurement'. Every organisation constantly buys goods equipment and services as a normal part of its operations. Sometimes the promoter will buy simply materials for consumption, such as raw materials, fuel or office supplies. Sometimes the promoter will buy spare parts or replacements or servicing/maintenance work for equipment plant or facilities. Sometimes contracts would be placed for minor or major modifications to equipment plant or facilities. Sometimes the purchase of major works or facilities will be required. Obviously the same *principles* must apply to all of these activities. They all involve the use of a contractor to supply some thing or some service to the promoter, and the contractor should be treated on the same general basis, whatever the supply. However, the *practices* will change. At some point the promoter will stop purchasing items of equipment or work and begin to operate on a project basis. That point will change from organisation to organisation, and will often depend upon the personnel involved and upon the circumstances of the case. This book is concerned with the overall principles and the practices that should apply when the project is involved.

During the initial phases of a project, discussions should take place under the direction of the project manager between the various departments and disciplines concerned to formulate, review and develop the different aspects of the project, culminating in the preparation of a coherent contracting plan for the project. It will often be necessary to prepare a number of plans to cover the detail of each contract to be let. (In particular 'user' departments should *always* be involved in drawing up the project plan.)

An overall schedule should be kept to monitor the critical dates in the preparation and letting of contracts over the whole project and the critical dates within the sequence of activities for letting each individual contract. As discussions proceed, attention will need to be given to the various implementation considerations including those factors that will influence the type and number of contracts, selection of possible contractors and method of operation.

The contracting strategy must address the choice of the number and types of contract that will best contribute to the success of the project bearing in mind the amount of the promoter's resources that would need to be committed. The types of contract chosen also affect the master plan because of the varying requirement for control and monitoring.

Competitive tendering would be the normal method adopted for letting contracts and the contract should be awarded to the most acceptable tender provided other criteria of comparability and acceptability are met. Often the most acceptable tender will be the lowest-priced tender. However, that will not necessarily be the case where time scale or, particularly in the process/equipment industries, the quality or performance of equipment may be more important than the simple price.

Number of contracts

The first step in contracting planning is to decide the number of contracts into which a project will be divided. One of the basic considerations must be the effect of the number of contracts on the promoter's management effort. More contracts mean more design risk and more interfaces for the promoter to manage and a greater degree of management involvement. Fewer contracts reduce this involvement but may increase the promoter's exposure to other risks.

The following principles should be used in determining the number of contracts.

(1) Each contract must be of manageable size and consist of elements that the contractor can control.
(2) The contract size must be within the capacity of sufficient contractors to allow competitive tendering. Occasionally, only one contractor may be capable of undertaking certain types of work and competitive tendering would not be possible. Such commitment should be minimised.
(3) The time constraints of the work may require parallel activities and this may mean that capacity restrictions necessitate separate contracts rather than a single contract.

In setting the number of contracts it is useful to list the elements of work and the contract phases from conceptual design to test and commission. These can then be considered in a number of different combinations to decide the minimum number of manageable contracts.

Tender stages

There are two stages at which the promoter and his project manager can control the selection of contractors: first, before the issue of tender documents and, second, during tender analysis/contract negotiation and placement up to contract award. Both evaluations are important but have different objectives:

Pre-tender. To ensure that all contractors who bid are reputable, acceptable to the owner, capable of undertaking the type of work and value of contract and competent to manage that type of contract.

Pre-contract. To ensure that the contractor has fully understood the contract, that his bid is realistic and his proposed resources are adequate (particularly in terms of construction plant and key personnel).

12.4 Contractor pre-qualification

As a general principle the promoter should always operate a pre-qualification procedure for project contractors. The choice is whether to adopt a full pre-qualification procedure specific to each contract or whether to develop standing lists of suitably qualified contractors for various sizes of contracts and types of work. Pre-qualification is never a totally objective matter, but subjective judgements should be kept to the minimum, wherever possible.

A full pre-qualification procedure may include:

- a press announcement requiring response from interested firms or direct approach to known acceptable firms;
- the issue by the promoter of brief contract descriptions including value, duration and special requirements;
- the provision of information by the contractor including affirmation of willingness to tender, details of similar work undertaken, financial data on number and value of current contracts, turnover, financial security, banking institutions and the management structure to be provided with names and experience of key personnel;
- discussions with contractor's key personnel;
- discussions with other promoters who have experience of the contractor.

The evaluation may be done qualitatively, for example, by a short written assessment by a member of the project manager's staff to narrow down the number of suitable contractors.

After the potential bidders have been interviewed and evaluated, the project manager should recommend a bid list for the promoter's approval. This should be a formal document that provides a full audit trail for the selection process. Specifically, it should discuss the reasons for inclusion and exclusion of each contractor considered and confirm that all of those selected for the bid list are technically and financially capable of completing the works satisfactorily.

It is the usual practice to pre-qualify about 1.5 times the number of contractors to be included at tender. This is often achieved by a combination of quantitative and qualitative methods as no standard procedure exists.

The EU Procurement Directives

All companies within the EU are subject to obligations, under the Treaty of Rome, not to discriminate on the grounds of nationality against contractors or suppliers of goods and/or services from elsewhere in the Union. Some organisations, primarily national and local government and state-owned or privatised organisations, are also bound by the EU Procurement Directives. These directives require that all procurement of works, equipment and certain categories of services above minimum defined values must be procured in accordance with procedures laid down by the directives. These procedures involve:

- publication of pre-information on procurement intentions in the Official Journal of the EU;
- notification of individual invitations to tender in the official journal;
- prescribed award procedures;
- stated criteria for rejection of unsuitable candidates;
- permitted proofs of economic, financial and technical standing;
- non-discriminatory selection of tenderers;
- publication of prescribed contract award criteria;
- publication of award details;
- debriefing of unsuccessful tenderers.

12.5 Contract documents

For a small amount of work it can be sufficient for a contract to consist of a drawing and an exchange of letters or a simple 'purchase order' issued by the promoter to the contractor and acknowledged by the contractor. The drawing can show the location and amount of work. The materials can be specified on the drawing and the completion date, price and other terms stated in a letter or the order.

For the supply of equipment a specification, describing the promoter's requirements in detail will generally be necessary, as no drawing or series of drawings can normally contain enough information.

If the contract is the result of a series of written or verbal interchanges or negotiations between the parties, the final contract must state all that has been agreed and replace all previous communications so as to leave no doubt as to what the terms of the contract are.

To avoid doubt on all but the smallest of projects the practice has evolved of setting out the terms of the contract in standard sets of documents. These can be lengthy, but will often be virtually identical for many contracts and so do not have to be prepared anew each time. The set of documents traditionally used in UK engineering contracts covers:

- agreement;
- general and special conditions of contract;

- specification;
- drawings;
- schedules.

A civil engineering or building contract would also contain a separate schedule of rates or bill of quantities.

The basis of the contract is established by the tender documents. The tender documents must provide a common basis on which contractors can bid and against which their bids are assessed. The issues between the parties are then clarified and settled by subsequent negotiation between the parties prior to the award of the contract. The precise terms of the contract are then defined by the contract documents.

The contract should be concise, unambiguous and give a clear picture of the division of responsibilities and legal obligations between the parties. Risks should be identified and clearly allocated.

12.6 Tender review

For larger projects particularly, a dedicated tender review team will be required by the promoter. Practice varies from industry to industry and organisation to organisation, but this might comprise a core of two to three people supported by specialists reviewing particular aspects of the different tenders. A typical core team would include the project manager, the lead designer or process specialist and a representative of the user department, with specialist support from other design specialisations, a contractual/legal expert and a quality assurance specialist.

The tender review criteria will depend upon the chosen contract strategy:

Lump sum. With this type of contract the objective of the review is to find the tenderer who offers the lowest price, best programme and yet meets the specification in terms of scope, quality, operability and economic maintenance. It will need to concentrate on identifying areas where tenders do not comply with the promoter's requirements. If the contractor is to have design responsibility, design capability will also be reviewed, if this has not been dealt with at the pre-qualification stage.

Reimbursable cost. The objective with this type of contract is to find the tenderer who has appropriate design and management skills and a project execution capability that gives the promoter confidence that he or she will meet the project requirements. The tenderer must also understand the promoter's requirements and offer a collaborative management team able to work with the project manager.

Quantities/rates-based and other contracts. These will fall somewhere between the above two extremes in terms of the promoter's risks, and the review team must make a judgement on which tender will provide the best value for money.

12.7 Tender evaluation

The project manager will open the tenders from the various contractors on a given date and time. A systematic evaluation of the tenders would include examination of:

- compliance with the contractual terms and conditions (and any qualifications made by contractors);
- technical correctness of tender;
- design or other advantages offered by the different contractor;
- correctness of bid prices (if errors are detected in multiplying rates by quantity);
- screening of bids for detailed analysis;
- pre-award meetings/negotiations (optional but usually essential);
- selection of the best bid and recommendation to the promoter for contract award.

Bid conditioning is a term sometimes used to define a process of reviewing all tenders so as to be able to compare like with like. Some organisations take a rigid stance and reject all tenders that fail to conform totally to the inquiry requirements. Others accept that tenderers may offer alternatives of genuine benefit to both parties and hence consider all submissions. This second approach is normal within the equipment/process industries, though rather less common in the civil engineering and building industries – for obvious reasons. In that event it is usual to require tenderers to submit a 'conforming' bid as well as variants to allow comparisons to be made. Where direct comparison cannot be made, exceptions must be carefully noted.

A misconceived tender based on an error or misunderstanding by the tenderer should be easily identified during tender evaluation. Exceptionally low bids are automatically rejected by some promoters as they suggest that the contractor has made basic errors or is desperate to obtain work at any price. However, in some circumstances a very low bid will be a strategy for 'buying' the job for the contractor's own valid commercial reasons. In the evaluation an attempt must be made to discover the contractor's philosophy. If the promoter is to accept a low price he or she will want to be sure that the project will be completed for this price. The pre-award meeting/negotiation is the ideal time to discover the motives of the contractor.

Contract award recommendations

It is essential that the review team put together a formal recommendation for award of contract. The report should:

- explain the background;
- summarise the recommendations;
- describe bid opening and the initial position;

- describe any bid conditioning process;
- give reasons for rejection;
- identify the tender recommended;
- summarise reasons for recommendation;
- set out the cost time and other implications for the project.

12.8 Typical promoter procedure

Sometimes the tender packages are prepared by an organisation employed by the promoter or by a consultant engineer or management contractor responsible for the design. In either case the procedures to be followed are similar although the titles of the departments and individuals concerned may differ between organisations. In the case of large projects such as those implemented offshore, the level of project management required often makes the services of a management contractor extremely cost-effective. For the sake of conformity the procedures will be described as if a management contractor has been employed by the promoter to coordinate the project.

The project manager controls the tender package preparation possibly as head of a department with duties solely relating to the preparation of contracts. The main purpose of the contracts department is to check:

- that no gaps or overlaps exist between the individual subject contracts;
- that all the particular requirements of the owner are included and covered in the subject contracts.

In particular, the contracts department will liaise and coordinate its work with the other departments with regard to matters such as:

- engineering/design;
- quality assurance/quality control;
- materials/procurement;
- construction;
- legal and insurance;
- planning and scheduling;
- costs and estimates;
- accounts and payment;
- performance;
- planning and other approvals;
- licencing;
- safety/environmental considerations;
- 'buildability/operability'.

These departments in turn should liaise with their counterparts in the promoter organisation.

The individual tender packages are usually developed from pro-forma/standard documents developed as standard for the project by selection of alternatives and/or amendments.

Draft formats of the individual subject contracts should be circulated to each department for review allowing sufficient time to incorporate amendments within the tender package. A standard procedure should be adopted for the reporting of comments and amendments to the contracts department. The final draft of the tender package is then sent to the promoter for formal approval prior to issue of tenders.

The contracts department will maintain a register of tender documents, including information such as:

- reference numbers;
- description of requirements;
- date of issue to the tendering contractors;
- tendering contractors;
- contractors' questions and answers;
- date of return of tenders and the date that tenders remain open to.

The procedures and pro-forma documents used by the contracts department in preparing tender packages should be set out in a document (sometimes referred to as a works plan).

The inquiry will often comprise or include the details given below.

Invitation to tender

This is in the form of a letter to tenderers written on behalf of the promoter. The letter simply invites the contractor to submit a tender for the performance of certain work. The letter lists the tender documents attached and requests the contractor to acknowledge receipt of the documents and their willingness to submit a tender. A 'form of acknowledgement' is usually included with the letter so that replies from all tenderers are set out in a standard way.

Instructions to tendering contractors

Instructions to tendering contractors inform the tenderers what is specifically required of them in their tender and usually comprise:

- tendering procedures;
- commercial requirements;
- information to be submitted with tender.

It will normally be made clear to tendering contractors that the tender submission should be based on the scope of works described in the contract documents, and that any permissible alternative tenders are to be submitted with a conforming tender as set out in the original tender documents so that the promoter may compare the two. Where the promoter will not consider alternative tenders this should be expressly stated in the instructions to tendering contractors.

Conditions of contract (articles of agreement)

The conditions of contract can be prepared by the contracts/legal department in consultation with the project manager and insurance, finance, shipping, and other disciplines. For an international contract, legal and other experts in several different countries may need to be consulted. They are prepared as a standard for the whole project and must be agreed with the promoter.

The conditions of contract are often based on an industrial standard or the promoter's standard. In either case it is likely that modifications will have to be made to suit the unique requirements of the project, the structure of the contract package and the philosophies of the contracting procedures. See Section 11.10 above.

The standard conditions will be modified as necessary for the individual contract packages.

The promoter and the contractor will usually be the parties to the contract, unless another arrangement is required for reasons such as project finance. A project manager may be appointed by the promoter via a separate contract of employment and to monitor the performance of the works.

Brief description of the works

The brief description of the works explains the overall requirements and parameters of the works to be performed. The description should be drafted with the aim of creating a broad appreciation of the works. The overall dimensions of the project should be stated and technical links between the elements of the contract package should be described.

The description should be kept brief and simple and not repeat detail already covered by the drawings, specification, programme or any other work element of the contract package. It should define any work that is not covered elsewhere in the contract package. The interface with other contract packages should be described without reference to the actual work contained.

This section may also be used to detail such things as the facilities that are required to be provided at site by the contractor.

Programme for the works

For larger projects the programme is usually prepared by the planning and scheduling department liaising with the contracts department. It should contain all key dates for the particular contract, including:

- award of contract;
- start of the works;

- completion of the works;
- intermediate key dates for particular elements of the work, where such elements are required to interface with work outside the scope of the subject contract;
- critical dates within the contract.

Indexes of drawings and specifications

These are usually prepared by the engineering department. The indexes should be correctly numbered to identify the latest numbers and revisions of drawings and specifications contained in the tender packages. The indexing system will then usually apply throughout the life of the project.

The format, but not the contents, of the indexes should be prepared as a standard by the contracts department in consultation with the engineering department.

Drawings and specifications

These are prepared by the engineering or quality control and assurance departments. Generally two copies of each drawing and specification are included in the tender package for each bidding contractor.

The terminology of the specifications must be consistent with that of the contract documents. The contracts department should review all specifications before issue to ensure the terminology is consistent. The contracts department should check the individual parcels of drawings and specifications against the indexes in the tender documents. The specifications and drawings issued with tenders (i.e. those listed in the index and specifications and drawings) should be given an identifying revision number.

Promoter provided items

Where items are to be supplied by the promoter they must be listed. This would normally be done by the engineering department in liaison with the contracts department. The following items are usually considered:

- descriptions of the items provided;
- delivery periods;
- specific storage requirements;
- explanation of any markings;
- details of returnable packaging.

Practical delivery periods should be stated against each item. It should be noted that such periods become contractual commitments and should be strictly adhered to.

Contract coordination procedures

The contract coordination procedures describe the administrative requirements for the implementation of the contract.

The procedures explain the day to day duties and responsibilities of the site management team and the contractor's site team. The procedures also detail the lines of communication.

Guideline procedures detailing the implementation of technical requirements can also be included but this is generally inadvisable. If they are included a check must be made for any duplications or contradictions with the specifications.

In compiling the coordination procedures the construction department will need to work closely with the various disciplines to obtain their requirements for the implementation of their specific responsibilities.

Form of tender

This is prepared by the contracts department. Practice varies widely between different industries.

A typical inquiry might comprise the sections shown in Table 12.1.

Review of the tender package

Following the assembly of the tender package a copy is sent to the department managers of the management contractor for review and comments.

Table 12.1   Preparation of a tender package.

Section	Pro-forma used as standard	Individual sections to be prepared
Invitation to tender	X	
Instructions to tendering contractors	X	
Conditions of contract	X	
Brief description of the works	X	X
Programme for the works		X
Index of specifications		X
Specifications	(format)	X
Index of drawings	(format)	X
Drawings	(format)	X
Owner-provided items	(format)	X
Contract coordination procedures	X	
Draft schedules	(format)	X
Form of tender	X	
Form of agreement		

Following the reviews comments if any should be incorporated where necessary. Reasons should be given by the contracts department for not incorporating any comment.

At least two copies of the reviewed tender package should be sent to the promoter for review and comments.

Collation and issue of tender packages

The final collation of the tender packages and the issue to the individual tendering contractors should be in accordance with the contracting schedule and is normally the responsibility of the contracts department.

Queries from tendering contractors

All queries from the tendering contractors during the tender period should be answered by the contracts manager. Any other contact between the tendering contractors and the promoter or other members of the staff during this period is to be discouraged.

Further reading

Ashworth, A. (2005) *Contractual Procedures in the Construction Industry*, fifth edition, Prentice Hall.

Boyce, T. (1992) *Successful Contract Administration*, Thomas Telford.

Wright, D. (1994) A 'Fair' set of model conditions of contract – tautology or impossibility? *International Construction Law Review*, **11**, 549–555.

Chapter 13
Project Design and Structure

Earlier in the book, the evolution and development of projects was examined. All projects originate in organisations that have ongoing business needs in either the profit or not-for-profit sectors. No project is an island as it interacts with the promoter's organisation, and external and internal specialists and contractors. Projects are often described as unique; yet more organisations are using projects as part of their core business activities. This chapter is concerned with the way projects are structured and the manner in which they integrate with permanent organisation structures.

13.1 Organisations

At the heart of all projects are parent organisations that have goals that when translated can become projects. Projects are either the final products for organisations, such as a road for a transport authority, or as an enabler for production, such as in a process plant project. The new organisation landscape that is emerging is more 'projectised'. In this organisation projects take place internally and externally and there is a greater need for integration of specialist project contractors and suppliers. The modern organisation is an open system that has to react to changes in its environment. The projects that operate in this environment need to be adaptable as well. However, a key part of the management of organisations now involves the management of projects and their interaction with the permanent structure.

Faced with the complexity of managing an organisation and its projects, it is therefore essential to adopt a systematic approach to managing these activities. At the heart of all managerial activity is the creation of an organisation structure to execute the managerial objectives and to establish how the people within it relate to each other. Organisation is about creating a control and communication system that allows management to achieve its objectives. It puts in place a structure that defines roles, hierarchies, communication, coordination and control mechanisms.

As organisations become older and larger, they tend towards bureaucracy, that is, they operate extensive systems with rules and regulations to manage the organisation. This approach is adopted to achieve control and stability. Smaller organisations tend not to have extensive rules and regulations and are more flexible in their approach. The most bureaucratic organisations tend to be less flexible, more predictable and better suited to stable rather than dynamic conditions. The more flexible, organic organisations respond faster and adapt more easily in changing and dynamic situations. Interacting with either type of organisation will have an impact on how the project organisation is developed. The design of the project organisation has to serve two masters, flexibility, speed and dynamism, and processes, planning and procedures. This should not only allow it to interact with existing organisation but also fulfil its own goals.

13.2 Building blocks of organisations

There is no universal system of organisation that suits all circumstances. Modern attitudes to organisation design refer to the 'contingency approach', which suggests that the most appropriate form of organisation is contingent on the influences on the organisation. Project organisations are temporary in nature, but often have to interact with permanent external and internal organisations. The manner in which these interactions take place will have an influence on what the project organisation can achieve. By understanding the influences on the design of interacting and sponsoring organisations, it is possible to orientate the project organisation accordingly.

Every organisation, large or small, is influenced by its approach to the system of authority. This is commonly referred to as the hierarchy of the organisation. Hierarchy relates to the number of levels of authority and control, and where decisions are made. The hierarchy is influenced by the senior management's attitude towards control. The closer the levels of control or supervision of subordinates, greater are the levels in the organisation. The classic approach to organisation design resulted in a pyramid-shaped hierarchy that is best illustrated in traditional organisation charts. The hierarchy in an organisation is a function of the senior manager's attitude to control.

The next element in organisation design is how the roles of the members of the organisation are defined. A rigid system of role definition and responsibility creates a sense of stability, in that, there is a place for everyone and everyone knows their place. Bureaucracy was founded on the principles of having this sense of order. This approach, on the other hand, may stifle flexibility, creativity and prompt decision making. More recent approaches to role definition tend to set more flexible parameters for the participants of the organisation. Roles are defined to allow for more autonomy and scope for individual innovation. This approach resembles the organic organisation form. The less rigid approach to role definition allows hierarchy to be broken down and places more faith in the people in organisations. However, role definition is made more complex as projects make increased use of external contractors and specialists. These roles are determined by contracts and relationships that are not always under the direct influence of the project manager.

A balanced hierarchy and sensible role definition do not necessarily create a good organisation. The design of the organisation should also facilitate coordination and communication. An organisation that does not coordinate its activities or communicate efficiently will face difficulties. Communication should be both vertical and horizontal, that is to say, communication should occur up and down hierarchies, as well as between units or teams at a horizontal level. The emphasis on the need for communication should be built into the early thinking in the organisation design. This preplanned approach is becoming more important as organisations use more external sources for achieving project goals. The design of the organisation should also serve a coordinating function. Traditionally, organisation design has tended to use division of work and specialisation as a means of creating structures. This is a sound approach where projects and products are simple, and high levels of coordination are not required. In a project that has complex processes and products the need for organisation design to help facilitate coordination is essential. There is a move towards creating multidisciplinary teams and cross-functional groups that interact more freely and are not constrained by specialist boundaries. In these circumstances, good communication and coordination have to be designed into the organisation.

Most of the design dimensions we have considered relate to the internal perspective of the organisation, but the most important influence on the organisation is how it relates externally. The customers, markets, regulators, shareholders, government, and so on can ensure success or create failure. Burns, in his work on organisations, showed the impact of structure on success or failure in turbulent environments. Peters and Waterman highlighted the importance of external orientation for organisations in their study of 'excellent companies'. It is difficult to identify individual factors that make organisation design more responsive to external conditions but there are certain characteristics that should be present. Decision making should be quick and decisive; there should not be a situation in which there is 'paralysis by analysis'. The structure should allow the company to be responsive to changing markets and customer needs. Role definitions should have loose and tight properties, that is, loose to empower staff to take decisions and be innovative, and tight to ensure that the organisation's goals remain a key focus. External orientation in organisation design cannot be built into the structure, but it develops because of the need to respond to external circumstances. The role of senior management in creating a response becomes paramount.

How does our understanding of these building blocks of organisation design help us in managing projects? They assist us in creating a structure for project organisations and help to improve understanding of interactions between projects and permanent organisations. Organisation design is contingent upon the influences on the firm; hence, there is no right answer. Probably the best advice on organisation design was given by Peters and Waterman when they suggested that structure should be simple, lean and easily understood by those who work in it.

13.3 The project as a temporary organisation

A key feature of a project is that it is temporary: it has a beginning and an end. A project can then be seen as an agency established by a parent organisation to achieve specific objectives that support the parent. During the life of a project, the structure of the parent organisation is likely to remain stable. For a project to perform its functions, the parent organisation has to appoint a project manager to manage the project. The project manager, in conjunction with the parent, has to develop structures, create control systems and develop information and data systems so that the activity can be monitored and aligned to the parent's objectives. The project manager needs to put in place information, control and communication systems to monitor the project objectives as well as the objectives of the parent organisation.

The project as a temporary organisation is also a useful agency for change. By its very nature, a temporary organisation is unlikely to be constrained by organisational structures designed to manage routine activity. The temporary nature of the project, its aims and delivery make change a more inherent feature within project delivery and has to be managed accordingly. The project takes on the form of an organisation in its own right with the project manager at its head. All organisations need resources to function, but a temporary organisation starts with no resources. The project manager has to ensure that the most effective resources are assigned to a project to allow it to deliver its aims. Resources are not just about money and equipment but about ensuring that the necessary skills and expertise are made available to the project. There is the potential for conflict between the resources already allocated to permanent structures within the parent organisation and the project. This interface between the project and the parent has to be carefully managed to reduce the prospect of less effective resources being assigned to the project. Although the project is a temporary organisation, it has to deliver its aims to meet the requirements of its parent organisation.

13.4 Organisation types

Considering dimensions of organisation design, there are a number of generic organisation designs that already exist. These organisation forms have been developed over time and are present in engineering organisations that operate today. Of importance are the factors that underlie their creation and how they might influence projects that interact with them. In engineering, and in many other fields, jobs have become more specialised. The growth in specialisation and expertise has had an influence on the way organisations develop. Specialisation is an extension of the principles of division of work and specialists tend to group together and form teams, sections, units and departments. In its simplest form groupings of specialists give rise to the functional structure, Figure 13.1. Functional organisations arise out of the principle of division of work, build on specialist skills and dominate most organisations through specialists. Expertise and information are contained within each specialist function, group or department. Projects operating within a functional environment are reliant on the cooperation of specialist functions and communication with other departments. Different specialist departments often guard their own expertise,

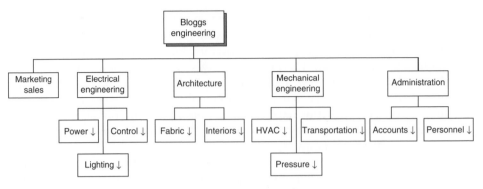

Figure 13.1 Functional multidisciplinary engineering firm.

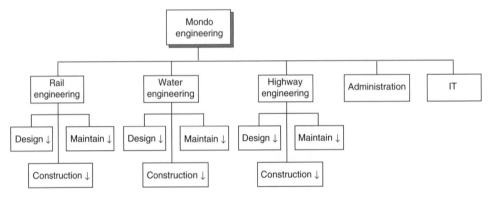

Figure 13.2 Divisional organisation form by type of project.

have their own specialist objectives and are not good at integrating with other experts. They may lack common interest and lack understanding of other specialists, and consequently communication and decision making suffer. It is this separation of interests that pose problems to projects. Individual specialist interests are promoted ahead of project goals and the potential to disrupt the project cycle is more likely. Decision making is based on ensuring functional performance. Projects operating in a functional environment require project managers to concentrate on integrating activities and communication with specialist departments.

Some companies try to overcome the particular problems of functional organisations by creating organisational structures that have more focus and facilitate communication between specialists. Organisations of this type design their structure by focusing on the specialist nature of the work or project rather than on individual expertise. This form of structure is known as the divisional structure Figure 13.2. The aim of the divisional form is to replicate specialist characteristics but to focus on a final product through its specialist type, size, location, customer, and so on. Participants at every stage of the production process are brought together to overcome the specialist niche mentality, to improve communication and to identify with the end product. Projects are easier to facilitate within a division as all the components within the organisation are geared to meeting the division's goals. Divisional organisations are predominantly found in large companies. Moreover, the entire focus of activity is on the division and decision making is based on divisional requirements. Projects are vulnerable when they operate between the boundaries of divisions as interdivision rivalry, poor communication and inefficient information exchange may exist. Projects developing in the divisional environment require project managers be aware of divisional and corporate priorities and how they may impact projects.

13.5 The matrix

Matrix management structures started to emerge in the 1960s, with the need to link project-orientated systems to senior management. The matrix organisation system is a mixed organisational form in which traditional hierarchy is overlaid by some form of lateral authority, influence or communication. Projects by their very nature are temporary and are difficult to integrate within permanent organisation design. This problem is further exacerbated in organisations that tend to manage largely by projects, which usually involves a balancing act between permanent specialist functions and temporary project structures.

The matrix structure, Figure 13.3, attempts to resolve this problem by imposing a temporary project structure across the permanent specialisms. The idea is to move groups of specialists to projects as they are needed. Projects fall under the control of project managers, resulting in the situation that specialists are

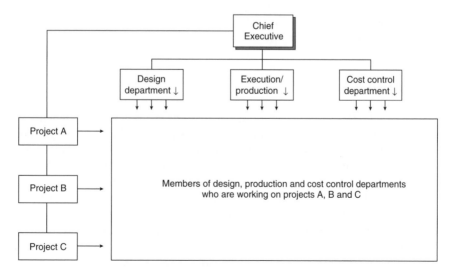

Figure 13.3   Matrix organisation structure.

answerable to two managers. Working in matrix requires careful role definition and a clear authority framework so as to prevent conflict. Three types of matrix organisations exist – the functional, the balanced and the project. In the functional form, the staff remains under the control of the functional manager, while the project manager works across functions. The project manager primarily plans and coordinates while the functional manager retains responsibility for specialist sections of the work. The balanced matrix is where the project managers and functional managers share responsibility for the project resources. In this approach there is sharing of decision making and control. The project matrix shifts the responsibility to the project manager who now has the responsibility of project delivery. The functional manager is responsible for assigning resources and technical support as needed. Most companies will adopt hybrid systems to suit their own organisations.

The personal skills of the project manager are important in matrix organisation for balancing the requirements of the project with the specialist functions of the organisation. These problems become more prominent with planning and resourcing constraints within the firm. Implementation within the matrix organisation is difficult particularly with roles and responsibilities. This represents an area of confusion and conflict between functional managers and project managers. The solution would seem to be to define roles and responsibilities, with tasks at project and functional levels. The drawback to this approach is the possibility of hundreds of tasks and duties for each role. Simpler, easily understood key roles and duties in functional and project areas would be a better solution. Another approach would be assigning priorities between projects and functional arrangements at the organisation level. A centrally coordinated system needs to be developed where staff are asked to give priority to either their functional or project roles depending on organisational urgency. This approach requires senior management to establish priorities and determine importance accordingly. The problems of role conflict are the greatest challenge to matrix organisations and have been instrumental in preventing its widespread adoption as an organisational form.

13.6 Networks

The complex markets of global business environments will demand the ability to deliver customised, high-quality, high-speed products. These products will be differentiated in form, function and service provision. The projects that source these products will not be self-contained. The project will be the hub around which a network of suppliers, contractors, specialists and customers orbit. As demand is shaped in the marketplace, clusters of shared activity will develop networks of suppliers, contractors and customers. Networks have the ability to develop more formal relationships, and over time alliances and frameworks may develop.

Networks are not hierarchical structures; rather each business within the project has its role defined by the contractual relationship it has with other members in the network. The project acts as the integrating mechanism for the network. The role of the project manager is to act as integrator and change manager, as well as to plan, monitor and control these relationships. The power of the project manager is derived from the nature of the contracts set up between the project sponsor and external specialists. These contracts determine the roles, responsibilities, tasks and risk of each network member on the project. The project manager is the coordinator, organiser and controller of these network members.

Networks are dependent not only on the companies coming together but also on the teams of people that operate within the network. The importance of the people and the manner in which they cooperate is important within networks. A more participatory culture with open communication and shared decision making is needed to facilitate project networks. All projects are agents of change with project networks having the need for a more prominent approach to change management. Integrating external organisations into projects requires a cross-functional approach, and bureaucratic organisations may hinder progress in managing change. As the trend towards networked organisations increases, the project manager will become more and more involved in managing projects that arise from networks or federations of specialist companies coming together for a project, see Figure 13.4.

A further development of the matrix organisation has resulted in the hypertext organisation. This term was suggested by Nonaka and Tekeuchi in their book *Knowledge Creating Companies*. This organisational form views bureaucracy and project teams as complementary rather than mutually exclusive. The metaphor that is used for such a structure comes from hypertext. Hypertext consists of multiple layers of text, while conventional approaches consist of one layer of text. Under hypertext, each text is stored under different layers or files.

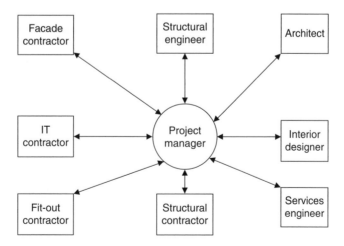

Figure 13.4   Network organisation.

It is possible to not only read through text but also to delve into different layers of detail and background material, thereby creating a different context. The hypertext organisation is made up of interconnected layers. The central layer is the main business systems layer in which routine operations are carried out and may have a conventional hierarchy. The top layer is the project layer, where multiple project teams are engaged in new projects. The project team members are brought together from the different units operating in the business system and are assigned exclusively to the project teams until the project is completed. The third layer does not exist as an organisational entity, but is embedded in organisational culture, corporate vision, technology, and so on. It is the layer that helps contextualise the other layers and provides the basis from which the organisation can learn and develop. The third layer also helps to focus the organisation externally, creating a more open system approach. The hypertext form of organisation is driven by the deadlines set by projects and resources are concentrated to achieve this. A team member in a hypertext organisation belongs or reports to only one part of the structure at any point in time, either the project or the business system. The vision, culture, technology and learning layer ensures that what is developed in the other layers is communicated to around the organisation. The key characteristic of the hypertext organisation is its ability to shift people in and out from one context to another.

13.7 Virtual organisations

The past decade has seen profound shifts in our understanding of organisations, and what it takes to be a member in an organisation. The contracts and spatial relationships between the organisation and its members have been redefined. A major redefinition of the open-ended employment contract has taken place with greater use of temporary flexible arrangements. New forms of working involve different combinations of contractual and locational variables. This creates a working environment that is more fluid and flexible without permanence. A new form of decentralised organisation, the virtual organisation, has now emerged. The virtual organisation is difficult to characterise in terms of work patterns, organisation structure, boundaries and physical form. They are temporary organisations that produce results without having form in the sense of traditional organisations. The virtual organisation defines roles in terms of the task now, rather than the role anchored by the organisation and codified job description. Time, space, the tasks and shifting group membership are the primary definers of responsibility. People are moved in and out of the organisation as and when they are needed. The virtual organisation is taking network organisations a stage further. They are networks of people activity brought together to fulfil particular tasks without the bounds of traditional organisation form. Virtual organisations maximise the use of people and knowledge rather than investing in the costs of permanent organisations. The use of virtual organisations on engineering projects will increase as we use more specialists and experts on a shorter time, task and role basis. The virtual organisation is an exchange network and will continue to develop as an organisational form as we improve communication technology. The primary constraint on virtual organisations is that it does not conform to the traditional approaches to management.

13.8 Multiple projects

Projects were traditionally managed independently of each other, but the multi-project scenario is becoming more prevalent. Multiple projects have to fulfil the strategic requirements in an organisation in a same way as an individual project. The management of multiple projects throughout their life cycle can be extremely challenging as projects start and terminate within different time frames. Each project has its own goal; yet there is a need to provide an organising framework to provide direction to a group of projects so that they can deliver the strategic aims of the organisation. Most organisations operate in a multi-project environment, as there is likely to be more than one project being executed at any point in time. These projects have to share resources according to the strategic direction the firm is adopting. There is a need to find the balance between the individual needs of a project and the long-term needs of the organisation and its supporting projects.

There is no one organisational structure that multiple projects environments can follow. In his paper 'Alternative approaches to programme management' Gray suggests that the multiple project environment is a nominal grouping of projects mainly on the basis of the interdependencies among projects, sub-projects or any kind of project type activity. The vertical and horizontal relationships between these elements are the key to proper structuring of multiple project relationships. Gray also suggests that there are three different approaches that organisations may adopt for a programme of projects: loose, strong or open structures.

The loose approach exists when projects are primarily managed individually with an overview from the parent organisation. Each project runs independently. The strong approach is where there is a coordinated central management without which the projects will not exist. A strong organisational structure is needed and may be formalised through creation of a 'programme or multiple project organisation' where a degree of commonality exists between projects. This approach works well when there is a need for structured control and monitoring.

The open model is a derivation of the open approach. It seeks to empower project managers and give them authority to make decisions for their projects. However, there is still a need for a coordination function for resources and integration with strategic aims.

As organisations become more involved in multiple projects, they have to adapt their structures to suit this changing environment. The hypertext structure gives an indication as to how organisations may have to evolve to accommodate projects within their structures.

Multiple projects create problems for organisational structures as they start individually at different times, consume resources at different rates and have individual performance requirements. They also have to come together in a coordinated manner to meet the aims of the organisation. There are a number of issues that need to be considered.

- Resource allocation has to be carefully managed to ensure that requirements are met for all projects.
- Careful consideration has to be given to overlapping and non-integrated projects in terms of resourcing.
- Roles and responsibilities between projects and line activities need to be addressed.
- Management support of projects is important.
- There needs to be careful monitoring of multi-project environment.
- Information flows are essential in multi-project environments to allow better decision making.
- As organisations have more projects they need to adapt existing structures to accommodate the requirements of projects.

13.9 The human side of structure

Organisation structure in itself is not the end but the means of achieving organisation and project goals. People have to interact within these structures to achieve success.

At the centre of the structural arrangements are leaders. The role of the leader in setting the style of management and the manner in which the structure is operated is crucial. There are the two extremes of leadership style, the centralised approach and the decentralised approach. There are supporters of both these views, but it is also important to consider which approach is suitable to the task in hand and whether the structure is also suited to the task. The traditional approach to leadership was to centralise decision making and exercise control through a pyramid-shaped hierarchy. This is regarded as a top-down approach to management. In clearly structured task environments the centralised approach works well, as there is a need for clarity of direction. More recent views with regard to control and leadership tend towards a decentralised and participative decision making. Organisations rely increasingly on specialists that come from all levels within the hierarchy and it is their knowledge and expertise that influences decision making. The more participative environment is better suited to dealing with complex problems where more ideas and opinions are needed. Decentralised leadership transfers the responsibility of leading to the individuals or groups involved in projects. The role of the leader is important, in that the leader's style of management can set the tone of the organisation, the culture of the organisation and the way it performs.

13.10 Project teams and empowerment

Engineering firms have also made increasing use of dedicated project teams. Establishing project teams within existing structural arrangements is always fraught with problems, and a clear relationship needs to be established between the project team and the rest of the organisation. Role definitions, reporting relationships, power and authority needs to be addressed. Equally important is the development of the project team. Projects are temporary, and hence project teams are temporary, which, however, have to achieve their objectives without the luxury of a second chance or prolonged learning curves. Project teams need to develop a sense of unity, purpose and identity fairly quickly to become fully functioning. It is common prior to the project team becoming directly involved in the project that some form of team-building exercise takes place. This allows the project team to bond, and these measures may also be backed up by locating the team together and using other symbols of a distinct project identity. The extent to which a project team is successful is not entirely about human interaction; it is also about the extent to which the team is empowered to undertake its task. Empowering a project team is about transferring responsibility and authority to the team. The key to empowerment is ensuring that those within the team feel a sense of ownership and control over the project. The greater the sense of project ownership, the more the individuals within the team are likely to work towards succeeding on the project. The transfer of the control of the project to the project team creates problems with the traditional structure, but these have to be overcome for project teams to succeed.

The people within structure have a great deal of influence on the extent to which the organisation can succeed. They are the source through which the organisation and project goals are transformed into reality.

13.11 Structure in collaborative relationships

The manner in which projects are carried out has changed over time. Early management thinking operated with the view that one company conducted all the activity required to produce a product. The view today is that a firm should be good at what it does but it cannot be good at everything. This focus on core competencies has led to the situation where collaboration between firms is more common. Collaboration takes many forms such as strategic alliances, joint ventures, partnerships and joint ownership. These forms of working have an influence on projects and on existing organisations. They influence hierarchy, roles, control and goal setting.

At the heart of all collaborative relationships is partnership. It is safe to assume that considerable thought would have gone into the use of a collaborative relationship and the selection of the right partner. This is the start of the relationship and there are a number of pitfalls that may damage the relationship. One of the first issues to consider is how the collaboration is structured and managed. Is the collaboration going to take on a conventional structural form, a network or a temporary project team? Each system will have merits and negative factors. Consideration will also have to be given to how integration is going to take place with the parent firms. Another factor that has an influence on collaborative projects is who leads the venture and from which firm does this person come. The question of bias towards one partner or the other may arise as a result of the senior project manager's decision. It is also possible to select an 'unbiased' external manager to head the collaboration, but there is the possibility that there will be no political support from parent organisations should the project start to fail. The leadership arrangement will ultimately be a compromise between the collaborating companies.

The control exercised by parent companies will also influence the structure of the collaboration. A parent that has a bureaucratic approach will expect the collaboration to conform to these ideals, while a parent with a more organic approach will be more flexible. Ideally the collaboration will be empowered to achieve the targets that have been set for it with minimal parent company interference. This is an idealistic situation; reality suggests that, because resources are involved, the parent companies will attempt to influence the collaboration. In collaborative ventures, the more powerful parent is also likely to exercise undue influence. It is essential not to forget the role of the parent as the initiator of the collaborative project. They have to be kept appraised of how the venture is progressing. Communication protocols should be established to ensure that issues relating to the collaboration get to the relevant levels within the parent company management structure. The parent company is an important stakeholder on collaborative projects.

Alliances, partnerships and joint ventures involve people from two or more groups of companies. Successful partnerships are not just about the contracts but also about the relationships. A good relationship goes a long way to making the project successful.

13.12 Summary

The management of organisations requires a systematic approach to create a sense of order and control. The lessons from this approach that can be adopted for projects are the following:

- clear definitions of the roles of the participants in the project;
- an established and clear relationship between the project and its sponsoring parent;
- an external orientation in setting up the project management structure;
- an understanding of how multiple projects relate to each other in terms of structuring, coordination and resourcing;
- an understanding of new approaches to organising projects and the management of these relationships;

- the realisation that more projects will require collaborative relationships;
- people and the way they are led is the key to making structure work.

Further reading

Belbin, R. M. (2003) *Team Roles at Work*, Butterworth Heinemann Ltd.

Edwards, P. and Bowen, P. (2004) *Risk Management in Project Organisations*, Butterworth Heinemann Ltd.

Gray, R. J. (1997) Alternative approach to programme management, *International Journal of Project Management*, **15**, 5–19.

Hickson, D. J. (1997) *Exploring Management Across the World*, Penguin.

Hofstede, G. (2004) *Cultures and Organizations – Software of the Mind, Revised and Expanded*, second edition, McGraw-Hill.

Nonoka, I. and Takeuchi, H. (1995) *The Knowledge Creating Company*, Oxford University Press Inc.

Reyck, B. D., Grushka-Cockayne, Y., Lockett, M., Calderini, S. R., Moura, M. and Sloper, A. (2005) The impact of project portfolio management on information technology projects, *International Journal of Project Management*, **23**, 524–553.

Turner, J. R. and Muller, R. (2003) On the nature of the projects as a temporary organisation, *International Journal of Project Management*, **21**, 1–8.

Chapter 14
Design Management

This chapter will concentrate on the stages of design where the greatest use of design resource usually occurs – concept through detail design. It is also at these stages, particularly concept, that the ability of the designer to determine the characteristics of the project deliverable is critical. The ranges within which cost, schedule and quality criteria will fall for the project are determined during design.

Effectively and efficiently managing design is clearly of fundamental importance in the overall project management activity. This chapter will first describe the aspects of creativity in design, then move on to the activities within the design stages and thereafter discuss the techniques used to manage the design work on a project.

14.1 Introduction

Designers occupy a tremendously important position in the project life cycle. The input to the design stages is usually a brief, statement of requirements or performance specification. But the input of the designer to the project does not begin at the receipt of this document, in whatever shape or size it arrives. The very creation of the input document is likely to have been moulded significantly by designers, and in some enlightened projects the designer may even have had an input to the business case development that leads to the initiation of the project in the first place.

Design input to the project does not finish at the time when the implementation of the design solution starts. As the building, facility, infrastructure work, process plant or other type of engineering project deliverable begins to be 'made', design input is needed to guide the implementation people as unforeseen problems arise. Designers will probably be called on during the commissioning stages to assist when the design solution is finally being put into operation. Throughout the operational life of the facility designers will be called upon to create solutions to new requirements not originally envisaged by the project owner, such as new production facilities to cater for increased market demand. When the facility has reached the end of its economic life, designers may well be called upon once again to advise and assist in the process of safely decommissioning and removing the facility.

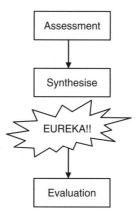

Figure 14.1   The assess–synthesise–evaluate model of creativity.

14.2 Understanding design

At the core of the ability to design is some element of creativity. By the nature of the concept, creativity is hard to define and hard to measure. It is often associated with the 'arts' as a generic term to capture all things creative. However, creativity is also fundamental in many other areas of work and life, such as science, engineering and business. Creativity within the context of design is difficult to isolate. There are many areas of industry and commerce where design activities are carried out. These include such diverse areas as creating new clothes fashions, the preparation of food and drink, graphic design, lithography, business strategy, engineering design, industrial design and architecture. The ways in which creativity manifests itself in these different design processes are many, varied and mostly not well understood.

A number of models have been developed that are used to represent the creative process, and perhaps the most commonly used one is identified in Figure 14.1. This is known as the assess–synthesise–evaluate model. At the assessment stage the person behaving in a creative way assesses a number of information inputs, some well known to the potential creator such as historical solutions to the particular problem that is trying to be solved, and others less well known and more fragmentary in nature. This period is often frustrating as attempts are made by the thinker to manipulate the better and less-well-known elements of information into a solution. There often follows a period of time, in the synthesis stage, when the person working on the problem consciously moves away from seeking a solution to the problem and works on some other activity or problem. At this stage the input information is being synthesised in the thinkers subconscious mind. This is usually followed by the 'Eureka' moment when the thinker becomes aware of what seems to be a major breakthrough in finding the solution. At times a fully formulated solution 'appears' in the mind of the creator, although this is more often illusory. The person working on the problem is then able to evaluate the solution, or a part solution, they have arrived at, contrasting it with the problem as they have defined it to themselves in the assessment stage of the process. The failings in the solution to meet the needs of the problem are then taken into the assessment stage of the next round of the process, when the three steps are repeated until the evaluation of the solution indicates that the problem has been solved.

There is clearly then some element of unpredictability in the process of creation. The period of synthesis between assessment and eureka can be the wink of an eye, days, months and has even been reported in some cases to be years!

There are also a number of different ways of thinking about the way the design process itself progresses, as distinct from the creative aspects of design. These models of design process suggest that the way design is approached can influence the degree of control that can be exerted over design and designing. Two models of design are commonly used.

- **Bottom-up, top-down and meet-in-the-middle**. The bottom-up style combines basic structures eventually producing a final output; top-down, conversely, works back from a desired final behaviour to sub-behaviours that are linked to components and their structures; meet-in the-middle design process style combines both bottom-up and top-down processes. Either method may be used according to the context of each particular aspect of the design.
- **The person-level methods**. With a depth-first strategy, an attempt is made to make work whichever hypothesised design is thought of first; several ideas may be pursued simultaneously until there is a convergence to a final solution; or a methodology is used in which designers know intuitively which design to select at an appropriate stage in the process.

At a less abstract level, design activity can be made more prescriptive, and processes are more and more frequently being captured in design process maps. The common processes modelled include determining functions and their structures, elaborating specifications, searching for solution principles, developing layouts, optimising design forms and dividing design work into realisable modules. Some element of planning is often incorporated, usually to enable the designers themselves to plan their work better.

Working within these explicit processes should encourage more creativity, as less attention needs to be paid by the designer to ensuring ad hoc processes are in place to meet the needs and constraints of the project for which the design is being carried out.

14.3 What design has to do

The design stages

Design is carried out at various stages of the project life cycle and is usually the predominant activity at the project front end. However, the activities that happen at each stage are quite distinct. Although it should be noted that there are

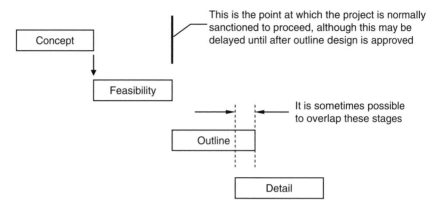

Figure 14.2   Typical design phases for an engineering project.

several different models of the design stages, the stages may generally be depicted as shown in Figure 14.2, and described below.

- *Concept.*   Various options are generated that will provide the design solution necessary to solve the engineering problem. The proposals are at a low level of detail, but are sufficiently well developed to be costed using a global estimating approach.
- *Feasibility.*   The feasibility of the various solution options are considered with respect to a number of criteria, typically including the cost to deliver the solution, the amount of time that would be needed to deliver the solution, the capability of the implementation people to deliver the solution, whether the solution is congruent with the technology strategies of the project participants and what environmental impact the solution will have. There are of course many more criteria that may be used to assess the design solutions proposed.
- *Outline.*   The concept design, which may or may not have been further extended during the feasibility stages, is now developed to the outline level of detail. The major work packages are defined in terms of form, function and delight. This may include process design, space planning, general arrangement drawings, system architecture, and so on. A decision will be made to progress the outline design to detail design stage.
- *Detail.*   The outline design solution that has been approved is now designed in detail. The work packages that make up the overall solution are broken down still further to the greatest level of detail. Individual components are designed, then integrated to form the work package. The work packages are then integrated to provide the overall solution.

Design does not only occur on projects to build new facilities and infrastructure; design is also carried out when facilities are extended or refurbished. These design projects can be huge undertakings, as when oil refineries have new process plant capability added to existing facilities, or they may be much smaller in scale requiring few designers to complete the design task, such as extension works on a private house. However, the stages of design are almost always followed as outlined in Figure 14.2. As with design work for new installations or infrastructure, larger refurbishment design projects usually have a project manager assigned to the design group.

The product and work breakdown structures

The processes required to define the work that is required to be done by the designers working on the project will result in a:

- product breakdown structure (PBS) – identifying the products, or components, that together form the design solution
- work breakdown structure (WBS) – identifying the work required to deliver the designs needed to produce those products or components, and what resources (human and machine) will be needed to undertake the work.

Some industries also generate an assembly breakdown structure (ABS) that shows how the products or components are assembled together to create the design solution.

The process used to tie the PBS and WBS together (and ABS, if used) is the creation of the organisational breakdown structure (OBS). This describes the way that the members of the design team will be organised to carry out the work to produce the designs required for the products identified in the PBS. The mechanisms of PBS, WBS and OBS are common to the project management method, and this is unsurprising as creating the design needed for an engineering project is a project in its own right. Design groups frequently have project managers assigned to them in an attempt to ensure the design work is delivered to the client according to schedule.

14.4　The role of design management

Design management is a relatively new area of professional interest. It has only been recognised as an important subject since the early 1970s, and postgraduate qualifications began to be offered in the 1980s. In fact, effective design management within the project management context is still not well understood.

The creative aspect of design work does not lend itself easily to being managed with the same mechanistic focus that can be applied to engineering projects when they move into the implementation stage. As the models of creativity and design discussed in the previous section of this chapter illustrate, the creative element of the designing process requires a period of synthesis that it is not always possible to 'force' – the subconscious mind needs to work on the problem. The amount of subconscious activity needed may be little or great. This can appear to make the time management, and hence cost, of design work an impossible task. Indeed, many designers resent the 'imposition' of a mechanistic management regime, as they feel this constrains their ability to design effectively. Despite this, there are a number of design management models that have proved to be effective.

Design management is defined in many ways, from the effective control of the flows of design-related information between the various project participants, through an approach involving control over the design process itself, to a more abstract view relating design to corporate strategy. As, within engineering projects, design can be considered to be a project in its own right, it would seem that project management techniques ought to be applicable to design management. However, managing the classic project management triple constraints of time, cost and quality is insufficient. As a fundamental aspect of design is the element of creativity and the difficulty this brings in terms of accurately estimating time to complete a design task, this must be considered in the management regime. Project management is not only about managing time, cost and quality though. It is about understanding the impact of the environment (in the widest sense) on the project. Part of the design project environment is the creative aspect of design work, and hence using a truly holistic project management approach to design management automatically allows for the somewhat unpredictable nature of the design task.

Designers may be classified by the industrial sector in which they are working such as aeronautical, naval architecture, chemical and food. Then the sector breaks down again, with specialisation in each, such as airframe, hydraulics, fly-by-wire and propulsion in the aeronautical sector. Thereafter the designers will fall into one of several professional disciplines such as mechanical, electrical, control or electronic, and all have their own professional institution to represent them and ensure that a universal standard is applied to the education of these design professionals. There is therefore a likelihood of designers in these disciplines acting in a 'tribal' manner. However, experience shows that this 'tribal' behaviour, if effectively managed, can lead to significant innovation in the design stages of projects and does not necessarily have a negative impact on the design organisation. These culturally driven misunderstandings can lead to a lack of predictability, imaginary and real, that for those attempting to manage the delivery of design can be difficult to deal with.

In addition to the difficulties of lack of empathy between different design disciplines, there is an equally difficult, culturally influenced misperception between designers and those responsible for implementing (in other words building or constructing) those designs. In engineering, as in most other fields of endeavour, the people charged with implementing the design solution have a mental attitude different from that of designers, creating a difficulty of identification between the designer and the manager. That design delivery should be managed must, however, be without serious question, from either implementation people or designers.

The inherent difficulties created by the difference of perception between designers and those charged with building the artefact or facility that the project was initiated to provide is not necessarily overcome by physically integrating designers and makers. The best solution to the problem of location of these two groups,

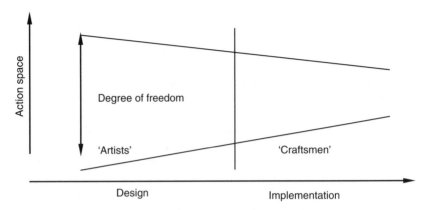

Figure 14.3   The contraction of the action space during the life of the project.

and the design of their overlapping work processes, is determined according to the circumstances of each project. Projects with nearly identical design and implementation processes, which may even be at the same geographical location, will often have completely different organisational arrangements.

As engineering designers and those who manufacture, produce and construct are in many cases educated together (and are unlikely to be firmly set on the course of designer or implementers at that time), these differences in perception must be formed, at least in some part, in the workplace that they do eventually move in to. One of the important exceptions to this 'co-educated' system is that of architects, and it is likely that the seeds of the antagonism that exist between the architectural profession and the constructors who build their designs are planted during their separate training.

This difference in mental attitudes between the two groups may be visualised as shown in Figure 14.3. The designer works in the area to the left of the diagram, where there is a great deal of freedom. This allows multiple perspectives to be taken on a problem and many alternative solutions generated. As the project life cycle moves from concept design, through feasibility studies, to outline and then detail design the degree of freedom gets smaller as the final design solution becomes clear. As the project moves into the implementation phases, the degree of freedom becomes increasingly smaller and the required actions by implementation people correspondingly more prescribed.

The need therefore of an understanding by the project participants of the cultural differences between design and implementation is important and is not necessarily intuitive. The manager of these design processes must be able to work effectively at the interface between 'artistic' designers and 'craftsmen' implementation people. The person managing design, seeking to influence the designer's environment within the project, must be able to understand both groups' perceptions of the project reality. This position between design and implementation allows the manager of the design organisation to have significant influence over the amount of design creativity within the confines of the project. This role can be seen as to some extent that of nurturing, or at least supporting, creativity within the design organisation. Sensitivity to the designers' need for autonomy and management focus on design (and hence project) objectives, is a prerequisite for a successful manager of design.

One of the fundamental requirements for effective management of design is the efficient flow of information between the participants in the project. This applies particularly to those that have an input to the design phases of the project, and a list of such participants would include at least:

- the client and appropriate groups within the client (such as their design department);
- the users/operators of the project deliverables – that may or may not be part of the client;
- the project manager;
- team leaders in the project team;
- the design manager;
- the lead designers;
- the design team leaders;
- sub-designers and appropriate team leaders within those groups;
- design approvals consultants acting on behalf of the client;
- design checking consultants;
- local authorities;
- statutory bodies.

The length of the list indicates the prospective complexity of the paths (channels) along which information needs to flow, not only in one direction (asymmetric) but in two directions, as feedback on the content of the information is vital. For instance, for designers to fully understand the brief they have been given requires that they question the brief to facilitate proper understanding of the brief writer's true meaning. Hence, the briefer and the briefed must have a two-way dialogue (symmetric). It also means that the communication path (or channel) must be duplex and not simplex, that is, information can flow in both directions at once (as in a telephone line).

It is important that an explicit communication strategy is developed, and the necessary channels between the project participants are established. Rules for the use of the channels must also be put in place. These rules will cover such issues as the medium that certain types of information should be transmitted in (e.g. paper, verbal, email or video), standard formats within each medium that should be adopted (e.g. how to lay out minutes of meetings), to whom within each channel the information should be passed and levels of authority on the information is passed. There are also many other considerations that are dependent on the circumstance of each project. The important thing to remember is that communication must be managed. Always consider that the information being transmitted is important, in terms of content, and that the correct person receives it.

The information being discussed in this context is that which facilitates design activity. There are also other types of information that are related to the monitoring and control aspects of the management of design, such as reports, scheduling diagrams (e.g. Gantt charts and dependency networks) and document-control processes. The control of these types of information is also important for the effective management of the design phases of the project and forms a significant part of the communication strategy.

14.5 Managing the project triple constraints

The management of time in the design phases is fundamental to the delivery of the design at the appropriate time to enable adherence to the project schedule. The first part of the chapter described the inherent difficulties in managing time in the design process. Delivering design information to those needing it, when they need it, is managed using several processes.

The concept of gates in the project life cycle is relatively new, and has in the main been learnt in engineering projects from the way in which new products are developed for the consumer market. A gated design process means that at certain points in the design phases, the design must 'pass through' a gate. The rules for passing through the gates must be established, as must the points at which the gates are positioned in the design process.

There are commonly two types of gates. They may be 'hard' or 'soft'. A hard gate is where the design cannot be progressed to the next stage if the gate is not passed. The design process may not move into the following stage until sufficient rework has been done to allow the design to pass through the gate. Soft gates are ones in which the design is allowed to progress to the next stage, even if not being accepted as 'compliant' (dependent on the gate's rules). However, a commitment must be made by the person responsible for the design (the design team leader or the design manager) to make changes to the design to ensure that it becomes compliant before the next gate. It is also possible to have 'fuzzy' gates, which are essentially a combination of hard and soft gates. In a typical fuzzy gate process, parts of the design may be progressed to the next stage (those that comply with the rules), while the non-complying parts must be reworked in the previous stage until they do comply.

Figure 14.4 shows where the gates are commonly positioned, although different organisations and different projects will position the gates at different places in the process. Figure 14.5 describes the different types of gates.

The process for scheduling design work must be based on knowledge of the individual deliverables (drawings, calculations, reports, specifications and other documents) from the PBS and WBS. Scheduling the creation of the deliverables can then be done from the dependency between the deliverables. Creating a dependency network and then a Gantt chart from the network, is the most effective way of building a schedule. Smaller design projects can be scheduled by creating a Gantt chart without first creating a separate dependency network. The duration of each individual task in the network is decided after consultation

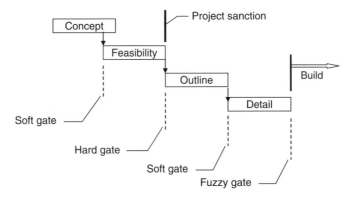

Figure 14.4 How a gated design process may be set up.

with those with knowledge of the work to be carried out. The difficulty with this type of scheduling tool is that the iteration that is a fundamental aspect of the design process cannot be modelled. The amount of iteration required between design tasks, or groups of design tasks, must be 'built-in' to the schedule in some way. This is not a very satisfactory way of scheduling, leading to continually adjusting the schedule as the iterative cycles in the design work unfold. There is one particular method of scheduling that can overcome this difficulty that involves a technique called dependency structure matrix (DSM). However, there is still no commercially available software application that can carry out DSM.

An aspect of planning in the design stages that is crucial to consider when estimating the time and manpower requirements is the degree of front-end loading that will be employed. Front-end loading refers to the practice of employing a significantly higher number of designers earlier in the design phases of the project than has normally been the case. The idea of doing this is to concentrate project resource at the point where the most effect in shortening the overall project time scale can be achieved. In broad terms the phases at which the biggest influence can be brought to bear on reducing uncertainty later in the project is at the outline design stage. However, the loading of extra resource is also done at the concept, feasibility and detail stages; see Figure 14.6. It must be remembered though that at the concept stages highly experienced designers will be employed to quickly identify the most advantageous possible options for solving the design problems, and there is very probably a scarcity of such design resource in the firm.

Putting extra resource on the work at the outline stage allows the development of the chosen design solution to be carried out much more quickly. The iterative cycles at this stage can be moved through rapidly and the final outline design solution can be articulated in a much shorter time. This means that the uncertainty that surrounds this stage of design (i.e. whether the outline design actually fulfils the promise of the concept chosen at feasibility) is removed from the process much sooner. Detail design can then be started with less risk that the outline design will have to be revisited (often meaning that the design process

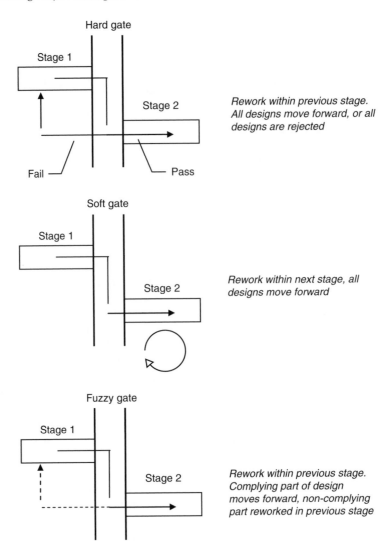

Figure 14.5   Types of gates.

has to be stopped while the implications of technical risk in the outline design are reassessed).

The cost to deliver design work is almost entirely the cost of the time of designers assigned to produce the design – essentially the cost of the human input to the process. There are overheads to consider such as the cost of the facilities required to carry out the design (the design office itself, power, design tools such as CAD and software packages), administrative support and management overhead. But the predominant cost is design man-hours required to produce the deliverables. This cost can be estimated using the information contained in the schedule. The time required to complete a design task can be multiplied by the hourly rate for

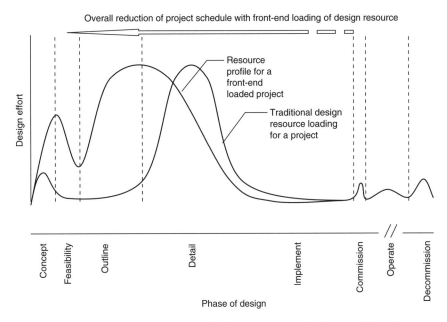

Overall reduction of project schedule with front-end loading of design resource

Resource profile for a front-end loaded project

Traditional design resource loading for a project

Design effort

Concept · Feasibility · Outline · Detail · Implement · Commission · Operate · Decommission

Phase of design

Figure 14.6 The design resource profile for a project with front-end loading.

the appropriate skill level of design required, producing the cost to complete that task. Overhead can be added per task, or added to the overall estimate once all the tasks have been costed.

Once the cost estimate to carry out the design work has been completed, the next stage is to consider the risks associated with the design tasks. These may include such issues as, How well known is the technology that is being designed? If the technology is mature, and there is great experience and knowledge in the design firm of working in this technology, there is probably little risk in this area. However, if the technology is new, or the designers have little experience of working in this technology, then the risk of overrunning the time to produce the design deliverable is high. Other risks may be associated with the lack of stability of the design brief (or user requirements if these were issued instead of, or in addition to, a brief) and the likelihood of changes to the brief from the client. There may be unknown risks from the construction process, classically in the construction industry, the actual ground conditions as found opposed to the expected conditions predicted from the geophysical investigation. There are a large number of risks that could impact the design schedule, and hence the cost estimate. These must be assessed and the appropriate contingency made in the schedule and cost estimate. It is most helpful if the design organisation, meaning all the firms with a significant input to the design process, are involved in the overall project risk management process. Often, risks to other elements of the project are not identified as having a possible effect on the design process. Design involvement in the risk management process can help to ensure these risks are picked up and incorporated into the design schedule and cost estimate.

Once the estimate has been produced and accepted by the management function of the design firms, the estimate is turned into a budget. This is an allocation of money to the manager of design to pay for the design work. At this time the firm becomes committed to the expenditure and the manager of design now has to provide accurate cost information to the accounting function to allow effective control of the business to be carried out. Cost control on larger design projects may be done using earned value techniques, and this methodology is discussed in Chapter 10. Often, design cost control is based on simple, and not integrated, cost reporting against scheduled progress of work.

It is during the design phases of the project that much of the quality of the ultimate project deliverable is established or enabled. There are a number of definitions of quality, but most of them in some way or another address the need to meet the requirements of the client.

Two clearly different quality processes must be encompassed in the design phases of a project. The first is to ensure high-quality design work, in and of itself. Doing so is not a straightforward process, but consists of a number of aspects. Accurately *capturing* the requirements of the client is fundamental. Putting in place a design *process* capable of developing an appropriate solution is essential. Ensuring that the solution developed *satisfies* the client's requirements is critical. Carrying out these activities effectively is dependent on approaching the tasks with a consistently good attitude to achieving high-quality design. The second process is about enabling the construction, building or manufacturing (the implementation) of the design to be carried out to high-quality standards. This entails designing within the process capability of the implementation phases of the project. Process capability is well established in manufacturing industries, where the production machinery's capability to make a component to the required accuracy must be considered in the design stage. This means, for instance, that designing a component to a machining tolerance not achievable (by the equipment used to make the component) is explicitly prevented by the design processes used. This concept of process capability is directly analogous to ensuring that the design for a structure is buildable *in practice* as well as in theory.

The most well-established and comprehensive quality system for the design phases of a project is known as quality function deployment (QFD), or the 'house of quality'. QFD can be considered to be a system for designing. It monitors the transformation of the client's requirements into the design solution to ensure that quality is inherent in the solution. To do this QFD integrates the work of people in the project's participant organisations in the following areas:

- briefing (to understand what the client's requirements are);
- engineering (to understand what technology is available to be used);
- implementation (e.g. manufacturing and construction – to understand the process capability);
- marketing (to understand the client's perceptions of the solution that satisfies the requirements);
- management (to understand how the processes to ensure quality can be put in place).

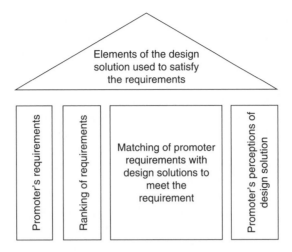

Figure 14.7   The quality function deployment matrix.

The primary set of considerations for the QFD team are given below.

(1) **Who** are the clients – in the broadest terms, that is, the users of the project deliverable, the owner and other stakeholders?
(2) **What** are the customer's requirements – which may or may not be explicitly stated in the design brief?
(3) **How** will these requirements be satisfied – including an evaluation at the highest level of abstraction, such as should the project actually build a road or a railway to meet the requirement to transport people from A to B?

The term 'house of quality' comes from the shape of the matrix used to capture the information generated in the QFD process; see Figure 14.7.

The client requirements are scored in order of their relative importance and ranked, after a weighting criterion is used. The 'roof' of the matrix contains the elements of the design solution that will satisfy the requirements of the client. The ability of these elements of the design solution to satisfy the requirements is then estimated using experience and judgement in the central core of the 'house'. The final part of the QFD exercise is to evaluate the client's acceptance of the design solution in total, and at the elemental level of the matrix, with regard to the client's requirements. It is important to understand that the QFD will not generate a design solution, but it will enable the quality of that solution in meeting the client's requirements to be monitored with rigour and accuracy.

14.6 Design liability

A key consideration in the management of design is the issue of the legal liability of the designer for defects in the solutions they create. Failure of the design solution may have many repercussions and affect many people. The designer may be liable under many different areas of the law – falling under the three main groups of contract law, tort and statute law. There is insufficient space in this chapter to make any attempt at a meaningful coverage of the issue of design liability, so a few commonly found issues will be touched upon.

Designers providing a design solution to a client with whom they have a contract are required to comply with the terms of that contract, usually to provide a design within a certain time, to an agreed-upon price. The contract will in addition specify technical details of the designer's work, against which it will be possible to determine whether the design solution fulfils the contract between the client and the designer. Failure to meet the terms of the contract is known as a 'breach of contract' and is pursued in the civil courts to enable the client to get recompense, usually in monetary terms, for the damage inflicted by the failure of the design. The case will hinge on the terms of the contract (unless overridden by statute law, for instance, where health and safety has been compromised).

The designer may also be held liable in tort, which is a non-contractual liability to third parties. When designers produce a design solution that is faulty, they may, in legal terms, be negligent – but only if they did not take reasonable care when carrying out the work. The definition of negligence is that the designer must use 'reasonable skill and care' in the execution of the work. Common law (under which most contract and negligence law falls) is based on judging a present case against the principles laid down in previous similar cases in which a 'precedent' has been set by a judge's ruling. There are appropriate precedent cases in most sectors of industry that relate to the negligence of designers. There are also two particular cases that relate to the term 'reasonable skill and care' of a professional (which is what a designer is considered to be). The first is known as 'Bolam's test', after a case in which the reasonable skill and care of a doctor, called Bolam, was tested in a negligence case. Bolam's test sets the precedent for the skill to be expected from an average person working in their field of professionalism. The second case is called *Roe v The Ministry of Health* (1954). This case established that a professional cannot be liable for negligence if following the normal practice of that time, even if subsequently that practice can be shown to be flawed. This has importance for designers working on innovative design solutions that have not been tried before. In these situations, despite taking reasonable skill and care, a design fault may occur that can only come to light after failure of the design – it could not have been predicted in advance. In this situation the designer is unlikely to be found negligent.

It is usual for designers to have in place professional indemnity insurance in the event they are found to have been negligent in the execution of their work for a client with whom they have a contractual relationship. This type of insurance is, however, expensive and does not buy a great deal of financial protection. For this reason, contracts for the supply of design usually have strictly limited liability clauses, with the designer (the supplier) often only accepting liability up to the value of the contract.

Another very important issue for designers is their responsibility for 'product' defects. In the area of product liability, most commonly found in products for the consumer market, the law does not allow for the professional designer to exercise reasonable skill and care. Liability in this case is 'strict'. This means that the designer is liable for any defect in the product that is there as a result of the design process. There is no issue of professionalism, or lack of it. If there is a defect in design, the designer is liable. In reality the designer's risk exposure due to design fault is often submerged beneath the other product liability issues of manufacturing, workmanship and materials liability.

Health and safety law is also an important consideration for designers. There are general industry-wide requirements for designers to produce safe design solutions, particularly (but by no means limited to) where the general public are exposed to the dangers of inadequate design. Many industrial sectors also have specific legislation to protect workers and the public from design failure, an example of which is the Construction (Design and Management) Regulations 1994 and 2007.

The importance of assessing and then controlling the design liability issues by managers of design is clearly of great importance. The design manager must be aware of the safety, contractual and insurance implications of the design project to ensure they are managed effectively.

14.7 Briefing

The brief is often considered the key document for a design project. Although most sectors of industry would recognise the term, some do not – software design houses are more used to working with a statement of requirements, although the basic concept of the document is similar. Simply stated, the brief is the method by which the client (external to, or within the same organisation as, the designer) tells the designer what it is that is required to be designed. This sounds pretty straightforward until one begins to articulate exactly what is wanted by the people commissioning the design. The briefing process can in fact be excruciatingly difficult. The main cause of the difficulty is in the nature of the way in which a solution to a problem is arrived at. The iterative and uncertain processes involved in arriving at a design solution were discussed earlier in this chapter. It may be considered that the ideal place from which a designer ought to begin working from is a pure statement of the requirements of the client (the requirements of a client are a clear and concise statement of the problem that the design is to overcome, completely devoid of any suggestion of the solution). Unfortunately, things are not so simple.

First, it is enormously difficult to state requirements without some reference to the possible solutions. The requirements capture process is only in recent years becoming better understood and reliable models for requirements capture are not always easy to find. Some are available, however, although they tend to be associated with specific industries such as automobiles and information systems. To carry out the process properly is often a costly and time-consuming exercise as the information required to generate a complete picture of the requirements must come from many people (the users, builders, suppliers, marketing, maintenance engineers and other stakeholders). The time and money to do the work may well not be available.

Second, just as it is becoming better understood how to create 'solution-less' requirements, some authoritative design experts are cautioning against the creation of such information at all. It is frequently found that the briefing process is distorted when the initial design concepts are presented to the client, and the client says 'I don't want that!' The initial brief is then adjusted by the client (hopefully with input from the designer) to reflect what type of solutions are not actually acceptable to the client. Sometimes this cycle can extend far beyond the original design period, costing the client much more money than originally envisaged to employ the designer, and can create great tension – and probably reduced motivation – in the designer or design group.

Third, to make the briefing process even more opaque, it is also being acknowledged that one of the cardinal rules of the briefing process can be legitimately broken. The rule of never changing any of the requirements is often considered to be inviolate and fundamental to the existence of the brief. This is frequently being challenged in some sectors, particularly in new product development, where the market for a particular product can be as short as a few months. Changing, or relaxing, a requirement in a design brief in these situations can mean a design solution is created that will get the product in the market earlier than competing products – and may mean the difference between the organisation surviving or failing.

The simple fact is that briefing is less about a repeatable process, although this may be desirable in many design project situations, and more about the interactions between the brief writer, those responsible for determining the requirements upon which the brief is based and the designer. Creating a brief from which the designer can most effectively work, and the client has most confidence of an appropriate design solution resulting, is a joint exercise. Frequently this is not the case. Many clients believe they alone are capable of creating the brief, even if they employ specialist design expertise to help them do so. The point is that the actual designer who is to provide the design solution has a big input to the brief because the way he or she works will impact the types of solutions that will be created, and this can then be reflected in the brief. Involvement in the brief writing will also enable the designer to far more completely understand what the client is likely to accept as a solution. The three parties together are also much better equipped to challenge the requirements definition feeding the brief, seeking sensible changes in the requirements based on information they all bring to the process. This means that less iteration between the client and designer is likely, although the design process iterations will still be required as part of the natural evolution of the final solution to the problem. The design outputs stand a far better chance of being delivered within the time scales envisaged by the client and at the cost expected.

One of the many checklists for creating a design brief is included here for reference. It was designed by the Department of Trade and Industry and published in 1989. It provides a useful basis from which to assess the needs of a brief in any industry.

- Value for money/lifetime operating costs
- Product uniqueness/superiority
- Selling price
- Performance
- Reliability
- Serviceability
- Maintenance costs
- Life expectancy
- Versatility
- Running costs
- Ease of operations/user appeal
- Ergonomics
- User friendliness
- Appearance
- Legislative and community factors
- Compliance with regulations and standards
- Safety
- Environmental impact
- Factors important to the manufacturer
- Time scales of development programme
- Unit cost of production
- Facility for future range expansion or product improvement
- Balance between in-house manufacture and brought-in items
- Other factors
- Size and weight
- Ease of transport and installation
- Cost of development programme
- External consultancy requirements and need for involvement of sub-contractors or suppliers
- Production levels envisaged and organisational implications
- Investment requirements – for stock, work in progress and production capacity.

(Department of Trade and Industry, 1989).

14.8 Interface control

A design solution very rarely has a single discrete product or is comprised of a single component. Far more usually, the solution comprises a significant number of components, and most often there are a significant number of components arranged in a number of sub-systems. For instance, a software program of any size at all will be made up of a number of modules of code that act together to create the program's functionality. A car engine consists of many sub-systems – the fuel supply, the oil supply, the cooling water, the electrical system and so on. This property, of almost any system, is not limited to engineering. Sophisticated drugs similarly comprise a number of systems of interacting components such as the material used to bind the 'active' parts of the drug together, the complex series of components that form the active' component and often a special outer containment coating which itself is a complex formulation. In almost all situations where designers work, they will be dealing with a design solution that comprises multiple sub-systems, each containing multiple components. The management of the interfaces between these sub-systems is often crucial to the effective creation of the solution. Some disciplines manage these interfaces better than others, and those that do it well and consistently are usually 'system'-orientated, for example, electronics, information systems and aerospace. Other sectors such as heavy engineering and civil and structural engineering are far less system-focused despite the self-evidence of the fact that they also create systems (a bridge is clearly a system of interacting sub-systems).

Effectively managing the interfaces between the sub-systems creates significant advantages in the overall management of the design process. Setting, and subsequently 'freezing', the interface requirements between sub-systems means that the designers of the sub-systems can then work on designing their part of the overall solution without further reference to those working on adjoining systems. Each sub-system design only needs to satisfy the interface constraints. If these are met by the sub-system, then the operation of the internal components in the sub-system is not relevant to other interfacing sub-systems (from which property the term 'black box' arises). Hence, the need for information to constantly flow between the designers working on the separate systems is removed.

The work of defining interfaces is not trivial. In a system-orientated design solution the 'architecture' of the overall solution (the way in which the sub-systems 'fit' together) is usually the responsibility of a senior and experienced engineer. The architecture determines the way in which the design solution is broken down into manageable sub-systems, and what, and at what level, the interface constraints are set. The degree to which the overall design is broken down, and the size of the sub-systems, is fundamental to effective system, and hence design, management. The crucial interface management issues are given below.

- Deciding to what level the overall design should be broken down into sub-systems – and therefore determining the number and 'positioning' of the interfaces.
- Ensuring that the setting of the interface constraints reflects the needs of the overall design solution – this means compromises will need to be made for individual interface constraints.
- Deciding on the tolerances that the constraints should have – if the constraints are too tightly specified, optimisation of sub-system design can be

reduced dramatically, and if too loosely specified, the overall design solution will probably perform poorly.

- Ensuring that the interfaces, and their constraints, are 'frozen' at an appropriate time in the design project's life cycle – freezing too soon will lead to sub-optimisation of the overall system as not enough is known about the system's properties, whereas freezing too late will prevent the designers from making the technical (and quite likely commercial) decisions needed to deliver the sub-system on time.

A schedule of the interfaces showing freeze dates and required delivery dates for sub-system designs is a valuable management tool. It will also be needed for configuration management – broadly, the process of ensuring that the system architecture is allowed to change in a strictly controlled manner.

14.9   Design for manufacturing

In most sectors of engineering, the majority of the cost in a project life cycle is incurred in the implementation stages – in construction when concrete is cast into its form in the position where it will remain, in mechanical and electrical engineering when the designed components are manufactured, and in electronics when the circuitry is assembled. The 'making' stage of a project typically accounts for between 75% and 90% of the total cost of the final project deliverable. Therefore, anything that reduces the cost of manufacturing or producing the components is worth pursuing. One of the biggest cost drivers in manufacturing is design that does not take account of the most cost-effective processes for making the components.

It is clear that it is vital that manufacturing specialists have significant input at the design stages. The process of bringing in this expertise to design is called design for manufacturing (DFM). In essence the design is optimised at the earliest stages of the processes that will be used to make the components. This is not an easy or comfortable approach to design for many designers and manufacturing specialists alike. Reference to Figure 14.3 reminds us of the fundamental difference in mental models between designer and implementer. Getting these groups of people to work together effectively is a key task of the design manager when DFM is being done. It is important to recognise that, to be most effective, DFM needs to be started at the earliest stages of design, when concepts are being generated for the various solutions to the design problem. There is little point in choosing a concept design to progress into detailed design work if the concept chosen cannot be supported by the existing capability of the organisation to make the components. At the least, DFM allows a logical debate to take place about trading off the costs of new manufacturing capability against the attributes of the design that can create extra value in the final product.

The success of DFM can be assured by recognising and acting on the realisation that differing cultures within design and manufacturing exist. The primary obstacle that this difference creates is that of effective communication. There are two key ways to improve communication between these two groups. The first is to plan for it to happen. This means identifying where in the design project life-cycle DFM will have most effect (invariably early on) and then ensuring appropriate DFM processes are created in time to be used most effectively. It also implies that DFM workshops and review meetings are built into the schedule. The second way to ensure communication happens is to ensure some common understanding of the issues faced by the two groups. For instance, it is far from obvious to designers that the manufacturing process capability required to make their design solution may not exist – particularly when an external client is manufacturing. However, this lack of knowledge of manufacturing capability is also frequently found when the design will be made in-house. Equally, manufacturing specialists are rarely aware of the specific reasons why a particular feature of the design is necessary to create added value to the client. These differences in awareness between design and manufacturing are natural and to be expected. It is up to managers of design to proactively manage the DFM process for the greater good of the client and, ultimately, the design organisation itself.

A related design management process is design for assembly (DFA). A major part of the 'making' cost for a design solution is the time needed for the assembly of the various components forming the overall product. In such industries as aerospace, power engineering, consumer electronics, and the manufacture of white and brown goods, assembly time is heavily influenced by the ease of assembly of the product that will be sold. Consequently, the specialists in assembly processes must be brought into the design process in the same way as the manufacturing experts are involved in DFM. Unsurprisingly, the differences in culture between the designers and assembly specialists are just as evident in the DFA process as for DFM. Communication between the two groups is facilitated in the same way as for DFM – plan to communicate and create a situation where common understanding can be gained.

The management processes of DFA and DFM also interact with the design solution. It is quite possible that the design of a component that has been optimised for manufacturing is very difficult to assemble, adding time (and therefore expense) to the processes that will deliver the final product. Conversely, a design optimised for assembly may be expensive, or even impossible, to make using existing manufacturing process capability. It is incumbent on the design manager to ensure that the correct trade-offs are made between designing for maximum client value, low-cost manufacturing and ease of assembly.

Further reading

Allinson, K. (1998) *Getting There by Design: An Architects Guide to Design and Project Management*, Architectural Press.

Cooper, R. (1995) *The Design Agenda: A Guide to Successful Design Management*, Wiley.

Department of Trade and Industry (1989) *Design to Win: A Chief Executive's Handbook*, DTI.

Eisner, H. (2002) *Essentials of Project and Systems Engineering Management*, second edition, John Wiley & Sons.

Hamilton, A. (2001) *Managing Projects for Success*, Thomas Telford.

Lawson, B. (2006) *How Designers Think*, fourth edition, Architectural Press.

Reinertsen, D. G. (1997) *Managing the Design Factory*, The Free Press.

Chapter 15
Supply-Chain Management

This chapter defines supply chain management drawing upon studies by academics and practitioners across a range of industries to identify the implications for the project manager. The chapter presents a module for supply chain management.

15.1 Introduction

Supply-chain management (SCM) is concerned with strategic internal and external activities in the supply and purchasing function of a firm. It focuses on the management of the sourcing, acquisition and logistics of the resources essential for the transformation process of an organisation to produce products or services from a network of firms that add value for its customers.

There is considerable ambiguity about terms and definitions and the scope of the field of SCM. This chapter presents the terminology used in the field, and looks at the concept of supply-chain strategy, including highlighting the influences on the purchasing and supply function. The notion of the 'world-class' organisation is introduced and subsequently implications are drawn from this for construction. The chapter introduces the idea of a supply-chain system compromising a project-focused demand chain generated by the client and the contractor acting as a multiple project-focused demand-chain hub that has to develop a supply-chain strategy to meet these different needs. Within this, the implications of different types of clients are explored and the concept of the project value chain introduced. A recently introduced procurement route initiated by the Ministry of Defence (MoD), termed prime contracting, is discussed. The chapter goes on to explore research conducted by the Warwick Manufacturing Group into SCM that builds on work related to the MoD's prime contracting initiative. The chapter also presents research work conducted on supply-chain scenarios in the constructional steelworks sector as an example of the potential impact of the newer procurement routes such as prime contracting and private finance initiative (PFI). It suggests that a new role will emerge in the industry, that of the strategic supply-chain broker, who will compete on the basis of core competencies in SCM. It also explores the possible restructuring within the sector that may occur due to the emergence of the broker role. Finally, the chapter concludes with a section that develops a framework for the supply-chain system in construction. This section also draws on empirical work conducted by Male in this area.

15.2 Perspectives on terminology

The 'traditional' model of purchasing and supply focuses on developing and retaining appropriate knowledge and skills in the purchasing area; hence, typical elements in the traditional model include a specialist department or section within a department dealing only with purchasing and the placement of orders with suppliers. It operates within a known hierarchical organisational structure, where paperwork systems dominated task activity prior to the advent of computers. Policies and procedures will have been established to deal with the enquiry and competitive bidding processes, order placing and contract management. 'Price' will figure strongly in purchasing managers' thinking to evaluate supplier performance, with decision making focused on securing the right quality, quantity, price, supplier and location and delivery at the appropriate time. The focus will also be on the individual transaction, although some repeat purchasing may occur. Finally, relationships with suppliers are predominantly of an adversarial nature and at arm's length, using competitive mechanisms for supplier choice. The traditional model is one in which purchasing acts as the interface between the firm and its suppliers under conditions of market-based competition and economic rationality.

Due to competitive pressures, however, a number of recent and significant trends have resulted in a need for organisations to become more effective and efficient with consequent influence on the way they are managed, including the purchasing function. Organisations, especially the larger ones, have to become more adaptable, responsive and flexible to changes in the business environment. Organisational layers have been removed, with an increased use of cross-functional teams and delegation of responsibility to lower levels in the hierarchy. Also, communication is horizontal as well as vertical, with manager roles changing from directing to facilitating. Incentives and staffing systems are being aligned to accommodate these changes. Managers have changed the manner in which they have approached the wider organisational context. This is also linked to a greater appreciation of the opportunities that can accrue from more cooperative ways of working between suppliers and customers resulting in cultural changes within firms. The consequence of this is that firms are more willing to consider working closely with suppliers and customers to create a more integrated production and supply process that goes beyond legally defined organisational boundaries. The impact has been that firms now have to work out their role and positioning within a wider configuration of organisations, with a consequent impact on organisational structuring. This has put the supply function clearly on the strategic and not operational agendas of firms and potentially opens up new roles for them, including associated impacts on the marketing function, with 'relationship marketing' coming to the fore.

This different way of thinking about supply and purchasing has resulted in a series of new terms being adopted and used to describe the domain. The *supply-chain* concept has emerged to have an underlying implication of a sequencing of interdependent activities that are internal and external to the firm, as a legal entity. This can encompass single location activities for fairly simple firms to multi-site, geographically dispersed locations, often located internationally.

The idea of the 'chain' has been extended to include analogies with rivers and 'upstream' and 'downstream' terminology has become infused into the language, often to include the concept of a 'supply pipeline'. Further extensions to the concept have included viewing a supply chain in network terms, seeing it as a series of connected, mutually interdependent organisations cooperatively working together to transform, control, manage and improve materials and information flows from suppliers to customers and end users. One implication of the supplier network is that the boundaries of different supply chains might overlap and particular suppliers might become nodes within a more complex web of patterns of suppliers. Typically, flowchart mapping techniques are used to understand these 'flows' within the chain or supply pipeline. The impact of lead times and individual cycle times is also encompassed within the analysis.

For the supply chain to work as an integrated system requires both the management of materials and information. It also involves managing the upstream and downstream business relationships between customers and suppliers to deliver superior customer value at less cost to the supply chain as a whole.

15.3 Supply-chain strategy

Product strategies provide the basis upon which a supply-chain strategy is built. Product strategies requiring different approaches to time, cost, quality and product innovation place different requirements on the supply chain and its structure and infrastructure. The structural features of the supply chain, as indicated in Figure 15.1, involve the fundamental physical activities of *make* – the production activity, *move* – the logistics activities, and *store*. The infrastructure of the supply chain comprises those features concerned with controlling the operation of the physical system and includes planning and control systems, human resource policies and communication strategies.

Design and technology strategies are concerned with decisions on which activities are carried out internally and which are carried out by external suppliers. Design strategies would also include *product* and *process* design. Decisions on design and make would encompass those to be undertaken either within the firm or by suppliers. Equally, it would entail related decisions on which design and technology competencies are to be retained internally and the extent to which supplier innovation would also be encouraged. Design strategy considerations could also include:

- the use of concurrent engineering principles;
- supplier involvement in design teams;
- product simplification and standardisation;
- the use of computer-aided design through the supply chain and the coordination of design processes and data when in use.

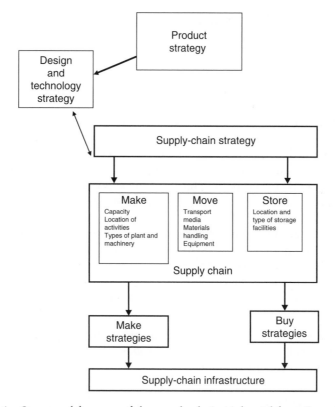

Figure 15.1 Structural features of the supply chain (Adapted from Saunders, 1997, figure 5.3, p. 150).

The extent of 'Make' or 'Buy' is one of the fundamental strategic questions that managers in a firm have to ask in SCM, that is, the scope of the activities undertaken within the firm versus those that are carried out by other firms external to the firm. Answering the Make versus Buy decision defines that part of the supply chain that is within the firm's direct control as opposed to that requiring the investment of resources to develop, enhance or retain internal capabilities. Subcontracting, in its various guises, is an extension of the make or buy decision.

Finally, the creation of sub-assemblies within an overall finished product, as part of a sourcing strategy, creates a tiered structure to supply chains. The final 'manufacturer' or 'producer' has a choice. They can decide to purchase and assemble all components and items and then sell the completed product. However, they may decide to break the final product down into major sub-assemblies and contract these out for manufacture. The final producer thus assembles a series of sub-assemblies. Tier 1 suppliers are responsible for producing major sub-assemblies. Tier 1 suppliers, in turn, will therefore source from their own suppliers – Tier 2 suppliers. It is possible that Tier 2 suppliers may well have components requiring the inputs of Tier 3 suppliers, and so on. The number of tiers within a supply chain creates its structure and shape.

15.4 The nature of the organisation

Saunders proposes four types of organisation, described below:

- Those in the primary sector, that is, they are involved in extracting a product or material that exists in nature.
- Manufacturing organisations that process, shape, form, transform, convert, fabricate or assemble materials, parts and sub-assemblies into finished or intermediate products for sale to external customers. Saunders acknowledges that within a heterogeneous group of firms there are sub-types that have an impact on the corporate, business and supporting manufacturing, purchasing and supply strategies. The on-site construction process is the equivalent to this type.
- Those in the tertiary sector – wholesale and retail. This type of organisation has no internal transformation process: it buys in from external suppliers and sells on to external customers.
- Those in the tertiary sector – service providers. The product received by the customer is intangible. Design team firms in construction are part of this sector.

Saunders has also explored SCM issues through the type of manufacturing organisation. Hill, quoted in Saunders, proposes that there are five process types of manufacturing organisation:

(1) project based;
(2) jobbing unit or one-off;
(3) batch;
(4) line;
(5) continuous processing.

This schema accounts for variations in volume, continuity and variety of products being manufactured; each will have a different impact on SCM requirements. Construction is clearly project based. In addition, factory layout has an impact. Saunders proposes four distinct approaches to factory layout.

(1) Fixed position, where all resources, workers, machinery and equipment, and materials are brought to a fixed location where the product is built. This typifies construction.
(2) Functional or process layout, where the factory layout is divided into separate areas or workflow areas specialising in a type of process.
(3) Line or flow process, where materials proceed in a fixed sequence through a series of processes, often dedicated to a particular product.
(4) Cellular manufacturing or group technology, where separate work centres are established with self-sufficient workers and equipment to make specialised families of parts or products.

15.5 The world-class organisation in manufacturing

Because of global competitive pressures, the manufacturing sector has concerned itself with developing the 'world-class organisation', one that has an international reputation for overall effectiveness and knows its core business well. This class of organisations will encompass four critical elements.

(1) A clear customer focus, normally through using flat organisational structures.
(2) Flexibility to provide rapid responses to customers and competitors.
(3) A programme of continuous improvement, which is the key to differentiating the firm from its competitors, achieving sustainable competitive advantage and promoting internal organisational learning. A programme of continuous improvement will normally involve the following elements:
   - Benchmarking against competitors, either in terms of products, functions or processes, using best practice as an extension of process benchmarking or strategic benchmarking in terms of the future direction of competitor firms. Benchmarking priorities in the supply chain revolve around an assessment of which processes within the chain are strategically important, have a high relative impact on the business and have a crucial impact on competitive advantage.
   - Corporate strategies that are designed to expand the organisation's knowledge assets.
   - Empowerment of employees and an incentive-led innovation policy.
   - The use of outsourcing where it adds value.
(4) The effective use of information technology to provide accurate and reliable information, to create the ability to respond rapidly and to differentiate the firm and provide a competitive advantage and greater knowledge, and hence an understanding of the market place.

Depending on the procurement route adopted – a project-focused demand-chain issue – Figure 15.2 indicates that the client and design team generates the major cost commitment for most projects but are only responsible for approximately 15% of the client's expenditure, primarily through design team fees. The contractor is responsible for the major element of expenditure, some 85% of the project cost, and yet, depending on the procurement route adopted, may be cut-off from a direct influence on client and design team thinking, and their commitment of cost in the early stages of projects. Hence, when looking at a project from a value-for-money perspective for the client, certain procurement routes preclude the contractor's knowledge and expertise from being accessed to the benefit of the project and potentially adding-value much earlier in the project process.

Also, in terms of roles and responsibilities within the distinct chains, the primary function and nature of the organisation will have an obvious impact. Design team firms are clearly tertiary-level service providers whose role can change depending on the procurement process chosen. Main contractors are also service

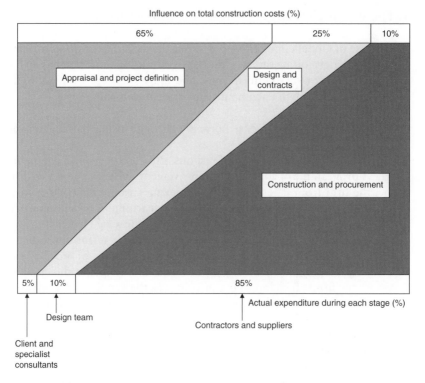

Figure 15.2   A schematic comparing the cost and expenditure during phases of a construction project (from Standing, 1999).

providers offering to manufacture end-products that are designed, purchased and made to order across numerous project-focused demand chains. However, the multi-project supply chain of the contractor has to handle not only this but a number of other critical influences, namely, multiple fixed-position 'factory' locations that require all resources to be brought together for the assembly process at each location. Equally, where contractors act as manufacturers, they will face considerable product diversity due to different client types, with their individual requirements and design team influences on this, coupled with the impact of the choice of procurement route on roles and responsibilities. In terms of the capacity and capability to influence their demand chains, regular, knowledgeable, volume-procuring clients are in a considerable position of power to influence their own project-focused chain. Some contractors who have the advantage of size will be able to influence their own multi-project supply chains.

Client influences

One of the important considerations for SCM in engineering is the impact that the client (or customer) has on the process (Table 15.1). Each client has distinct

Table 15.1 Client/Promoter and demand impacts on the construction supply chain.

	Client type									
	Private sector						Public sector			
	Knowledgeable			Less knowledgeable			Knowledgeable		Less knowledgeable	
SCM demand response	Consumer clients: Large owner/ occupier	Consumer clients: Small owner/ occupier	Speculative developers	Consumer clients: Large owner/ occupier	Consumer clients: Small owner/ occupier	Speculative developers	Consumer clients: Large owner/ occupier	Consumer clients: Small owner/ occupier	Consumer clients: Large owner/ occupier	Consumer clients: Small owner/ occupier
Unique			✓			NA			NA	NA
Customised		✓	✓	✓	✓	NA			NA	NA
Process	✓	✓		✓		NA	✓	✓	NA	NA
Portfolio	✓		✓			NA	✓	✓	NA	NA

The ✓ denotes that this is the probable occurrence. NA indicates no occurrence, and a blank cell indicates a possible but unlikely occurrence.

requirements and value systems, driven by their own organisational configurations, business and/or social needs for a project, the external environment to which they have to respond and the manner in which they approach and interface with the construction industry. The client commences the process of procurement and brings a demand chain together through a project process and requires a completed product – a facility of some type – to meet a need.

A number of distinguishing characteristics can be applied to clients. They separate into *public* or *private* sector clients. The public and private divide provides different pressures on the project-specific demand chain. Public sector clients are now driven by public accountability and best value. Private sector clients are much more homogeneous, with influences ranging from the impact of share holder value on projects to time-to-market considerations and ownerships considerations due to private limited company (plc) or family business status. Clients also differ in their level of knowledge of the industry and can be characterised as of two types.

(1) **Knowledgeable clients** will generally have a structured approach to project delivery, often encapsulated in some form of project delivery manual or set of procedures or guidelines, including the manner in which they will have developed the project brief for the industry. They will treat the supply chain and its members as 'technicians' to delivery a project or projects to meet their business or social need. They will employ either internal or external project managers to act on their behalf as the interface with the industry and they will tend to be innovative in the manner in which they approach procurement. The knowledgeable clients will generally be the volume procurers of services and be demanding of the supply chain.

(2) **Less knowledgeable clients** will often have limited or minimal in-house expertise and knowledge of the operations of the industry. They will rely on the design team to brief the project and will often approach the consultants first, depending on the type of project. This type of client will have limited, if any, appreciation of the complexities of engineering and will tend to be directed onto a traditional procurement path because of their initial point of contact with the industry.

Clients, promoters or customers to engineering projects can also be classified by type.

- **Large owner/occupiers,** who will use facilities as part of their ongoing corporate strategic plan to meet a business or social need and will normally have undertaken an intensive study of their project needs.
- **Small owner/occupiers,** who will often react to change and will be driven to approach the industry because their existing facilities are inadequate in some way.
- **Developers,** who view facilities as a method of making profit, and will trade the asset to achieve this or see it as an investment to generate profit and look for business opportunities and available sites to ensure a quantifiable return.

A fourth dimension to client characteristics is the economic demand placed into the industry in terms of volume – frequency and regularity – and the extent to which standardisation may exist from project to project in terms of parts, processes and design.

(1) **Unique** projects have the distinctiveness of technical content, the level of innovation or are leading-edge projects that push the barriers of the industry's skills and knowledge to the limit. With this type of project, there is limited, if any scope, for efficiencies in process or standardisation and repetition. Typical SCM tools and techniques suggested by Croner (1999) include:
   - use of competitive tendering coupled with strong pre-qualification and post-tender negotiation processes.
   - control over product delivery exercised through specifications and forms of contract and quality assurance processes.
   - a reliance on good professional advice.

(2) **Process** projects can occur where the client has repeat demands for projects and a high degree of standardisation is possible through the volume placed into the industry. Efficiencies can occur from standardisation of design, components and processes. There are many similarities to the manufacturing sector assembly lines. MacDonalds, the fast-food restaurant client, is an example of this type. Typical SCM tools and techniques proposed by Croner (1999) include:
   - use of forward planning and demand forecasting techniques.
   - rationalisation and consolidation of suppliers by spend.
   - use of strategic alliances, joint ventures and partnering with suppliers using non-contractual forms of agreement.
   - use of performance management and continuous improvement, quality circles, total quality management, just-in-time and inventory management and lean supply systems.

(3) **Portfolio** projects are those in which clients have large and ongoing spends across a range of project types. However, unlike the process approach, this type will involve a diverse range of needs in terms of technical requirements, degree of uniqueness or customisation and content, but regular spends will permit long-term relationships with some suppliers. Clients involved in this type would be the Defence Estates Organisation (DEO) of the MoD, BAA and London Underground, and typical SCM tools and techniques suggested by Croner (1999) include:
   - clustering of suppliers.
   - use of forward planning and demand techniques.
   - agile and flexible supply agreements normally using some type of 'framework agreement' or 'call-off' contract arrangements using schedules of rates and partnering philosophies.
   - use of the learning organisation philosophy and supplier innovation, benchmarking and continuous improvement.

The next section explores the concept of the project value chain.

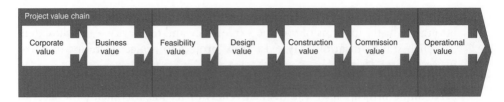

Figure 15.3   The project value chain (from Standing, 1999).

15.6 The project value chain

Value chain activities are the basic building blocks from which an organisation creates value for the customers of products or services. The project value chain forms part of an organisation's value chain as project activities are superimposed on the organisation's normal operating activities. This leads to the concept that a project adds value to the organisation through its own processes. The project value chain concept developed by Standing is set out in Figure 15.3.

There are two primary transition points in the project value chain. The first is the decision to sanction the engineering and the second is the handover of the completed facility into the operational domain. There are also other transitional points as different organisations become involved. Discontinuities can occur resulting from changes in values due to the influence of those organisations involved and a different focus being applied to the project.

Standing's project value chain framework – Figure 15.4 – is subdivided into three distinct value systems:

- the promoter value system that creates the demand chain;
- the multi-value system, involving parts of the demand chain and main contractor's supply chain;
- the user value system.

The client value chain is concerned with a project to be constructed to meet a business objective, or perhaps a social objective or a combination of both depending on the type of client. The decision to build stage is the point at which the client effectively outsources the 'business project' to the construction industry in the form of a 'technical project' to meet that need. The problem becomes one of ensuring the alignment of the different organisational value chains involved in the project process to form a holistic value-driven project-focused demand chain working to the benefit of the client. For example, a more generic arrangement is set out in Figure 15.5, indicating levels of complexity that can creep into the project value chain, as a supply network.

The next section looks at the impact of the procurement system on the project value chain.

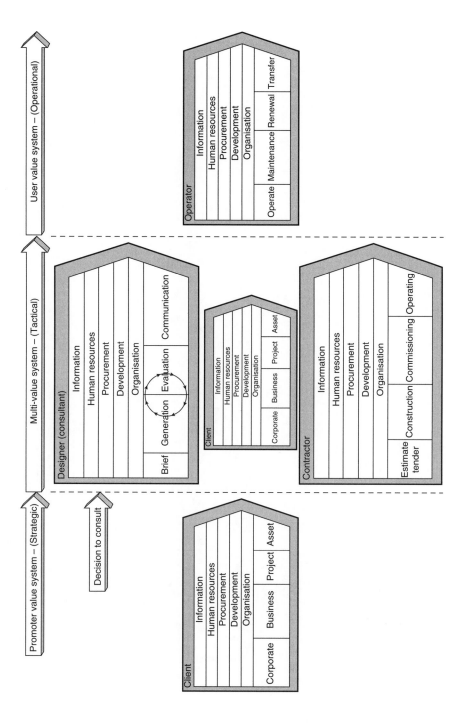

Figure 15.4 Alignments in the project value chain (adapted from Porter, 1979 and Standing, 1999).

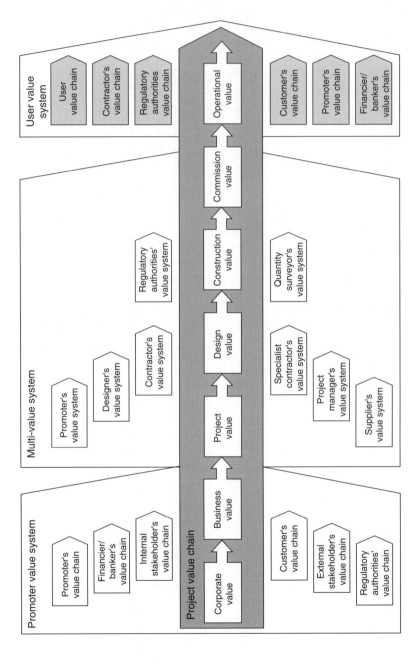

Figure 15.5 Typical value systems and value chains that impinge on the project value chain (from Standing, 1999).

15.7 Procurement and the project value chain

Provided the project value chain remains in alignment, it is a series of inputs and outputs that create value for the client. Each value transition should be adding value until the complete project forms an asset for the client's organisation to meet a corporate need. Complexity is added to the project value chain when other value systems impart skills and knowledge into it or create barriers to its effective operation as an integrated system on behalf of the client. One of the most important strategic decisions that impacts the project value chain is the choice of procurement route, which can act as either an enhancer or barrier to value creation and improvement. Single-point responsibility for the whole delivery from concept to operation for the client comes to the fore and should, in theory, create the capability to maintain the integrity and alignment of the project value chain.

Figure 15.6 provides a schematic overlay comparing some of the major procurement systems with the project value chain concept. Schematically, those procurement systems at the top of the diagram provide more opportunity to maintain the integrity of the project value chain as an increased number of discrete activities come under one umbrella organisation for single-point delivery. While the project remains within the client value system, value should be maintained internally; although once transferred into the multi-value system through procurement, there is a potential for loss in client value.

The contractor-led procurement systems, where the contractor offers a one-stop-shop service to a greater or lesser extent has the potential for greater integration within the project value chain, depending on the method of tender. Profession-led design procurement systems involve additional interfaces and provide more opportunities for disruption to the project value chain. Under profession-led systems, any change by the contractor must have the client's approval as they are the only party that can sanction change – none has a mechanism permitting a change to occur to the design by the contractor.

The management forms of procurement lie somewhere in the middle of the schematic. They permit increased involvement of engineering knowledge earlier in the process but essentially they are profession-led routes with a consequent increase in the number of interfaces.

Turnkey procurement has similarities to PFI, but unlike the latter, turnkey procurement does not have the additional requirement and liability for operating the facility. The contractual positioning and role of the designers will alter the impact on the project value chain. If the contractor employs the designers in-house, then there should be increased alignment in the project value chain. However, under turnkey, where the designer is independent of the contractor, another value system is imposed.

The traditional construction procurement route, at the bottom of the schematic, is probably the most disruptive to the project value chain as single-stage competitive tendering occurs at the transition point between design value and construction value. Two-stage tendering, overlaid onto the traditional procurement route, does, however, have the capability of bringing construction

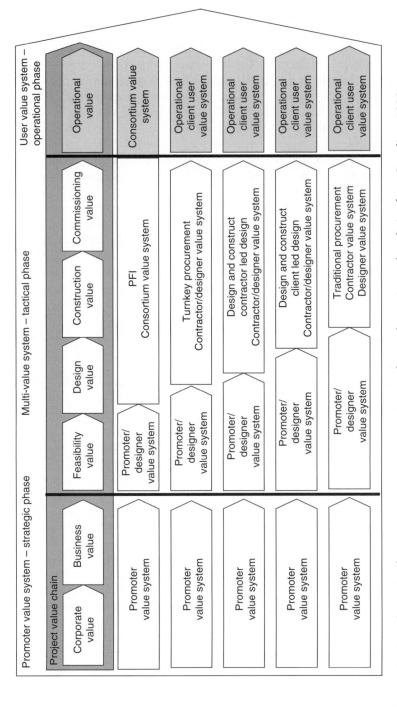

Figure 15.6 Schematic of procurement systems superimposed over the project value chain (from Standing, 1999).

expertise into the project much earlier. Therefore, there is also an interaction between procurement method and choice of tendering strategy in terms of impact on the project value chain.

To summarise, the choice of procurement route is a strategic decision made by the client and/or its advisors that has a fundamental impact on the demand chain. It has the capacity to assist or hinder the transfer of value through the project process. The use of value management and value engineering are methodologies for aligning or realigning the project value chain.

15.8 Prime contracting

There have been two major factors that have caused the MoD to change its approach to procurement. The first is the abandonment of Compulsory Competitive Tendering in the public sector in favour of the Best Value regime that came into effect on 1 April 2000. This requires government client to think in terms of the life-cycle cost of designing, building, operating and maintaining a facility. The second is Rethinking Construction or the 'Egan Report', which builds on the Latham Report and has had a major impact on government thinking at all levels. The government is committed to implementing the principles identified in the Egan Report, including partnering, SCM and continuous improvement. The DEO of the MoD has taken the initiative with its *Building Down Barriers* project to adopt a radical approach to procurement and has embraced the principles of SCM and strategic alliancing within its new procurement process called *prime contracting*. The prime contractor has single-point responsibility for the design and building, or design, building, operating and maintenance of a facility. The distinguishing characteristic of prime contracting from other procurement routes is seen as *the* requirement for single-point responsibility for the total process, from concept to operation. There are, however, some similarities with projects delivered under PFI, which also provide single-point responsibility. However, the primary differences are that no financing is required by the prime contractor and there is a limit to the maintenance and operating period.

The prime contracting initiative was established in January 1997 and ran until late 2000. It is jointly funded by the Department of the Environment, Transport and the Regions (DETR), the DEO and the MoD. The initiative has three primary objectives:

(1) to develop a new approach to construction procurement – prime contracting – based on supply-chain integration;
(2) to demonstrate the benefits of prime contracting in terms of improved value for the client and profitability for the supply chain through two pilot projects;
(3) to assess the relevance of the new approach to the wider UK construction industry.

There are two main themes to the Building Down Barriers project: undertaking pilot projects using the new procurement route and a research-and-development theme involving developing the new procurement process and a 'toolkit' to support it, and then evaluating the pilots and toolkit.

The DEO, in conjunction with the Tavistock Institute, the Warwick Manufacturing Group (WMG), AMEC Construction, John Laing Construction, British Aerospace Construction Consultancy Services and the Building Performance Group, utilised two pilot projects where prime contractors were given full responsibility for the design, construction and maintenance for a 'proving period', including an additional focus on operational and through-life costs. Consultant and trade suppliers became part of the prime contractor's supply chain. The intention is that the MoD's £2 billion per annum procurement regime will be let under the prime contracting route, whereby the number of supplier firms will contract by approximately 90% and the value of its main construction contracts will increase to £100–200 million, with benefits accruing in the longer term across multiple projects.

Prime contracting has already come under close scrutiny by architects, who have already expressed concerns that the prime contracting route will further undermine their professional standing since the drive will be towards repeat solutions. It has also been viewed as institutionalised Design and Build.

There are five phases in the whole life of a prime contracting project. These are set out in Figure 15.7.

The core conditions envisage prime contracts being divided into two main categories: first, the 'one-stop-shop', where the contract period can range between 5 and 10 years, and second, major stand-alone capital works, where the prime contractor will design and construct an asset with a through-life compliance period of approximately 3 years.

Both prime contracts are based on the core principles below.

- Using SCM as a fundamental underpinning to prime contracting.
- A commitment to collaborative working.
- Use of open-book accounting.
- Fraud prevention and detection policies adopted by the prime contractor.
- The use of an integrated project team (IPT) as a critical success factor (the IPT commences initially with the client and then incorporates the prime contractor and his supply chain, on appointment).
- Seeking innovative solutions to improve value-for-money and continuous improvement.
- Use of output specifications; for one-stop-shops, the range of services will include:
  - capital works including refurbishment;
  - planned maintenance;
  - reactive maintenance;
  - soft facilities management services – cleaning, catering, space planning, and others, depending on the type of prime contract.
- The use of best practice in value engineering and through-life costing as a mechanism to provide better value for money and providing a more transparent view of the cost of ownership.

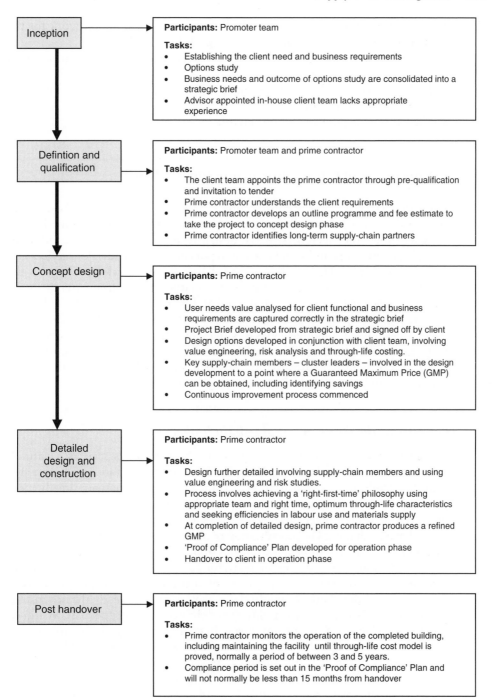

Figure 15.7   Plan of prime contracting.

- Warranting fitness-for-purpose for all design and construction work for the intended purpose as set out in the strategic brief.
- The use of a maximum price target cost arrangement, including procedures for sharing 'pain and gain'.
- TUPE (Transfer of Undertaking from Previous Employer) arrangements using the MoD's Code of Practice.
- The use of a dispute resolutions board.

15.9 The operation of future construction supply chains

In future it is likely that the construction supply chain will be much wider than envisaged currently and will include, on a more regular basis, funding bodies, designers and other members of the design team, and a 'general' contractor, specialist contractors and suppliers of components and providers of services and facilities managers. They will offer and deliver a total service package to the client from concept through to use. Extrapolating from current trends in the industry associated with design and build and more importantly PFI and prime contracting, the optimum solutions project for improving the delivery of multi-storey steel-framed buildings envisages that *strategic supply-chain brokers* and *phased, seamless teams* will emerge in the industry to become the norm. The emergence of, and pressures for, the strategic supply-chain broker and phase, seamless team are outlined in the following section.

The emergence of the strategic supply-chain broker role

The emergence of the strategic supply-chain broker role is seen as a natural, direct, consequence of the UK construction industry having a long and engrained record and culture of adversarial contracting. Through initiatives already mentioned, such as the Latham Report, the Egan Report, M4I and the Construction Best Practice Programme, there is a clear determination by policy makers that things have to change. The globalisation of markets, especially in the service sector, coupled with governments' strategies of public sector privatisation in many countries will also increase the pressure for change and attract investment and foreign firms to deliver new types of combined public–private sector projects. The concept of the strategic supply-chain broker assails directly the adversarial culture of the UK construction industry and the need for integrated SCM argued for in numerous government initiatives. Regular procuring clients are already pressurising the industry for a 'one-stop-shop' service, coupled with cost and time certainty in the delivery, and they are very vociferous in their demands of the industry to meet their expectations as customers.

Newer procurement routes such as PFI and prime contracting, as well as more integrated, team-based routes such as management contracting and construction management are attempts at drawing together design and construction interfaces and responsibilities. PFI and prime contracting also incorporate the operational phase. The broker role is, therefore, an organic extension of these procurement options and industry-led initiatives. The branded product is a response to rethinking construction, using a core philosophy of a fully integrated design and delivery process that ensures that value, quality and speed of delivery are 'designed in' from the outset. Use is made of standardised pre-designed buildings and components enabling engineering solutions, systems and architectural component interfaces to be identified and resolved early. In addition, at the heart of the branded product are IT/CAD systems that enable work to proceed concurrently. PFI, prime contracting and developments are precursors to the emergence of the broker role, one that will take responsibility for the contracted-out strategic, tactical and potentially the operational delivery of an asset for a client. The key is that the broker is able and willing to take the risk away from the client as a one-stop-shop supplier of skills, expertise, product components and knowledge, for a price. Optimum solutions' research proposes that brokers will compete on their expertise using their supply-chain base.

In the industry of the future, it is predicted that two main forms of brokers would emerge, namely, the *volume brokers* and the *innovative brokers*, based on the type of demand that will exist in the industry and the requirements of knowledgeable and less knowledgeable clients. Using the Pareto principle, approximately 80% of all projects can be categorised as 'routine' and would act as one of the drivers for working with the same project teams. This would form the basis of volume delivery within the industry and volume brokers would emerge over time to develop a brand reputation for timely, regular delivery to cost at an appropriate quality and functionality. Approximately 20% of all projects can be categorised as technically demanding, innovative or leading edge. The current management forms of procurement would be the drivers behind team delivery for these types of project. Brokers emerging from this domain would develop brand reputations as innovative brokers. The requirement here would be for the phased seamless team to be put together to deliver innovative solutions. Brokers operating under these conditions would have a pool of leading-edge supply-chain members that would be niche providers to the industry. The next section discusses the phased, seamless team.

The phased, seamless team

Contractually based relationships highlight the dangers of adversarial interactions developing when things go wrong. They also tend to be much more linear, with one task or sequence of events following another, with different organisational relationships procured at a particular point in time to suit a particular project programme. This may militate against securing the right skills at the right time regardless of position within the supply chain. Underpinning the operation of the supply-chain broker and construction supply chain is a *phased, seamless, team*. Seamless teams would pre-exist around a particular supply-chain broker. Members of these teams, through long-term working relationships, would be familiar with each other's working practices, including a full appreciation of the requirements for appropriate and timely information exchange to enable key project interfaces to operate effectively throughout the supply chain. The seamless team would operate on the basis of a moral and psychological contract founded on trust. Its operations will not be contractually based, and rewards and incentives would be based on performance indicators tied into client-focused value and not cost-driven service delivery. Teams would, through common goals, work in a cooperative, solutions-oriented culture, where pooling and sharing of information would occur through a combination of face-to-face problem-solving meetings supported by information technology. As the project progresses, the team would operate with seamless handover processes and procedures, as some new members enter and others leave the team due to changing task requirements. To support this, expectations, roles, responsibilities, skills and competencies would be known and fully understood among team members. Under a system where a psychological and moral 'contract' based on trust is the founding principle of a relationship, deeper levels of commitment can occur and the principles behind the seamless team would permit concurrency of inputs and skills for the benefit of the project and product, buildability, and decision making to take account of all aspects of the total process. Interfaces would be owned and information flows improved as they are internal to the team. This would lead to savings in both time and money. The seamless team would also provide greater opportunity for more continuity of work and learning.

15.10 Summary

This chapter has proposed that the supply-chain system in engineering comprises two components, the demand chains of numerous clients, generated through the procurement route adopted, and the supply chain of the main contractor, which has to respond to a diverse range of client types and hence demand chains. SCM is now on the strategic agenda of firms and is seen no longer as an operational issue. It involves thinking about activities within the supply and purchasing function that are internal and external to the firm. Supply chains in engineering can be classified under three main headings.

- **Professional services.** These suppliers provide a combination of skills and intellectual property into the process, typically comprising the designers and other professional consultants. Those supplying professional services operate under numerous professional institutional umbrellas. In this chapter the delivery of professional services, a tertiary-level provision, falls within the domain of either the client's project-focused demand chain or the main contractor's supply chain, depending on the procurement route adopted.
- **Implementation and assembly.** This on-site construction process is highly skills-focused and also forms part of the service sector of the industry. It comprises a range of different types of contractors either with overall responsibility

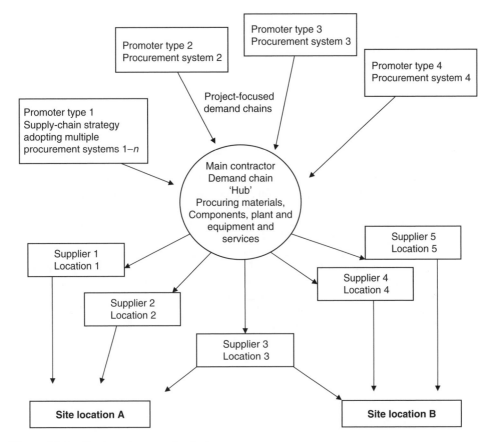

Figure 15.8 Engineering supply-chain system.

for the management of the process or for supplying inputs into the process. It involves fitting, installation, assembly, repair and on-site and off-site labouring. In this chapter it is also viewed as a manufacturing process.

- **Materials and products.** This comprises the materials, products and hired plants involved in the on-site process. These form part of the main contractor's supply chain and will involve the make, move and store activities to and at a particular site location.

Figure 15.8 sets out a schematic for the supply-chain system in engineering, from which supply-chain strategy is developed by the main contractor or, as this chapter has indicated, also by clients to the industry.

References

Porter, M. (1979) How competitive forces shape strategy, *Harvard Business Review*, March/April.

Saunders, M. (1997) *Strategic Purchasing and Supply Chain Management*, second edition, Financial Times & Pitman Publishing.

Standing, N. (1999) Value Engineering and the Contractor, Unpublished PhD thesis, University of Leeds.

Further reading

Aitken, J. (1998) Supply Chain Integration within the Context of a Supplier Association, Unpublished PhD thesis, Cranfield University.

Christopher, M.(1998) *Logistics and Supply Chain Management: Strategies for Reducing Cost and Improving Service*, second edition, Financial Times & Pitman Publishing.

Construction Task Force (1998) Rethinking Construction Report, available at http://www.constructingexcellence.org.uk/pdf/rethinking%20construction/rethinking_construction_report.pdf

Croner's Management of Construction Projects, July 1999.

Croner's Management of Construction Projects. Special Report No 11, 17th May, p. 3.

Gatorna J.L and Walters D.W. (1996) *Managing the Supply Chain, A Strategic Perspective*, MacMillan Business, Basingstoke.

Holti, R., Nicolini, D. and Smalley, M. (1999) *Building Down Barriers – The Concept Phase*, Interim Evaluation Report.

http://coconet.co.uk/supplychain/prime_process/prime_process.html, pp. 1–2.

http://www.coconet.co.uk/supplychain/supplychain.html, pp. 1–2.

Chapter 16
Partnering

This chapter examines the impact of team-based supply chains and partnering on project management performance. It starts by exploring team working and then discusses relational contracting principles that underpin partnering and other collaborative working arrangements. The key areas of supply-chain integration, partnering definitions and types, making the relationship work and the benefits of partnering, will be addressed so that readers can identify actions within their own project. Finally, a framework for partnering best practice is described.

16.1 Team working

Competition in the business world has led to the need for increased productivity, and one way that companies have tried to meet this need is by using teams. However, the construction industry worldwide has been criticised for its fragmented approach to project delivery and its failure to form effective teams, resulting in reduced project delivery efficiency. The Egan Report (2002) suggested that process and team integration were necessary if the industry in the UK were to become more successful. On the basis of the report, an 'integrated toolkit' was developed by the Strategic Forum for Construction (2002) to aid the move away from fragmentation and adversarial relations.

Team working is characterised by mutual trust and openness, where problems and risks are shared and resolved collectively by the integrated project team. It is the starting point on which relationships with other parties should be based, and it applies as much to the internal relationships between members of the client's in-house project team as to the working relationships between members of the client organisation and those of the supply team.

The Office of Government Commerce (2003) noted that for team working to succeed there should be the following.

- Team-building workshops led by an independent facilitator to help the team to understand and adapt new ways of working.
- Training and empowering of project staff to work jointly with others as an integrated project team to identify opportunities to do things more efficiently and to solve problems together.
- Commitment by the team to ongoing review and evaluation of performance.

- Aligned goals (with commitment to those goals).
- Post-project reviews with the principals of all the parties to identify and disseminate the lessons learnt.

Ultimately, the success of a project depends on the people involved. It is the enthusiasm, commitment and desire to succeed that will determine the outcome of the project. Particular attributes required of individuals include continuously searching for improvement; actively listening to and learning from others; and identifying which personal attributes we need to improve, to allow us to work better with others.

When selecting a project team the personal attributes of the individuals should be taken into account. Where it appears that, even with training, an individual is unlikely to be able to work constructively as part of the team, such a person should not be selected. To prevent confusion and misunderstanding, the roles and responsibilities of each individual should be clearly set down at the outset.

Team working is defined as working together as a team for the mutual benefit of all by achieving the common goals while minimising wasteful activities and duplication of effort. When problems or difficulties are encountered all parties should work together as a team to overcome those problems rather than blaming each other. Team working is the starting point on which relationships with others should be built and is equally important internally as externally. It does not replace proper and appropriate management structures but is a pragmatic way of working together to meet the project objectives. It should encourage greater openness and earlier involvement of contractors and suppliers.

The benefits of team working include

- understanding each others' objectives;
- use of collective brain power to find solutions;
- reduced numbers of personnel to monitor progress and prepare claims;
- reduction in correspondence;
- reduction in the duplication of effort;
- elimination of 'man to man' marking;
- improved working environment and cooperation, not conflict;
- enhanced reputation of individual when associated with successful projects;
- reduction in issue escalation.

A team-working culture can be encouraged by holding workshops and by arranging for team members to undergo team-building training. Co-locating is also believed to have a positive impact on team working. Where that is not practicable, attention needs to be given to establishing a common virtual environment, through intranets and other such mechanisms.

16.2 Supply-chain integration

The term 'supply chain' is imported from the manufacturing and service industries. The supply chain encompasses all those activities associated with processing from the raw materials to the end product or service for the client or customer. This includes sourcing and procurement, production scheduling, order processing, inventory management, transport, storage and customer service and all the necessary supporting information systems. With regards to construction, the supply chain for the delivery of a product includes:

- the main contractor appointed by the client;
- general and specialist contractors employed by the main contractor to assist in carrying out the works;
- suppliers of materials and products to be incorporated in the works, including any independent hauliers that may be hired to deliver the supplies or remove waste from the site;
- suppliers of professional services, including consultant engineers and architects employed to produce designs.

In the manufacturing and service industries, the supply chain stops at the point of delivery to the customer, who will have the choice between different services and products without having any input or interference in their design or production method. However, in construction, the client is the main consumer determining the production method and the requirements that lead to the design of the product. Hence, the client is considered as an integral part of the supply chain. The integrated supply team will invariably include the client and those involved in the delivery of the product, including manufacturers and specialist suppliers. Integrated teams involve assembling the right skills at the right time in the project delivery process irrespective of their position in the supply chain.

The integrated team can add value and create new capability by working in a team to achieve more than what can be achieved by working independently. The entire supply chain must be integrated to manage risk and apply value management and engineering techniques to improve buildability and eliminate waste from projects. This process should reduce through life and operational costs, leading to greater certainty of project time and budgeted costs, fewer accidents and more sustainable construction.

An integrated team includes the client and those involved in the delivery process who are pivotal in providing solutions that will meet the client's requirements. It comprises the client's project team and the supply team of consultants, contractors and specialist suppliers. It brings together the design and construction activities and involves the valued inputs of all parties of the supply team.

Integration is the merging of different disciplines or organisations with different goals, needs and cultures into a cohesive and mutually supporting unit. Integrated approaches demand that individuals from various organisations work together to achieve common attainable project goals through the sharing of information. This means that different company processes and organisation cultures have to be aligned in a collaborative manner. Integration is often recognised as a continuous process with the objective of improving team culture and professional attitudes.

In construction, integration refers to the introduction of working practices, methods and behaviours that create a culture of efficient and effective collaboration by individuals and organisations. It promotes a working environment in which information is freely exchanged between the different participants. The team brings together various skills and knowledge, and removes the traditional barriers between those with responsibility for design and construction in a way that improves the effective and efficient delivery of the project.

The delivery team in a construction project can be described as 'fully integrated' when it:

- has a single focus and objective for the project;
- operates without boundaries among the various organisation members;
- works towards mutually beneficial outcomes by ensuring that all members support each other and achievements are shared throughout the team;
- is able to predict, more accurately, time and cost estimates by fully utilising the collective skills and expertise of all parties;
- shares information freely among its members such that access is not restricted to specific professions and organisational units within the team;
- has a flexible member composition and is therefore able to respond to change over the duration of the project;
- has a new identity and is co-located, usually in a given common space;
- offers its members equal opportunities to contribute to the delivery process;
- operates in an atmosphere where relationships are equitable and members are respected;
- has a 'no blame' culture.

Creating an integrated team brings immediate benefits through a reduction in the manpower resources allocated to the project. It also offers the possibility of efficiency gains through having single-point accountability and more transparent processes than might otherwise be achieved. The benefits of an integrated team, compared with a traditional supply chain, are many.

- Teams work together, but a chain is only as strong as its weakest link – integration of organisations leads to fewer links in the chain and consequently fewer weak points.
- Direct communication between team members reduces the possibility of distortion of the message through multiple handling along the chain.
- A team will work to support and develop a weaker member – this is more cost-effective than re-tendering when one team member is struggling.
- As the team shares its knowledge, it will develop a bond of trust, which will lead to more learning.
- Team learning will help individual team members in their personal development.
- Shared learning, shared knowledge and shared understanding encourages, enables and supports better communication.

16.3 Relational contracting

Although the various team working approaches such as partnering and alliancing are clearly different from each other, they share a common core element of mutual cooperation to varying degrees. All such approaches can be effectively explained by the term *relational contracting* (RC), which may vary from post-contract partnering to vertical integration as if all the project participants belong to a single organisation. RC considers a contract to be a relationship among the parties, encourages long-term provisions and introduces a degree of flexibility into the contract on the basis of understanding each other's objectives. More relational and performance-oriented rather than purely priced-based contractor selection also encourages an amicable RC environment and more collaborative team work and higher productivity. These concepts can be extended throughout the supply chain to build a single coalescent team and to target optimal project performance. What is critical are the convergent culture and realigned attitudes of this coalesced team, which are expected to be generated and nurtured through appropriate change initiatives.

RC regards a contract as a present promise of doing something in the future that has dynamic ongoing states of interrelated past, present and future. A contract is a projection of exchange into the future and involves communication and commitment of a future event. But present promise affects the future by limiting choices that would be available during contract execution. Moreover, not all future events can be perceived or quantified at the time of contracting because of uncertainty and complexity. Therefore, contracts should be flexible so as to adjust for future events and effectively address the uncertainties as and when they arise. RC defines the working relationship among the parties who do not always adhere strictly to the legal mechanism offered by the written contracts. The parties themselves govern the transactions within mutually acceptable social guidelines. Thus, the relationship itself develops obligations between the parties.

While the contract establishes the legal relationship, the partnering process attempts to establish working relationships among the parties through a mutually developed (and agreed-upon) formal strategy of commitment and communication. It attempts to create an environment where trust and team work prevent dispute, foster a cooperative bond to everyone's benefit and facilitate the completion of a successful project.

Construction project teams comprise various hierarchically and interlinked parties possessing diversified skills, knowledge, and professional and organisational culture. As a result, complex relationships exist within project teams, which, if not managed effectively, can adversely affect the project performance. RC is an approach for managing such complex relationships between the members in construction contracts. The foundation of relational contracting is based on recognition of mutual benefits and win–win scenarios through more cooperative relationships between the contracting parties.

The principles of RC underpin various approaches including partnering, alliances, long-term contracting and other collaborative working arrangements and better risk-sharing mechanisms. As practised in the construction industry in a variety of forms, the core of RC is to establish the working relationships between the parties through mutually developed formal strategy of commitment and communication aimed at win–win outcomes for all parties. Although alliances in construction may vary from partnering arrangements to strategic commercial relationships, these are based on a mechanism that can improve interorganisational relations and project performance. However, partnering may be regarded as a strategic arrangement between clients and contractors over a series of projects, or only a short-term single project, with the aim of reducing costs and improving efficiency.

RC is characterised by the subordination of legal requirements and related formal documents to informal agreements in commercial transactions, such as verbal promises or partnering charters. This mode of governance calls upon all parties to (i) recognise the positive gains from maintaining the business relationship, (ii) transcend the hostility and (iii) overcome the uncertainties associated with unforeseen events in order to improve overall efficiency through motivation and improved attitudes. Disagreements are then negotiated towards solutions that do not jeopardise the relationship between the parties. Such objectives and approaches also provide an ideal framework for the joint management of risks that cannot be foreseen or clearly allocated to one party at the outset.

The aim of RC is to generate an organisational environment of trust, open communication and employee involvement. This is achieved through the creation of a project culture to fulfil the function served by a corporate culture in longer-lasting organisations. Although the benefits of partnering are numerous, the implementation of the partnering process is hard work. Changing old habits and building trust do not magically happen. RC is not a panacea, but rather a philosophy that must be tailored for each situation to which it is applied. Therefore, organisations considering RC should assess their business objectives, analyse the role of RC in helping them to achieve those objectives and determine an appropriate style of collaboration to implement. Once a culture of trust and cooperation is developed, transactional activities can be more cost-effective.

Supply-chain integration on an RC basis is often achieved using framework agreements, these are interpartner cooperation plans to coordinate two or more partners pursuing shared objectives, and therefore satisfactory cooperation is fundamental. It provides the parameters to all transactions and allows a reliance on litigation should the laws of the framework agreement environment be breached by any of the parties privy to the agreement.

The typical elements of a framework agreement are legal requirements, purpose, objectives, enabling elements, confidence in partner cooperation, ownership, balance and structure, disputes, termination and items common to all complex agreements (conditions, warranties, etc.). A typical framework agreement can contain a large number of conditions, warranties and implied and expressed terms, from which ownership balance and structure, knowledge and learning transfer systems, the control ingredient of confidence in partner cooperation and a number of enabling elements can be derived.

16.4 Partnering

Partnering has been successfully applied in the construction industry over the past decade. Originating from the USA, it has subsequently been widely used in the developed economies including Australia, Canada, the UK and, increasingly, in many other parts of the world. The concept of partnering emerged out of the realisation that the traditional forms of contracting for construction projects were often failing to deliver results that were satisfactory to either or both clients and contractors.

In the UK, the partnering concept has been given impetus by a number of government commissioned reports notably the Latham Report (1994) and Egan Report (1998). Both reports advocate the use of partnering and collaborative working arrangements to deliver client's objectives and value for money. One of the main recommendations proposed in the Egan Report, '*Rethinking Construction*', is to promote partnering between the construction industry participants and particularly members of the supply chain. It suggested the building of long-term relationships such as those formed from partnering instead of the traditional method of competitive tendering for the procurement and delivery of projects. The aim is to improve the relationship between the contractor and employer and overcome problems of adversarial relationships and the culture of blame that characterised the construction industry. Following on from the Egan Report, the 'best value' regime was introduced in April 2000 to replace the Compulsory Competitive Tendering for the award of works and services contracts. The former is now used for the delivery of services by local authorities and other public bodies.

Partnering is based on the idea of alignment of values and working practices by all members of the supply chain to meet the customer's real needs and objectives; although partnering may take many forms, they all seek to directly address the issue of adversarialism. The parties jointly share the risks by allocating them to the parties best able to manage them, thus creating a culture of trust. The parties communicate openly, discuss problems jointly and propose solutions and work towards continuous improvements to the benefit of the project. Subsequently, a win–win culture is created, thereby improving the relationship between the stakeholders. Success is rewarded and failure is penalised, which provides incentives to the parties to improve performance over critical factors of safety, quality, schedule, revenue, asset availability, production loss, operation and maintenance and capital costs. As trust and open communication exist between partners, the owner can reduce costs by downsizing its own project teams, particularly in the areas of procurement, contract administration and insurance. Partnering encourages team work and innovation where partners work collaboratively to develop more effective solutions that deliver better value in terms of whole-life costs. It facilitates the development of supply-chain strategies that encourages the modernisation of packaging and purchasing arrangements, helps suppliers reduce the number of contracts and facilitates the integration of procurement. Partnering also facilitates the early involvement of contractors and suppliers, allowing them to have their input into design process and participate with their knowledge and expertise to improve the deliverables of the project and advise on risk identification and management. The skills of qualified technical and management personnel of the partners will be utilised in lessening the resources constrains. Furthermore, savings and financial rewards can be used to fund additional investment, research, new technology, education and training.

The aim of partnering is to change the commercial style in which contracts are managed. It therefore concentrates on 'soft issues' such as attitude, culture, commitment and capability rather than 'hard issues' of price and scope of work. This requires a change in the organisation culture and the key stakeholders to become committed to the partnering values and the success of the project. The commitment should be established from senior management so as to spread the essence of collaborative working values within the organisation.

Although team working and partnering are not synonymous, they share some key characteristics, and forming project-based interorganisational teams can provide an opportunity for the related parties to explore each other's intentions and the possibility of developing a strategic partnership. However, even if the critical mass of forming stable, long-standing alliances has not been reached, it should not prevent project participants from striving for a high level of interorganisational team work in an individual project. In other words, project participants do not need to demonstrate a long-term commitment before they can work as a team for a relatively short period, although this is instrumental to team building. In the situation where a formal partnering agreement does not exist, interorganisational team work can act as an effective and flexible mechanism, promoting cooperation among project participants.

There are many definitions of partnering, some of which are as follows:

Partnering was defined by the Reading Construction Forum as:

> ... a management approach used by two or more organisations to achieve specific business objectives by maximising the effectiveness of each participant's resources. The approach is based on mutual objectives, an agreed method of problem resolution and an active search for continuous measurable improvements.

The US Army Corps of Engineers defined partnering as:

> ... the creation of an owner–contractor relationship that promotes the achievement of mutually beneficial goals. It involves an agreement in principle to share the risks involved in completing the project and to establish and promote a nurturing partnership environment.

The Construction Industry Institute Partnering Task Force defined partnering as:

> A long-term commitment between two or more organisations for the purpose of achieving specific business objectives by maximising the effectiveness of each participant's resources. The relationship is based on trust, dedication to common goals and an understanding of each other's individual expectations and values.

Expected benefits include improved efficiency and cost-effectiveness, increased opportunity for innovation and the continuous improvement of quality products and services.

According to the Association of General Contractors, partnering is:

... a way of achieving an optimum relationship between a customer and a supplier. It is a method of doing business in which a person's word is his or her bond and where people accept responsibility for their actions. Partnering is not a business contract, but recognition that every business contract includes an implied covenant of good faith.

The Egan Report (Egan 1998) defines partnering as:

Partnering involves two or more organisations working together to improve performance through agreeing mutual objectives, devising a way for resolving disputes and committing themselves to continuous improvement, measuring progress and sharing the gains.

As a concept, partnering is very difficult to capture in a short definition and that is why a number of definitions are presented. The Egan Report definition does identify the key elements that must be present for partnering to exist and is the one that the author usually recommends. Partnering provides the construction industry with a fundamentally different approach to team work. This revolutionary management process emphasises cooperation rather than confrontation, and it calls on the simple philosophy of trust, respect and long-term relationships.

Partnering should not be confused with other good project management practices, or long-standing relationships, negotiated contracts or preferred supplier arrangements, all of which lack the structure and objective measures that must support a partnering relationship. These other forms of team-based supply-chain arrangements are not being ignored, but rather it is suggested that the adoption of partnering proper will lead to greater benefits.

The main difference between partnering and other forms of procurement is the shift in the commercial route by which projects are procured. By changing the procurement and completion process, parties can reap benefits of cost saving, profit sharing, quality enhancement and time management. In the search for partnering relationships, clients are gradually altering their procurement criteria. There is a shift from 'hard issues', such as price and the scope of work, towards 'softer issues' that revolve around attitude, culture, commitment and capability.

The key to partnering is that it starts at the outset of a project, and is applied to the selection of the design professionals, consultants and contractors/suppliers. Partnering arrangements are flexible, and may be formal or informal, depending on the size, complexity and scope of the project. What is essential, however, is an understanding among all parties involved that they are committed to a plan to anticipate problems and deal with any as they arise. Figure 16.1 illustrates this process.

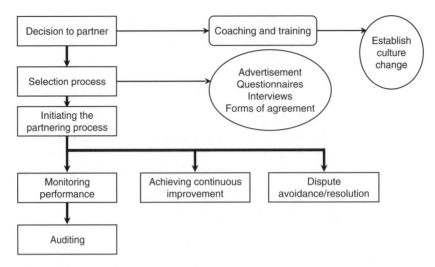

Figure 16.1   The partnering process (from ECI, 1997).

Mutual objectives

Despite widespread belief that these are obvious to all involved, team members typically do not understand other members' objectives very well at the beginning of the project. Developing a shared understanding of the mission and clear, detailed objectives are essential to a successful project. By getting people together to clarify what is important to each of them, and what their expectations for themselves and each other are, many suspicions, false expectations and lurking issues are dissipated. The agreed-upon mutual objectives must be constantly kept under review through meetings and effective communication. The principle that adversarial attitudes waste time and money must be mutually accepted.

Partnering is a tried and effective technique for placing the responsibility for anticipating problems early and on those best equipped to effect their resolution. The partnering concept has been proven as a powerful participant-driven method of anticipating problems that might be encountered on a project. The Centre for Public Resources' Legal Program Construction Dispute Committee stated that a goal of partnering is to 'mould a new synergistic bond between contracting parties in the industry to prevent disputes before they surface'.

Issue avoidance

The primary focus is on prevention and early resolution of issues. Partnering is seen as a method of settling conflict and disputes before they become destructive and costly. Early intervention is critical, because as issues develop into disputes, positions become hardened, formal claims are made and costs go up. In partnering, the issue resolution process is formally designed in 'kick-off' sessions between the partnering members so that everyone has a clear understanding of what the process is and agrees to use it. It is based on seeking 'win–win' solutions and not parties to blame.

Continuous and measured improvement

This ongoing process involves a fundamental understanding of customers' requirements, identification and aiming for best practices, development of new outputs and processes, implementation of new changes and the evaluation of the effectiveness of the improved process. The evaluation and feedback measures not only ensure that the objectives of the project are met, but ensure that they also aid in planning future projects. They identify and quantify what worked, what did not and how things could be improved next time. Measurement also keeps the partnering team focused on the terms of the partnering agreement.

Incentives should encourage the parties to work together to eliminate wasteful activities that do not add value to the client and to identify and implement process improvements, alternative designs, working methods and other activities that result in added value. Incentives should not be given for merely doing a good job, nor should they be made for improvements in performance that are of no value to the client. Incentives can be built into the bidding contract or into the non-binding partnering charter. Performance targets on which incentives are based must be measurable and they should not only cover cost and time but also safety and quality.

Further to these elements, the '*Seven Pillars of Partnering*' (Bennett and Jayes, 1998) establishes that the following issues must be addressed.

- **Strategy.** Developing the client's objectives and how consultants, contractors and specialists can meet them on the basis of feedback.
- **Membership.** Identifying the firms that need to be involved to ensure all necessary skills are developed.
- **Equity.** Ensuring everyone is rewarded for their work on the basis of fair price and profits.
- **Integration.** Improving the way the firms involved work together by using cooperation and building trust.
- **Benchmarks.** Setting measured targets that lead to continuous improvements in performance from project to project.
- **Project processes.** Establishing standards and procedures that embody best practice based on engineering
- **Feedback.** Capturing lessons from projects and task forces to guide the development of strategy.

16.5 Forms of partnering

Although collaborative arrangements can take many forms, partnering can be either project-specific or strategic.

- **Project partnering** involves the main contractor and the client's organisation working together on a single project usually after the contract for the project has been awarded.
- **Strategic partnering** involves the main contractor and the client's organisation working together on a series of projects to promote continuous improvement.

A variation of project partnering exists that is suitable for the public sector – this is referred to as post-award project-specific partnering. This type of partnering is used for contracts that undergo the normal competitive processes, but for which the intention to adopt partnering approach throughout the project is declared during the tendering process. As the name suggests, the partnering arrangement is entered into after the contract has been awarded. Here the concept of partnering is applied under the main contract for a particular project. The partnering application is detailed as part of the project contract document and both parties agree to overlay their formal contract with a partnering arrangement.

Partnering should provide clients with greater assurance that value for money is being achieved because (i) competition still applies – partners are still appointed competitively, adopting any of the recommended procurement routes; (ii) clear targets have to be agreed to – clearly measurable targets for improving building quality, delivery times and achieving cost reductions have to be agreed upon between the parties and arrangements have to be made for sharing efficiency gains to ensure that both parties benefit. The contractors in particular should be incentivised to come out with innovative and cost-effective solutions; open-book accounting is needed – contractors have much to gain from partnering arrangements in terms of better communication and sharing of information, which means that they have greater certainty over final costs and profit levels, opportunities to improve profit margins through meeting efficiency targets and sharing cost improvements with clients, and reductions in litigation costs due to fewer disputes. In return, clients need greater assurance as to the level of costs incurred by contractors and the efficiency savings they achieve and pass on to the client. To do this, an open accounting policy should be agreed to so that clients have reasonable access to the contractor's financial record and cost information so as to have confidence in the reported improvements in efficiency and performance, and there is commitment to continuous improvement – both clients and contractors have to be committed to a continuous programme to deliver measurable improvements in value for money and quality. The objectives of such a programme need to be clearly defined and agreed to from the outset. A common approach adopted by many organisations is to agree to a partnering charter that typically identifies the common goals for success, sets out a common resolution mechanism for reaching decisions and solving problems, defines the targets to demonstrate continuous measurable improvements in performance and sets out the incentive arrangements where these are not covered within the formal contract.

Partnering in different forms has the potential to improve the value for money of construction and is likely to be most appropriate in the following circumstances:

- on complex projects where user requirements are difficult to specify;
- for organisations wanting similar facilities repeated over time, giving scope for continuous improvements in cost and quality;
- for projects where construction conditions are uncertain, solutions are difficult to foresee and joint problem solving is necessary, for example, where the land is contaminated;

- for individual projects or series of projects where there are known opportunities to eliminate waste and inefficiency from the construction process.

16.6 Establishing the relationship

Partnering is generally established through a structured, facilitated process, normally consisting of organising workshops to bring the participants together. The workshop is designed to provide a positive partnering atmosphere. Its purpose is to establish commitment, trust, mutual goals and objectives among all the members of the partnership. The length of the workshop will depend on such variables as the complexity of the contract, experience of the participants in partnering, the number of partners and the time needed for team building.

In most cases, a facilitator-directed partnering workshop will accelerate the successful implementation of the partnering effort. The facilitator is a neutral person who helps the partners to get organised from the outset of the process. The facilitator helps to develop and lead the partnering workshop, and is instrumental in having the parties design their charter, identify potential problems and develop a conflict escalation procedure. The facilitator may also deal with any scepticism or bias brought to the workshop and keeps the team focused on the partnering process. Keeping the facilitator involved maximises the benefits to the partners by keeping them on the partnering path and by increasing the facilitator's knowledge of the programme, contract requirements and unique contract administration issues. The workshop may entail both individual and joint sessions with the facilitator and, generally, will be at least 2 days in duration. During the workshop, the essential ingredients of the partnering arrangement are drafted.

- The roles and responsibilities for each partnering participant.
- Specific programme issues and concerns, with an action plan developed for each.
- Conflict management techniques and specific procedures to resolve conflicts among stakeholders.
- Approach to resolving disputes.
- Metrics for the assessment of accomplishments.
- Techniques designed to improve team communications.
- The partnering charter.

At the end of the initial partnering workshop, the participants must identify the goals, develop a plan to achieve them and draw up a partnered agreement, or charter, to commit to them. The partnering workshop is just a starting point. Periodic reviews at regular intervals are critical to success. These reviews may involve the assessment of the partnering relationship and follow-up workshops. One reason why it is beneficial to keep the facilitator informed during contract performance is to enhance his or her involvement in follow-up workshops if they are required. Follow-up workshops should also be considered when major players in the partnering process are replaced, so as to ensure that new participants are knowledgeable about, and committed to, the process, and if there is a breach of the charter or conflict escalation procedure, or some other indication that it is necessary to re-affirm the process and remind participants of the need for their consistent commitment.

The partnering charter is the focal point of the relationship and the blueprint for success. There is no single approach to drafting a partnering charter. Some of the general matters that should be addressed in the charter, according to ECI, are listed below.

- A statement committing the parties to abide by the aims of partnering.
- An expression of intent to communicate freely.
- Acknowledgement that problems may occur during the currency of the contract.
- Acknowledgement that relationships may deteriorate owing to disputes being left unresolved.
- Commitment to monitoring the partnering process performance in accordance with the procedures set up at the workshop.
- Commitment to a safe project with a high-quality outcome.

Although not legally binding, the charter is a 'social contract' about how the team members have agreed to work together. The charter constitutes the first visual instrument of shared partnering commitment. It states each goal in specific terms, which are measured periodically. During the course of a project, joint evaluations of the charter keep its original objectives in place and underscore their continuing importance.

The charter is not intended to be a contractual document, nor does it supersede the contract. The contract establishes the legal relationship between the parties, while the charter is concerned with the working relationships. It is in effect a statement of how the parties intend to conduct themselves. The charter represents the common commitment of all the participants to the partnering process.

The partnering contract should not be allowed to become a barrier to the participants. From a purist standpoint, the contract is not an important part of the partnering process. Nevertheless, a formal agreement needs to be established between all parties. The contract, however, should promote and complement the partnering process, and not contribute to an adversarial relationship. The success of the contract as a working document depends on the commitment to fulfil the mission statement and the charter supporting that statement, as well as a statement of expectations and objectives. Flexibility should also be built into the contract to allow for change that may be a result of the continuous improvement or review process.

16.7 Making the relationship work

When entering a partnering agreement, there are a number of contractual issues that need to be considered.

- What effect the established benchmarking will have upon the improvements strived for by the parties.

- What redress an injured party can claim if the agreement is breached.
- Whether it provides recompense in the event that either party invokes the termination provisions.
- Whether there are any guarantees that the consultants and contractors will be appointed for any or all of the individual projects.

No matter how committed management and team participants are, the partnership will not run itself. To track, care for and feed the process, one individual, designated by each partner, must assume responsibility. This 'champion' will play a powerful and influential role in the process, and will generally be at the project management level. The champion will oversee the project, reinforce the team approach, overcome resisting forces, participate in resolution of issues escalated to their level, celebrate successes, maintain a positive image for the project and provide the leadership to ensure that the partnering process moves smoothly throughout performance of the contract. The champion will also communicate with senior management officials to keep them apprised of partnering efforts and to solicit their continuing commitment. Champions are more than figureheads. They must play a vital role in initiating and energising the partnering process for those in the team and implementing the tools developed at the partnering workshop.

16.8 The benefits of partnering

Partnering relationships offer advantages and opportunities specific to the individual members of the project team as well as the opportunities and advantages shared by each. The benefits associated with partnering arrangements are given below.

Effective utilisation of personnel resources may be the most important benefit to the owner, in terms of both staffing requirements and available expertise. The client may also benefit from increased flexibility and responsiveness in terms of added skills and resources available from other parties, from the presence of a diversity of talents not usually found in a single company, which will improve on delivery, and from reduced costs associated with contractor's or consultant's selection, contract administration, mobilisation and the learning curve associated with beginning a project with a new contractor or consultant. Other benefits to the client will be the reduced dependence on legal counsel, the development of a team for future projects and more control over possible cost overruns.

Partnering provides the design team with the opportunity to refine and develop new skills in a controlled and low-risk way. This occurs because new methods or approaches may be required to meet owner project requirements. Through partnering, the design team will benefit from the involvement of contractors during budgeting, development of the team for the future projects and optimal use of the design team's time.

Although a partnering relationship will not make a specific guarantee of workload, partnering implies a clear intent to maintain an active fundamental organisation. The long-term, non-adversarial aspects of partnering means that revenues may be more stable and the potential for the claims or litigation process is significantly reduced. The contractor may also benefit from increased opportunity for value-engineering involvement to provide value for money, faster decision-making process and more effective time and cost control.

Other benefits will include formation of teams for future projects, increased opportunity for financially successful projects, reduced dependence on legal counsel and the possibility of faster payments.

As with the other team members, the benefits that manufacturers and suppliers stand to gain through partnering include approval of their product recommendation, a voice in the design intent, involvement in coordination with other project trades and the possibility of repeat business. Other benefits are a better chance for quality in product installation and increased opportunity for financially successful projects.

Of all the potential benefits resulting from partnering relationships, perhaps the one that will have the most impact on the construction industry is improved project quality. An effective partnering agreement will improve project quality by replacing the potential adversarial atmosphere of a traditional owner–contractor–consultant relationship with an atmosphere that will foster a team approach to achieve a set of common goals.

Within this atmosphere of cooperation and mutual trust, companies can jointly determine and evaluate approaches to designing, engineering and constructing the project. By becoming partners in the project, team members can work together to achieve the highest level of quality and safety. The close, team-working relationship between the parties can provide an environment that encourages finding new and better ways of doing business. An effective partnering relationship will encourage partners to evaluate technology for its applicability to quality improvement for the project.

The partnering relationship also encourages the companies to identify major obstacles to the successful completion of the project and develop preventive action plans to overcome those obstacles before they impact schedule or cost resulting in many benefits.

- **Empowerment.** With collaborative working arrangements, the team members are empowered and driven to complete a successful project, which leads to more efficiency, innovation, and less waste and duplication. Subsequently, everyone can concentrate on meeting the project goals.
- **Early award.** Collaborative working facilitates the early award of construction contracts because it puts the participants on a single agenda. Consequently, suppliers and contractors can make input into the design and the delivery schedule before detailed drawings are complete. They can help to identify risk and suggest how it can be managed, reduced and eliminated. Their input can also improve buildability of the project and reduce the risk of allocating inadequate time for key milestones and overall completion. Finally, early letting of contracts also enables the advantages of long-led procurements to be transferred to the contractor.

- **Improved communication and problem-solving skills.** Project participants are encouraged to communicate effectively and share information about the project and any problems that may arise. With these improved communication skills, problems can be jointly solved in the mutual and best interest of the project. This will result in a project delivered on time, within budget and to the required quality.
- **Less resource constraints.** In traditional contracting arrangements, each management layer oversees the performance of the layer below it in the project's hierarchy. However, in partnering, integrated, co-located, multifunctional project teams are established and empowered to meet the project goals. Thus, collaborative arrangements can facilitate more efficient use of qualified technical and management personnel and other limited human resources.
- **Reduced risk.** Because risk is jointly evaluated and fairly allocated, the partners do not take on unmanageable risk. Moreover, contractors and suppliers tend to come up with more realistic schedules, resources and pricing on partnered projects.

According to the National Audit Office (2001), the benefits of partnering include the client and contractor working together to improve building design, minimise the need for costly design changes, identify ways of eliminating waste and inefficiency in construction process, replicate good practice learned on earlier projects and minimise the risk of costly disputes.

16.9 The limitations of partnering

Although research has shown that partnering arrangements offer many significant benefits to all parties partaking in such arrangements, a balance evaluation of partnering also requires the identification and assessment of its potential limitations. Some of the disadvantages of partnering are given below.

Direct costs

Available evidence indicates that partnering arrangements can reduce overall costs significantly. However, additional expense is incurred in providing training for each entity separately and running joint workshops, including the cost of employing an independent facilitator. Further considerations include additional managerial costs in finding partners and negotiating partnering agreements, setting up joint systems with partners, monitoring the progress of the arrangement and evaluating its performance.

Complacency

There is the possibility of a general complacency on one partner subverting the relationship. The employer's work procured through a partnering arrangement can come to be regarded as already won, and especially if margins on the work are keen, the work may become less exciting, requiring a lower level of commitment than newly acquired projects. Moreover, teams can become stale, thereby diminishing rather than promoting efficiency and motivation.

Commercial blandness

Partnering can limit the scope for a firm to perform well in the marketplace. Thus, it is much more difficult in the context of a partnering arrangement for an employer to take advantage of a contractor's underpricing, or a contractor to ignore a client's favourable distorted view of the marketplace. It is also more difficult for the client to explore fresh relationships with new and dynamic potential partners. It is envisaged that partnering relationships can suffer where participants perceive the loss of opportunity.

Exposure to changes in the market

If the employer ceases to deliver the workload in accordance with projections, the contractor may find himself or herself suddenly having an empty order book and, by reason of his or her reliance on work generated through the partnering agreement, may have become sufficiently distanced from the general marketplace to readily make good the deficit. This situation may arise not only because the client has a genuine difficulty in making work available, but in case the relationship falters for other reasons because of which the employer ceases to regard the partnering arrangement as beneficial and wishes to seek new partners elsewhere. Moreover, in a boom market, the client may discover that the contractor is reluctant to honour a long-term low-return arrangement in circumstances where more attractive opportunities exist elsewhere. This can subsequently erode the contractor's enthusiasm for future projects.

Career prospects

A possible disadvantage of strategic partnering is that it can sometimes lead to reduced career prospects for those involved. Employees may regard themselves as sidelined into a static part of their organisation, isolated from the main commercial impetus of the business. This concern must be addressed by partnering entitles – individuals must be seen to be regarded as valued for their contribution to the partnering arrangement and career structures should be maintained or enhanced.

16.10 Summary

Under a partnering arrangement all parties agree from the beginning to a formal structure that will focus on creating cooperation and team work, so as to avoid adversarial confrontation. Working relationships are carefully built, based on mutual respect, trust and integrity. Partnering can provide the basis for a 'win–win' approach to problem solving.

Partnering depends upon an attitude adjustment, where the parties to the contract form a relationship of team work, cooperation and good-faith performance. It requires the parties to look beyond the bounds of the contract and to develop a cooperative working relationship that will promote their common goals and objectives. The aim is not to change any contractual responsibilities, but rather to focus on what really makes the contract documents work: the relationships between the participants.

References

Bennett, J. and Jayes, S. (1998) *The Seven Pillars of Partnering; A Guide to Second Generation Partnering*, Centre for Strategic Studies in Construction, The University of Reading.

Egan, J. (1998) *Rethinking Construction*, Department of the Environment, Transport and the Regions.

Latham, Sir Michael (1994) *Constructing the Team*. Final Report of the Government/Industry Review of Procurement and Contractual Arrangements in the UK Construction Industry, HMSO, London.

National Audit Office Modernising Construction. Stationery Office. HC 87 2000-01, The Stationery Office. 11 January, 2001.

Partnering in the Public Sector, European Construction Institute, ISBN 1 873844 34 4.

Strategic Forum for Construction (2002) *Accelerating Change*, Rethinking Construction Ltd.

Further reading

Office of Government Commerce (2003) *Achieving Excellence in Construction Procurement Guide*, HM Treasury, London.

Bower, D. A. (2003) *Management of Procurement*, Thomas Telford.

Holti, R., Nicolini, D. and Smalley, M.(2000) *The Handbook of Supply Chain Management: The Essentials*, CIRIA Publication C546 and Tavistock Institute.

Partnering in the Team, Construction Industry Board, ISBN 0 7277 2551 3.

Chapter 17
Private Finance Initiative and Public–Private Partnership

This chapter describes the concession contract. In this form of procurement the contractor effectively becomes the promoter and, in addition to performance, finances the project and operates and maintains the project over a period of time to generate sufficient time to provide a commercial return.

17.1 Concession contracts

There has been a growing trend in recent years both in the UK and overseas for principals, usually governments or their agencies, to place major projects into the private sector rather than the traditional domain of the public sector by using concession or (Build, own, operate and transfer) (BOOT) (see Section 17.2 for definition) project strategies. The adoption of this form of contract strategy, often referred to as a concession contract, has led a number of organisations to consider its implementation for different types of facilities, on both a domestic and international basis and by speculative or invited offers.

Privatised infrastructure can be traced back to the eighteenth century when a concession contract was granted to provide drinking water to the city of Paris. During the nineteenth century, ambitious projects such as the Suez Canal and Trans Siberian Railway were constructed, financed and owned by private companies under concession contracts.

The transfer element of a BOOT project implies that, after a specified time, the facility is transferred to the principal and cannot be considered as real privatisation. In a BOOT project, however, ownership of the facility is retained by the promoter for as long as desired and is therefore more consistent with the concept of privatisation.

In the late 1970s and early 1980s, some of the major international contracting companies and a number of developing countries began to explore the possibilities of promoting privately owned and operated infrastructure projects financed on a non-recourse basis under a concession contract.

The term BOT was introduced in the early 1980s by the Turkish Prime Minister Turgat Ozal to designate a 'build, own and transfer' or a 'build, operate and transfer' project; this term is often referred to as the Ozal Formula.

17.2 Definition of BOOT projects

A 'build, own, operate, transfer' (BOOT) project, sometimes referred to as a concession contract, may be defined as:

> a project based on the granting of a concession by a principal, usually a government, to a promoter, sometimes known as the concessionaire, who is responsible for the construction, financing, operation and maintenance of a facility over the period of the concession before finally transferring at no cost to the principal, as a fully operational facility. During the concession period the promoter owns and operates the facility and collects revenues in order to repay the financing and investment costs, maintain and operate the facility and make a margin of profit.

Other acronyms used to describe concession contracts include

FBOOT	finance, build, own, operate, transfer
BOO	build, own, operate
BOL	build, operate, lease
DBOM	design, build, operate, maintain
DBOT	design, build, operate, transfer
BOD	build, operate, deliver
BOOST	build, own, operate, subsidise, transfer
BRT	build, rent, transfer
BTO	build, transfer, operate
BOT	build, operate, transfer

Many of these terms are alternative names for BOOT projects, but some denote projects that differ from the preceding definition in one or more particular aspects, but which have broadly adopted the main functions of the BOOT strategy.

17.3 Organisational and contractual structure

A typical BOOT structure illustrating the number of organisations and contractual arrangements that may be required to realise a particular project is shown in Figure 17.1.

The key organisations and contracts include those listed below.

> **Principal:** is responsible for granting a concession and the ultimate owner of the facility after transfer. Principals are often governments, government agencies or regulated monopolies.
> The structured contract between the principal and promoter is known as the concession agreement. It is the document which identifies and allocates the

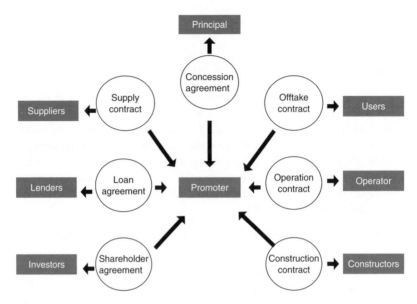

Figure 17.1   Organisational and contractual structure for BOOT.

risks associated with the construction, operation maintenance, finance and revenue packages and the terms of the concession relating to a facility. The preparation and evaluation of a BOOT project bid is based on the terms and project conditions of the structured concession agreement (SCA).

Promoter: the organisation that is granted the concession to build, own, operate BOOT and transfer a facility. Promoter organisations are often construction companies or operators or joint venture organisations incorporating constructors, operators, suppliers, vendors, lenders and shareholders.

The following organisations and contracts may be included within the BOOT project strategy.

> **Supply contract.** Contract between the supplier and promoter. Suppliers are often a state owned agency, a private company or a regulated monopoly who supply raw materials to the facility during the operation period.
>
> **Offtake contract.** In contract-led projects such as power generation plants, a sales or offtake contract is often entered into between the user and the promoter. Users are the organisations or individuals purchasing the offtake or using the facility itself. However, in market-led projects, such as toll roads or esturial crossings where revenues are generated on the basis of directly payable tolls for the use of a facility, an offtake contract is not usually possible.
>
> **Loan agreement.** The basis of the contract between the lender and the promoter. Lenders are often commercial banks, niche banks, pension funds or export credit agencies who provide the loans in the form of debt to finance

a particular facility. In most cases one lender will take the lead role for a lending consortium or a number of syndicated loans.

Operations contract. Contract between the operator and the promoter. Operators are often drawn from specialist operation companies or companies created specifically for the operation and maintenance of one particular facility.

Shareholder agreement. Contract between investors and the promoter. Investors purchase equity or provide goods in kind and form part of the corporate structure. These may include suppliers, vendors, constructors and operators and major financial institutions as well as private individual shareholders. Investors provide equity to finance the facility, the amount often determined by the debt–equity ratio required by lenders or the concession agreement.

Construction contract. Contract between the constructor and the promoter. Constructors are often drawn from individual construction companies or a joint venture of specialist construction companies. Constructors can sometimes take and have taken the role of promoters for a number of BOOT projects both in the UK and overseas.

17.4 Concession agreements

In BOOT contracts, a concession agreement is used as the basis of the contract. The concession agreement is the one that forms the contract between the principal and the promoter and is the document that identifies and allocates the risks associated with the construction, operation and maintenance, finance and revenue packages and the terms of the concession relating to a facility over the lifetime of the concession before transfer of the facility to the principal.

The statutory concession agreement is adopted when governments are required to ratify a treaty that may lead to legislation and consequent concessions. The terms of a concession granted under statute may usually only be altered or varied by the enactment of further legislation. Under this form of agreement, the promoter (concessionaire) would be required to enforce his or her rights by making an application to the courts for judicial review of a principal's (government's) actions. In the case of concessions granted under statute, third parties may potentially have the right to apply to the courts to enforce provisions of the concession that the government does not wish to apply. A statutory concession agreement was adopted for the Channel Fixed Link Project.

The contractual concession agreement is often adopted when one government organisation enters into an agreement with a promoter to undertake a specific concession. In the contractual agreement either party may amend or relax the terms of the agreement. Under such an agreement any breach of the concession by the principal would entitle the promoter to damages, and in some cases specific performance of the terms of the concession. In the contractual concession only the parties to the concession can enforce the terms. A contractual concession agreement was adopted for the North–South Expressway in Malaysia.

In some cases the concession agreement is a hybrid form of both contractual and statutory elements. This form of agreement is often adopted when a principal or promoter requires an element of legislative control and the benefits associated with the contractual agreement. For example, if planning consent is considered a major requirement for implementation of the concession, then a statutory element may be incorporated into the hybrid form to cover this requirement.

17.5 Procurement of BOOT project strategies

BOOT projects are procured by invited tender from the principal or by a speculative bid from a promoter group to an individual principal. In the case of an invited bid many elements of the risk may be determined by the terms of invitation. In the case of a speculative bid the promoter will need to approach the principal to determine his or her obligations under the terms of the concession agreement. About 60% of BOOT projects are awarded as the result of speculative tendering, but within the EU this percentage is much lower.

Speculative bids

A speculative bid is one by which a promoter approaches a principal with a proposed scheme considered commercially viable by the promoter and requests the principal to grant a concession to the promoter to build, own and operate a facility for a defined period of time before transferring the facility to the principal. A speculative bid is for a concession usually undertaken by the principal and requires the promoter to prepare a concession agreement as the basis of the bid. Many BOOT projects begin when promoters approach governments privately to propose a much needed project which government finds difficult to finance from the public sector budget.

In projects involving new transport infrastructure projects, a speculative bid is considered as one in which the private sector promotes an innovative project from concept to meet a perceived market need.

Invited bids

An invited bid is one by which the principal invites a number of promoters to bid, in competition for the privilege of being granted a concession to operate a facility normally undertaken by the principal or one of his or her organisations. An invited bid is often a solicitation of bids based on a speculative proposal made open to tender.

In the case of invited bids for transport infrastructure projects, a scheme will have been through a public enquiry. The public sector seeks the support of the private sector to design, build, finance and operate a route or where the government has defined the transport corridor and requires the private sector to select the actual route, assemble the land, design, build, own and operate the project on the basis of a time-limited concession. In the UK the practical consequence of privately financed projects requires each project to be authorised individually by an Act of Parliament. In invited bids such as the Dartford River Crossing, a hybrid bill procedure was adopted to authorise the projects go ahead and the invitation involved:

- the identification by the government of a corridor for the proposed route;
- a competition for the financing, building and operation of a road serving the corridor, inviting bids from private companies;
- the promotion by the government of a hybrid bill to authorise the road, the tolls, land acquisition and arrangements for the concession.

Competitive bids for BOOT projects should follow the normal procedure to that of awarding public works projects, and ideally the government would identify the project and define the project specifications, nature of government support, the proposed method of calculating the toll or tariff, the required debt/equity ratio and other parameters for the transaction.

The government would then invite preliminary proposals with the winner selected on the basis of normal competitive criteria such as price, experience, track record of the promoter or on the basis of side benefits of the host country.

17.6 Concession periods

Concession periods are sometimes referred to as a lease from government. The concession period normally includes the construction period as well as the operation and maintenance period before transfer to the principal. However, in the case of the Shajiao Power Plant constructed in China the concession period of 10 years excludes the construction time. In the Macau Water Supply Project, a concession period of 35 years included the refurbishment of existing plants rather than construction of new facilities and permitted the principal to repurchase the rights if deemed necessary. This agreement also permits the concession to be extended by mutual agreement of the parties, which in effect constitutes an addendum to the contract. Provision for the extension of the concession period, in this case the operating period, may be included in the concession agreement to protect a promoter against the principal's default on his or her contractual obligations that may result in projected returns not being met. Typical concession periods range from 10 to 55 years when granted by governments under a BOOT initiative.

In infrastructure projects the concession period is often longer than in industrial facilities. Concession periods of between 10 and 15 years may be successfully financed but principals need to accept that the project economics must be strong enough to bear the enhanced depreciation rate over such short periods as well as the return required on the capital investment. In the Dartford Crossing Project the concession period was a maximum of 20 years, which could be terminated at no cost to the principal as soon as surplus funds had been accrued by the promoter to service all outstanding debt.

The concession period should be sufficient to:

- allow the promoter to recover his or her investment and make sufficient profit within that period to make the project worthwhile;
- but not allow the promoter to overcharge users when in a monopolistic position having already recovered his or her investment and made sufficient profit.

The commercial viability of a concession contract and the difficulty and uncertainty of predicting revenues over long periods of time is a major obstacle to many promoter organisations. If there is flexibility in the concession agreement to adjust the concession period, then this may reduce the promoter's risk and allow predicted revenues to be achieved.

17.7 Existing facilities

In a number of concession contracts an existing facility is included as part of the concession offered or requested. This may be a requirement of the principal in an invited bid or offered as an incentive by a promoter in a speculative bid. In some industrial/process facilities, however, a promoter may only agree to tender a particular project provided he or she is given operational control of an existing facility that may affect the performance of the facility to be tendered.

The operation of an existing facility often guarantees the promoter an immediate income that may reduce loans and repay lenders and investors early on in the project cycle. The commercial success or failure of an existing facility must be considered by the promoter at the bidding stage so as to determine the success or failure of the proposed concession.

Principals can influence pricing mechanisms by making existing facilities available to promoters, which are capable of earning revenues during the construction period. In the case of the Sydney Harbour Tunnel Project revenues generated by the existing bridge crossing are shared between the principal and the promoter, which enables the promoter to generate income to service part of the debt prior to completion of the tunnel.

Assets capable of producing earnings that can be used to pay capital costs, debt service and operating expenses are a familiar feature of BOOT projects.

A concession to operate a section of existing highway generated us $16 million for the promoter of the North–South Expressway in Malaysia. The concession to operate existing tunnels as part of the Dartford Bridge Crossing concession offered an existing cash flow but required the promoter to accept the existing debt on those tunnels. An existing concession also formed part of the concession agreement for the Bangkok Expressway – this arrangement required the operation and maintenance of an existing toll highway with generated revenues being shared between the principal and the promoter.

17.8 Classification of BOOT projects

BOOT projects may be classified on the basis of the method of procurement, the type of facility, the location of the facility and the method of revenue generation.

Speculative/invited

In the case of a speculative bid, the promoter will determine which costs and risks should be borne by his or her own organisation and those to be borne by the principal and other parties involved. In an invited bid, the principal will determine the concession and the costs and risks to be borne by the promoter under the terms of invitation.

Infrastructure/industrial/process

The components of each of the packages of an infrastructure project will contain costs and risk levels different from those considered for an industrial or process plant. An infrastructure project may require large capital expenditure during construction but may operate on a small budget. A process or industrial facility, however, may require a low capital expenditure but a high operating budget over the operation phase.

The number and types of risks associated with a BOOT project may often be determined by the type of facility and the number of contracts and agreements to be included. Infrastructure facilities may often be considered as static or dynamic facilities. A road or fixed-bridge facility may be considered as static as the facility offers no moving parts and requires no input of raw materials or power; usually the static facility will have a smaller operation and maintenance cost than a dynamic facility. A light transit railway facility will require a major power source during the operation phase, which will result in operational costs being higher than that of a static facility.

Domestic/international

The location of a project will determine the host country's political, legal and commercial requirements, which will be a major factor in project sanction. In the case of a domestic project the promoter will often be aware of the country's requirements and have access to local financial markets. In international projects promoters may need to carry out in-country surveys to determine risks associated with meeting the requirements of the concession and determine how revenues may be repatriated to service loans. In effect each international project will be determined by the constraints of the host country government.

Market-led/contract-led

One of the major risk areas associated with BOOT projects is the generation of revenues, which often leads to market-led revenues being far more uncertain than those based on pre-determined sales contracts. The commercial risks to be considered in a BOOT project are often determined by the revenue classification.

The demand for a toll road, which depends solely on revenues from users, may be much lower than forecasted at the feasibility stage, which may be due to increased costs of fuel or a reluctance to pay tolls by users. In the case of a water treatment plant, revenues will be contract-led, and provided demand is met and an effective price variation formula takes into account inflation, a promoter may consider the risk associated with revenue negligible compared with other risks.

In a number of market-led projects, promoter organisations will often seek contract-led revenue streams to reduce the risks associated with revenue generation. In a toll road facility, promoter organisations may approach haulage contractors and enter into take-or-pay agreements for the use of the facility, in effect providing lending organisations with a guaranteed source of revenue, which will reduce the risks associated with the finance package.

In summary, the number of organisations, contracts, data and resources required to meet the project and the major risk areas will be determined by the classification of the project.

17.9 Projects suitable for BOOT strategies

BOOT projects are a means of meeting the needs associated with population growth, such as housing, water sanitation and transportation; industrial growth, such as power, infrastructure and fixed investments; tourism and recreation, such as airports, hotels and resorts; and environmental concerns, such as waste incineration and pollution control. Developing nations are receptive to the idea of funding such projects under a BOOT strategy, which will often reduce the capital and operating costs and reduce the risks normally borne by the principal. Provided sufficient demand exists for these projects, revenue streams can be identified and the commercial viability determined by promoters and lenders.

The two most fundamental constraints on project development are economics and finance. In a BOOT project the promoter must cover operating expenses, interest and amortisation of loans and returns on equity from project revenues. However, promoters often consider the suitability of a project based on global market forces and the commercial viability of a project, which affect the profitability rather than the facility itself.

Any public service facility that has the capacity to generate revenues through charging a tariff on throughput may be considered suitable for a BOOT strategy provided suitable financing can be achieved. The most successful BOOT projects will be those in the small to medium range up to US$500 million because private sector equity requirements for such projects are usually obtainable.

Tolled highways, bridges and tunnels, water, gas or oil pipelines and hydroelectric facilities are considered suitable projects, because a private economic equilibrium is obtainable. However, subsidies are often necessary for high-speed train networks, light-rail trains and hectometric transport because prices paid by users are often low and governments generally prefer to control prices.

The characteristics of BOOT projects are particularly appropriate for infrastructure development projects such as toll roads, mass transit railways and power generation and as such have a political dimension of public good, which does not occur in other private financed projects.

17.10 Risks fundamental to BOOT projects

The two types of risk fundamental to BOOT projects, elemental risks and global risks, are defined below.

(1) Elemental risks are those risks that may be controlled within the project elements of a BOOT project.
(2) Global risk are those risks outside the project elements that may not be controllable within the project elements of a BOOT project.

Many of the global risks are addressed and allocated through the concession agreement with elemental risks retained either by the promoter or allocated through the construction, operation and finance contracts. The author has developed an SCA based on the terms of the concession and the project conditions specifically for BOOT contracts. This SCA is used as the basis for risk analysis, bid preparation and bid evaluation.

There are two phases when risks associated with financing BOOT projects occur – the construction phase and the operation phase. These two distinct phases are considered as:

- the pre-completion phase relative to construction risks;
- the post-completion phase relative to operational risks with the first few years of operation having the major operation risk.

Promoters are exposed to risks throughout the life of the project, which may be summarised as:

- failure at several stages of the project;
- failure in the later stages of the project when considerable amounts of money have been expended in development costs;
- failure of the project to generate returns or the opportunity to recover costs.

Risks associated with market prices, financing, technology, revenue collection and political issues are major factors in BOOT projects. Risks encountered on BOOT projects may also include physical risks such as damage to work in progress, damage to plant and equipment and injury to third-party persons and theoretical risks such as contractual obligations, delays, force majeure, revenue loss and financial guarantees.

The major risk elements of a BOOT project may be summarised as:

- completion risk: the risk that the project will not be completed on time and to budget;
- performance and operating risk: the risk that the project will not perform as expected;
- cash flow risk: the risk of interruptions or changes to the project cash flow;
- inflation and foreign exchange risk: the risk that inflation and foreign exchange rates affect the project costs and revenues;
- insurable risks: risks associated with equipment, plant commercially insurable risks;
- uninsurable risks: force majeure;
- political risk: risks associated with sovereign risk and breach by the principal of specific undertakings provided in the concession agreement;
- commercial risk: risks associated with demand and market forces.

Demand risks associated with infrastructure projects are much greater than those facilities producing a product offtake because an infrastructure project is static and cannot normally find another market, whereas a product may be sold to a number of offtakers through the life of the concession. Facilities producing an offtake bear the risk of product obsolescence, and competition usually leads to market risks dominating especially when operation and maintenance costs are high and concession periods short.

17.11 BOOT package structure

A BOOT project structure is a highly sophisticated structure requiring the full participation of all the parties involved in identifying and allocating the relevant project risks and responsibilities and an appreciation of the political, legal, commercial, social and environmental considerations that have to be taken into account when preparing BOOT project submissions. The process of developing a BOOT project is immensely complicated, time consuming and expensive.

The major components of a BOOT project include the agreement to:

build: design, manage project implementation, carry out procurement, construct and finance;

operate: manage and operate plant, carry out maintenance, deliver product or service and receive offtake payment;

transfer: hand over plant in operating condition at the end of the concession period.

The number of components and their timing over the concession period need to be identified at an early stage of a project. This should be addressed in a format that can be utilised to identify the obligations and risks of each organisation involved in the project so that an equitable risk allocation may be determined.

BOOT contracts may be determined by four major packages described below.

> **Construction package** containing all the components associated with building a facility, normally undertaken in the pre-completion phase and may include feasibility studies, site investigation, design, construction, supervision, land purchase, commissioning, procurement, insurances and legal contracts.
>
> **Operational package** containing all the components associated with operating and where applicable owning the facility and may include operation, maintenance, training, offtake, supply, transfer, consumables, insurances, guarantees, warranties, licences and power contracts.
>
> **Financial package** containing all the components associated with financing the building, and in some cases the early stages of operation and may include debt finance loan, equity finance loan, standby loan agreements, shareholder agreements, currency contracts and debt service arrangements.
>
> **Revenue package** containing all the components associated with revenue generation and may include demand data, toll or tariff levels, assignment of revenues, toll or tariff structures and revenues from associated developments.

This structure incorporates all the components of a BOOT contract into discrete packages over both the pre-completion and the post-completion phase of the concession period. The type of facility, its location and revenue realisation would effectively be contained in one of the packages. Having identified and allocated components into the four packages, a promoter organisation can then determine the risk associated with each package and how such risks would be shared. The package structure provides a rational basis for financial appraisal of BOOT contracts, for the allocation of risk within the concession agreement and contractually between the parties concerned and for the structure of the tendering process.

17.12 Advantages and disadvantages of BOOT projects

The BOOT project may offer both direct and indirect advantages for developing countries, those being:

- promotion of private investment;
- completion of projects on time without cost overruns;

- good management and efficient operation;
- transfer of new and advanced technology;
- utilisation of foreign companies resources;
- new foreign capital injections into the economy;
- additional financial source for priority projects;
- no inroads on public debt;
- no burden on public budget for infrastructure development;
- positive effect on the credibility of the host country.

The introduction of new technologies, project design and implementation and management techniques are considered as advantageous to developing countries, the disadvantages, however, being host country constraints and financial market constraints.

A major advantage of a BOOT project is the financial advantage to a government as its off-balance-sheet impact does not appear as a sovereign debt. The advantage to an overseas principal is that he or she does not need to compete for scarce foreign exchange from the state purse, there is a specific need for the project and risks are transferred to the promoter. The most important attractions to governments of Asian developing countries is off-balance-sheet financing, transfer of risk, speedy implementation and an acceptable face of privatisation. The involvement of the private sector and the presence of market forces BOOT schemes ensure only projects of financial value are considered.

There are six main arguments for concession projects.

> **Additionality.** This would offer the possibility of realising a project that would otherwise not be built.
> **Credibility.** This would propose that the willingness of equity investors and lenders to accept the risks would indicate the project was commercially viable.
> **Efficiencies.** The promoters control, and continuing economic interest in design, construction and operation of a project will produce significant cost-efficiencies, which will benefit the host country.
> **Benchmark.** BOOT projects are useful to the host government as a benchmark to measure the efficiency of a similar public sector project.
> **Technology transfer and training.** The continued direct involvement of the project company would promote a continuous transfer of technology, which would ultimately be passed on to the host country. A strong training programme would leave a fully trained local staff at the end of the concession period.
> **Privatisation.** A BOOT project will have obvious appeal to governments seeking to move its local economy into the private sector.

There are three main arguments against concession projects.

> **Additionality.** Commercial lenders and export credit guarantee agencies will be constrained by the same host country risks whether or not the BOOT approach is adopted.

Credibility. This benefit may be lost if the host government provides too much support for a BOOT project resulting in the promoter bearing no real risk.

Complication. A BOOT project is a highly complicated cost structure that requires time, money, patience and sophistication to negotiate and bring to fruition. The overall cost to a host government is greater than traditional public sector projects although proponents of the BOOT approach argue overall costs are less when design and operating efficiencies are taken into account and compared with public sector alternatives

Although there are a number of advantages and benefits associated with BOOT projects, very few BOOT proposals have reached the construction stage. A review of BOOT schemes by an EC Commission concluded that there were three key problems associated with BOOT projects – availability of experienced developers and equity investors, the ability of governments to provide the necessary support and the workability of corporate and financial structures.

The risks associated with BOOT projects are far greater than those considered under traditional forms of contract as the revenues generated by the operational facility must be sufficient to pay for construction, operation and maintenance and finance. The uncertainty of demand and hence revenues, cost of finance, length of concession periods, levels of tolls and tariffs, effects of commercial, political, legal and environmental factors are only a small number of risks to be considered by promoter organisations.

17.13 The origins of private finance initiatives

The private finance initiative (PFI) was a policy born out of a series of privately financed projects beginning with the Channel Tunnel project in 1987. Other transport-orientated projects followed swiftly: Second Severn Crossing, Dartford Bridge, Skye Bridge, Manchester Metrolink and London City Airport.

A privately financed public sector is not a new concept – concession contracts have been utilised in the past with French canals and bridges privately financed in the eighteenth century and the railways in the UK in the nineteenth century. Toll roads were common in the eighteenth and nineteenth centuries both in the UK and USA, where a turnpike system saw owners collect money from users of the roads.

Many of these projects 'floundered' by the beginning of the twentieth century, with most major infrastructure projects having a government-financed structure. The trend back toward concessions started to emerge during the mid-twentieth century in the USA, which corresponded to a growth in the financial markets and their ability to finance such complex projects.

At present there is interest in concession contracts throughout Europe including toll roads in France and Spain and power stations in Italy, Spain and Portugal. The pressure on public finance is not restricted to Europe – there is growing interest in the use of private finance all over the world, especially in the emerging economies in Asia.

17.14 The arguments for privately financed public services

The underlying principle behind the introduction of the private sector has many dimensions, the obvious one is a pure private finance case where a facility and service is provided at zero, or minimum cost, to the public sector. An example of this is the Second Severn Bridge, held up by many as the perfect PFI project. Secondly, it is the public sector's exploitation of the private sector's ability to design and manage more efficiently. The public sector is characterised by substantial cost overruns and poor management skills, utilising PFI may reduce these inefficiencies.

There is a pan-European downward pressure on public capital budgets, which has forced governments to rethink their procurement strategies, with PFI-type schemes becoming more popular throughout Europe. There is particular emphasis on reducing public expenditure with regard to satisfying the criteria for monetary union under the Maastriht agreement. However, it must be noted that the targets set under Maastriht are subject to the European Union's standard measures. Therefore, reducing the public sector borrowing requirement (PSBR) through PFI will only go towards meeting this target if the PFI project 'adds value'. Hence, the need to cut, or restrict, public expenditure is not just a British preoccupation but a global objective. The rationale behind concession contracts and private finance is to achieve the following objectives.

- To minimise the impact of added taxation, debt burden on the finances of governments.
- To introduce the benefits that might accrue from the use of private sector management and control techniques in the construction and operational phases of the project.
- To promote private and entrepreneurial initiative in infrastructure projects.
- To increase the range of financial resources that might be available to fund such projects.

The reduction in public borrowing and direct expenditures are among the factors for adopting concession contracts.

Arguments against the concept include those listed below.

- Efficiencies may well be available in the private sector, but it is argued that such management efficiencies might not materialise, and if they do, who benefits?
- Construction efficiencies are also visible, with construction programmes reduced due to a lack of procedural constraints from government, but the overall time from conception to operation will be far longer than a conventional project. This is due in part to the processes involved in the evaluation and negotiations required by such projects.
- Financing charges are higher for the private sector than for public bodies, which reduce some of the efficiency benefits that may, or may not, occur from private sector involvement.

- Owing to the complexity of concession contracts, governments, and their agencies, have to spend large sums in advisory fees for lawyers and financiers.

17.15 PFI in the UK

PFI health care

There has been massive interest in the provision of services for the health sector, which has attracted much negative publicity and a conspicuous absence of signed deals. According to the Private Finance Panel (PFP), there had been 47 signed contracts for PFI service provision for the Department of Health by the end of 1997 – these range from clinical waste incineration to information systems to office accommodation.

A sticking point with high-value PFI hospitals is that trusts do not appear too realistic in their requirements, and consortiums responding to output specifications spend large sums of money preparing bids, which ultimately are not affordable. This problem arises because of the intensive level of borrowing required by the consortium at the early stages of a project and subsequent debt servicing. These costs are passed on to the trust who find them too high.

PFI prisons

The prison sector has been the beneficiary of modest PFI success, with privately financed prisons. Design, construct, manage and finance (DCMF) prisons were a progression from privately operated facilities; these existed in three prisons across the UK including HMP Doncaster and Wolds Remand Prison. These prisons are still operating successfully, showing 'significant cost benefits' over existing public sector best performers. Alongside this was the contracting out of prisoner escort services, which are provided by the security company, Group 4. The natural course of action follows that to expand the prison number capacity through building more prisons, potential private sector operators should input into the design, construction and financing of the asset. The existing links within HMPS and the experience with private operators meant that it was well suited to PFI treatment – there was an existing culture within HMPS to working alongside private organisations.

In addition to 'contractors turned operators' bidding for the project, there was the arrival of the US operator Wackenhut Correction Facilities, who joined as consortium members. This track record and the existing privately operated prisons in the UK meant there was already in place a valuable benchmark on which to evaluate the bids. The success of the Pathfinder projects was aided by the government's desire to see PFI realised, and the help they provided reflected this. In the Fazakerley and Bridgend prisons, the government acted as insurers as a last resort if the operators could not obtain commercial insurance. The follow-up projects, particularly in Lowdham Grange, have encountered problems due to the government's refusal to repeat this. This potentially terminates the concession agreement with Premier prisons unless they can reassure the insurers and point to their record of riots and disturbances in Doncaster. It was felt by HMPS that the success of Fazakerley and Bridgend meant they could transfer greater risks in forthcoming projects; they felt that commercial insurance could incentivise the operators to maintain an exemplary record when it came to discipline, as no claims bonuses could be earned.

Local authorities

Local authorities have been granted £250 million of help to cover advisor's fees to get the PFI process up and running in this sector. Local authorities have been looking at projects varying from small-scale heating projects to high-value transport interchange schemes. Along with other sectors there has been doubts expressed on the viability of smaller-scale projects. Some contractors have introduced a lower limit for bidding for PFI projects; one major contractor has determined a minimum of £5 million, which could effectively prevent such local authority schemes from getting off the ground. Examples taken from other countries could show the way forward for schemes such as district heating systems. A project in Germany has seen a developer award a 10-year concession to a small contractor to finance, supply and maintain a heating system to two apartment blocks. The interesting points about this project were that the capital cost amounted to slightly in excess of DM 15,000 and the project had a net present value (NPV) of just DM 2860 over 10 years at a 10% discount rate.

17.16 Bidding and competition

This section provides a brief explanation of the EC Procurement Rules and how they are applicable to PFI. These regulations are applicable to local authorities and utilities; hence, they have a significant impact on the way facilities and services are procured under PFI. The procurement strategies to which they apply are:

- works contracts;
- works concession contracts;
- subsidised works contracts;
- service contracts;
- design contests.

It should be noted that service concession contracts are defined as 'contracts under which a purchaser engages a person to provide services and gives him the right to exploit the provision of these services to the public' and are not subject to the rules. However, even though PFI is concerned with the procurement of services rather than assets, it still falls under the regulations applying the works projects. The reason for this inclusion is the typically high level of capital expenditure in the construction phases of projects; it is deemed that the payments for services are taking into account this expenditure, and thus PFI projects are treated from the standpoint of a capital works project.

The procurement rules mean that the procedure should follow one of three routes;

- open procedure;
- restricted procedure;
- negotiated procedure.

The open procedure allows everyone who has expressed an interest in the project to bid for it. This type of approach is not suitable for PFI projects as the cost of bidding is too high to allow mass bidding. More widespread is the use of the restricted procedure, where bidders express an interest and pre-qualify to bid for a project. The last type is where the preferred bidder negotiates the terms of the contract, but usually after winning in a competitive process. The final two methods must take place in an environment where competition is satisfactory, meaning that there should be a number of bidders in the competition. This number should not be normally less than three in the case of restricted procedure and five under the open procedure.

Public authorities are usually limited to using the first two methods and can only use competition and the negotiated approach in the following circumstances:

- if the pricing is not possible for the works and services;
- if specifications for services cannot be precise such that the open or negotiated procedure could be utilised.

The preferred bidder will win the right to negotiate the contract terms, but negotiation can be conducted directly without the need for tenders to be submitted. Competition is not always necessary; in some cases a negotiated non-competition procedure is allowable if:

- technical or artistic reasons mean no one else can do the job;
- exclusive rights exclude other bidders;
- it is a works concession contract;
- goods are manufactured for certain research, experiment, study or development purposes.

If a negotiated approach is to be used, it is important that it adheres to the following principles:

- transparency;
- objectivity;

- non-discrimination;
- equality of treatment;
- measurability.

In addition to the method of procuring a service, the contract duration may be subject to regulation under Article 85 of the Treaty of Rome. This addresses the concern that an excessively long contract period may distort and restrict competition; this is particularly problematic when considering the length of contract periods required to make the financing of concession type projects.

17.17 Output specification

One of the guiding philosophies of the PFI is that the public bodies allow the private sector the freedom to determine how they are to provide a service to meet the relevant specification. To give the private sector this freedom, the client has to be very careful in the way they express their requirements. This is achieved through the drafting of an output specification. This puts public bodies in an unusual situation as they have to be the authors of a service provision document rather than an asset provision document. Although this may seem straightforward, it is sometimes difficult to separate the service from the facility, and this leads to preconceptions of how a service should be run and from what type of facility. Such presumptions should be avoided at all costs as they remove the freedom of the private sector to innovate and produce alternative solutions.

The PFP supplies the following examples:

- a computer system is not an output; the information service it supplies is;
- a particular office building is not an output; a supply of services accommodation is;
- a school gym is not an output; regular access to sports facilities is.

In order to define needs successfully the approach must be carefully thought out and biased towards services provision. Clear specifications should comply with the following according to PFP guidance:

- a clear statement of requirements, concise, logical and unambiguous;
- a statement of what you want, and not how it is provided;
- allow for innovation from tenderers;
- provide all information a tenderer needs to decide and cost the services he or she will offer;
- provide equal opportunity for all tenders to offer a service that satisfies the needs of the user and may incorporate alternative technical solutions;
- solutions to be evaluated against defined criteria;
- indication of information needed in order to monitor that the service is being procured in compliance of the contract.

The issuing body must define its core requirement for the project, as it is important that these are clearly and accurately expressed. For example, a local authority may want a new school, but a consortium may develop a bid that includes other facilities such as a leisure centre and adult education centre. Most PFI projects can be negotiated as this allows bidders to develop and refine the specification, but there will come a point where the specification must be fixed, thus allowing bidders to develop the best way of achieving the requirements.

17.18 Financing public–private partnerships

Increasingly the projects that the principals wish to realise are not sufficiently robust to be procured by total private finance funding. This gap between commercial financial analysis and social cost–benefit analysis can only be closed by some type of public sector involvement.

The public sector has four main mechanisms for participation, which operate within a strict hierarchy, as shown in Figure 17.2.

Essentially and in sympathy with the concept of encouraging the private sector to take responsibility for project risk, the public sector wishes to make the minimum contribution sufficient to reduce the investment risk to levels acceptable to the private sector.

As the first stage the public sector can consider additional risk sharing. Risk is already shared through the concession agreement, using the basic principle that the party best able to control or manage a risk should take responsibility for that risk. For risks that cannot be controlled by either party, it is usual for the public sector principal to take responsibility because the risk will have to be paid for only if it occurs, rather than a premium paid to a contractor to cover the possibility of it occurring. Therefore, by assuming responsibility for some additional risk, this may enhance the robustness of the project cash flow, and hence attract investors and lenders at an acceptable rate of interest.

Once the risk sharing option has been exhausted and further responsibility for risk is unacceptable to the public sector, the option of additional equity can be considered. Often this might take the form of offering the promoter the use, the revenue and the operation and maintenance of existing facilities. This stream of revenue at the critical stage of the project can be valuable in closing

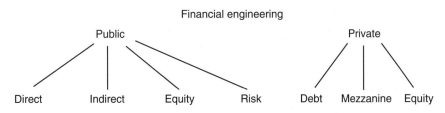

Figure 17.2   Public–private financial engineering.

the gap between debt finance and the equity provision that the promoter is able to raise.

An alternative mechanism was adopted for the new Athens airport at Sparta. When originally proposed as a private sector project, the debt financiers required equity funding in excess of the levels that any of the tendering consortia were prepared to offer. The Greek government then suggested a deferral of the project and during this period a hypothecated tax on all airline tickets would be raised. This funding was then utilised as a second source of equity to bridge the gap between the level required by the lenders and the equity raised by the promoter.

Typical indirect options include tax holidays, grace periods and soft loans. A tax holiday is a predetermined period of time over which the promoter will not be liable for tax on the concession. Conventionally, each project is treated as a 'ring-fenced' investment and, in the UK, is liable to be taxed on its operating profits in the same way as any other commercial enterprise. In a number of UK infrastructure projects a tax holiday of 5–10 years is likely to be most beneficial and after that period the returns diminish.

Grace periods and soft loans can be used separately or jointly to assist the financing of the project. A grace period is a fixed length of time before which a loan has to be repaid. Obviously, the concession contract has the maximum capital lock-up towards the end of the construction phase, and if a loan can be effectively extended without interest, this can be extremely beneficial to the cash flow of the concession. As most infrastructure constructions take between 2 and 3 years, it is a grace period of between 5 and 7 years that is at most benefit. Once the infrastructure is completed, the initial debt financing carrying high interest rates comparable with the risks associated with construction can be repaid and the debt refinanced at lower rates of interest commensurate with the known risks of operating similar infrastructure projects.

A soft loan is a loan offered at rates below the commercial market rate. The public sector loses some of the return that could have been gained, but assists the promoter by reducing the debt burden. The range of these options can be quite considerable, e.g. consider the financial implications for two concession consortia, one with a public loan at commercial rate over 12 years and another with the same loan but with a 10 year period of grace and interest at a quarter of the commercial rate.

Finally, if no other course of action is available, the public sector can provide direct financial investment. This may be necessary on projects perceived to be high risk or projects with a marginal cash flow. Examples can be found of the M1 Vienna to Budapest Toll Road in Hungary, where the National Bank of Hungary, the Nemzeti Bank, made a contribution of 40% of the project value in Hungarian Florints. Similarly, most light-rail transit schemes in the UK and in many other countries are financially non-viable without a major financial grant from the public sector, often between 75% and 90% of the capital cost.

Further reading

Akintoye, A., Hardcastle, C., Beck, M., Chinyio, E. and Asenova, D. (2003) *Public Private Partnerships: Managing Risk and Opportunities*, Blackwell Publishing.

Bult-Spiering, M. and Dewulf, G. (2006) *Strategic Issues in Public-Private Partnerships*, Blackwell Publishing.

Merna and Owen (1998) *Understanding the Private Finance Initiative*, Asia Law and Practice.

Merna, A. and Smith, N. J. (1996) Private finance and public partnership. In *Projects Procured by Privately Financed Concession Contracts*, volume 2, first edition, Asia Law and Practice.

Smith, N. J. and Merna, A. (1997) The private finance initiative, *Engineering, Construction and Architectural Management*, special issue, **4**.

Chapter 18
Project Stakeholders

This chapter will identify the issues facing the project manager during the project's life cycle because of its interaction with its internal and external environment. All projects operate as open systems and have a variety of influences operating on them. There is no longer the complete and unrestricted acceptance of new projects. Increased concern about the impact of new projects on our daily lives, society, the environment, and so on means that we are starting to adopt new perspectives on activities and actions that influence our lives. Society is taking a greater interest in its 'stake'.

18.1 Stakeholders

Projects have always been subject to scrutiny by a variety of interests. The classic definition of a stakeholder was provided by Freeman who suggested that 'A stakeholder in an organisation is (by definition) any group or individual who can affect or is affected by achievement of the organisations' objectives'. Extrapolating this idea, groups or individuals who influence and are influenced by a project are the project's stakeholders.

Identifying and analysing stakeholders is a simple way to acknowledge the existence of multiple constituencies of the project environment. The main insight of stakeholder identification is that project managers might pay some attention to those groups that were important to the success of the project. By adopting an open system's perspective, we can advocate that stakeholders have some sense of participation in the affairs of the project.

Stakeholders are categorised as primary and secondary stakeholders. Primary stakeholders are those groups that are directly associated with the project. These may include, among others, the project sponsors, clients, the project management organisation, project financiers, consultants, contractors and subcontractors. All these groups directly influence and are influenced by the project. These groups stand to gain or lose the most from a project.

Secondary stakeholders are those groups that may not have a direct involvement on the project but may be influenced by its outcomes. Prominent secondary stakeholders may include, among others, the government, other regulatory authorities, community organisations, non-governmental organisations (NGOs), trade associations and interest groups. These groups are concerned about the extent the project encroaches on their areas of interest. Our interest in stakeholders stems from the influence that stakeholders have and their potential to either promote or disrupt projects.

18.2 Stakeholder identification

Generally, identifying potential stakeholders in a project should be a major part of the strategic management process carried out at start-up. This is a potential problematic area for the project manager as the number and nature of stakeholders on a project will vary with the life of the project. The point at which the project manager joins the project will also influence perceptions of stakeholders. It would therefore make sense to carry out the identification throughout the life of the project. Stakeholder management is an important part of the strategic management process of an organisation as stakeholders have the ability to influence the organisation. Taking a project as the equivalent of an organisation, it follows that stakeholder management should form a major part of the strategic management process at the start-up phase of a project.

Project stakeholder management is the process of dealing with the people who have an interest in the project with the aim of aligning their objectives with those of the project. A systematic approach to the project stakeholder management process is shown in Figure 18.1.

The first step in the process is to identify the potential stakeholders on the project. The next step is the gathering of information about the stakeholders. This is then followed by identifying the stakeholder's likely mission. For example, if Greenpeace is identified as a potential stakeholder, gathering information about

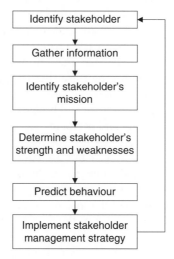

Figure 18.1   Project stakeholder management process. (Adapted from Cleland and King, 1998).

how they work and what aspect of the project is likely to attract their attention provides the basis for determining what their mission could be. The next stage involves the determination of the stakeholder's strengths and weaknesses. The important issue in stakeholder identification is trying to understand what the 'stake or interest' is and the extent of the stakeholder's power on the project. This stage helps the management to understand how much the stakeholder can actually influence the project, what avenues they have available to make their needs known and how far they will go to see them through. Based on this, it then becomes possible to predict what the behaviour of the stakeholder will be and how it will affect the project goals and objectives. Once this is known, the project manager can now devise and implement stakeholder management strategies, which should aim at mitigating negative stakeholder behaviour. The loop shows that the process is a cycle as the stakeholders in a project will vary from one stage in the life of the project to the other, and sometimes strategies may have to be revised. Stakeholder management is not a fixed process and will alter as stakes change over the life of the project.

18.3 The client

The client is the most influential project stakeholder. For our purposes, the client is the owner of the project and the group that stands to gain or lose the most on a project. The owner of the project must provide clear direction and timely decisions, and must assist the project management team to drive the project to a successful conclusion. From this definition of what a client is, it follows that if all clients fulfilled their obligations as project owners, there would be very few if any problems facing the project manager at any stage in the life cycle of the project as a result of client activities and conduct. This, however, is in the ideal world and, more often than not, problems arise within projects that arise from the relationship between the client and the project manager.

Defining who the client really is is essential – a project team will tend to recognise its client as the body that has the authority to approve expenditure on the project, set its schedules, set the form that the project has to take and assess its performance. The project arises out of strategic need of the client and in many cases sets the timing, performance and budget. On a project, such as the Olympic Games stadium, the timing and performance of elements of the project are determined when the announcement for hosting a games are made. The budget and funding are outlined at the start but are subject to the vagaries of time. On an Olympic Games project, the client is notionally the National Olympic Association; yet in almost all cases, the host government is always a key influence on the client. Although the Olympic Association is the notional client, the host government is just as important in the decision-making process. The motivation of the client is also important in that a private sector client is driven by commercial criteria, while a public sector client has other requirements such social accountability. However, the key project performance areas of cost, time and performance remain the same for both sectors.

Recent developments across all industries in both the UK and the rest of the world have seen an increasing popularity for partnerships on projects. Partnerships are usually complex organisations made up of individual companies each with their own separate aims and goals although they may share one or more common goals for which the partnership is formed. What this means for the project manager working for a partnership is that conflicting ideas may be presented, rendering decision making extremely difficult. Another problem is that, if the decision-making structure of the partnership is not clearly defined, there are likely to be problems with all aspects of the decision-making process. In partnerships identifying the key influences are important. Partnerships are also susceptible to the vagaries of partners' varying strength, internal project agendas and influence on the project.

The solution to the problem of defining who the client is lies with the project manager and his team making an effort to understand the client organisation. An effort must be made to identify the organisation's aims and the importance of the project. Consideration must be given as to who has the authority to make decisions within the client organisation and the speed at which these decisions can be made. Once this understanding exists, then the project manager can tailor project activities so that they are not in conflict with the way in which the client organisation works. For example, if it is known that approval for funding takes a minimum of 4 weeks to go through, the schedule of works must take this into account to make sure that it is realistic and workable.

It can be extremely difficult to tailor project systems to suit those of the client organisation. A highly bureaucratic client organisation may make timely decision making difficult, and yet this could be a key factor to the success of the project. In this case, it may be worthwhile for the client to set up a management team within their structure for the sole purpose of dealing with the needs of the project. This approach would quite easily work, as it would not require the complete overhaul of the whole client organisation structure just for the sake of a one-off project. Besides that, it gets rid of the resistance to change, which is one of the main obstacles to successful strategic project management. However, all bureaucracies are tied up with rules and regulations and any new decision-making structure within the organisation would have to fit into this environment. Project structures that emerge from a bureaucratic client will follow the parent organisation. External projects may take an independent structure provided reporting and control elements are agreed to with the client. Internal projects will tend to conform to the general structure within the parent organisation.

The client value system – projects are not independent silos for the client, but part of a larger business system so much so that its failure may affect more things than the obvious. For instance, the failure of the Wembley stadium to be completed on time has resulted in the client suffering commercial losses owing to the cancellation of major sporting events, concerts, conferences and other associated events while still having to meet payments to the project financiers. The commercial reputation of the client is also damaged with its financiers and other commercial partners. It is therefore important that the project manager and his or her team clearly understand what the project objectives mean for the client. The problem starts with the fact that sometimes clients are not so keen on divulging such information as part of their competitive strategy. Even when clients are willing to do this, it is not easy to do so for highly complex projects surrounded by a lot of uncertainty. This is clearly a problem for the start-up process as a lot of information required to start off effectively may not be available or, even if available, may be unclear.

Part of the start-up process should involve the client, ensuring that the project management team are aware of the link between the project and the client organisation's business strategy. The project manager has to be aware of the implications of project failure. This should be done to a level that does not undermine the client's competitive strategy but at the same time allows the project team to understand the importance of the project objectives. This means that the team will have a better understanding of the project's role in trying to satisfy their client's needs.

The project brief is essentially a document that spells out what the client requires from the project. The client's decision to construct is in response to changes in the organisation's business strategy and environment. The organisation's strategy, in either the public or private sector, will determine project deliverables in terms of timing, budget and performance. Changes in the political, economic, social, technological and legal environment can be extremely dynamic and influence the direction of the client's business strategy goals. On projects that are dependent on changing technology such as security systems or informations and communication technology (ICT), they are subject to the vagaries of change. Clients often expect the latest changes in technology to be incorporated into these types of projects, which results in many of the original brief ideas having to be reinterpreted. This means that the client may request for changes during all stages in the life of the project. Some changes may be subtle while others may be so big that they render earlier plans of work invalid. Project planning has to become more flexible when there is a dynamic project environment and greater potential for change.

Traditionally, when the client made the decision to construct, a brief was developed with the help of professionals like architects or engineers. The project manager was therefore not involved when certain important decisions were made. This was a limitation in that the project manager did not get a chance to clearly appreciate why the client wanted what the brief spelt out. The need for better project performance has led to earlier involvement of the project manager. In-house project managers for knowledgeable clients are more closely involved in earlier stages of the project. This approach has been carried through more widely in the project development process with earlier project manager involvement. The transfer of risk to the project manager has also necessitated earlier involvement. However, project managers are expected to adapt irrespective of whether they have early involvement or not, or whether they are internally or externally employed.

Another common problem with the project brief is that it sometimes fails to clearly spell out what is required even though it may initially appear to do so. This is a case of poor communication as the brief fails to specify what it was supposed to mean to the project manager. Poor communication may also plague other decisions if they are not relayed appropriately to the project team or to the client. This may result in expensive changes at later stages in the life of the project. This will affect project start-up as well. Making changes will inevitably require some decision making on the part of the client and other stakeholders.

In general, the client must make sure that their brief is as complete as possible and that the channels for any changes are made clear to the project manager. Regular meetings to discuss progress on the project can help to improve the communication between the client and the project team. Briefing does not stop after start-up. There has to be interaction between the client and the project manager throughout the life of the project. This creates the opportunity for change to be managed and to test the feasibility of changes before they are implemented. The integration of the client and the project team is the most significant factor in the success of the project and requires understanding and skill in its design. The implication here is that there is a need for the client to be involved with the team, but this integration must be designed such that it does not undermine the authority of the project manager. Furthermore, team building must be made a part of the project as early as possible in its life.

18.4 Contractors

The contractors who execute projects are the primary stakeholders. They have a significant commercial stake in the project. If a project does not achieve the desired outcomes, the contractor can face financial ruin. Losses accumulated from a single project can result in the bankruptcy of the contractor. Cost overruns from the Cardiff Millennium stadium and New Wembley placed their constructors under serious financial pressure. Contractors are the executors of projects and they assume a significant amount of risk to deliver the project.

The contractors and the other members of the production team interpret the client's requirements to deliver a physical product. The relationship between those who set the requirements, those who design the requirements and those who turn the requirements into reality is important. The client and the contractors share many of the risk attributes on the project. If the project is not delivered in time, on budget and to the required performance specifications, then both parties tend to suffer losses. The emergence of a long-term relationship between clients and contractors on engineering projects has also changed the dynamics of the stakeholding on projects. In a single project, the stake is relatively short term and project orientated. However, in longer-term relationships such as programmes or frameworks the stake relates to the ongoing interaction across multiple projects and activities. The stakeholder relationship between contractors and their clients are project-specific, but can move towards a longer-term corporate stakeholder relationship.

18.5 Financiers

Very few significant projects are self-funded. In most cases, self-funded projects are relatively small in scale. In all projects there will always be a financial case to be made irrespective of whether the project is self-funded or not. On an internal project, financial controllers still require due diligence over the financing of projects. This financial case has to be robust, or else the project is not likely to proceed.

In projects that are externally funded, the role of the financial provider is significant. A financial stakeholder is a primary stakeholder in the project. The financial stakeholder has a commitment to the project from its earliest stages of concept, through execution and, in most cases, when the project becomes operational. Financial stakeholders may be providers of equity, debt or subscribers to bonds that support the project or its parent company. The decision to provide financial support to a project is based on assumptions related to the project's ability to be within cost, to the schedule and to meeting its performance requirements. The next stage will be assumptions about demand, revenues, costs, reliability, deliverables, and so on. When a project does not perform, the financial stakeholders face the biggest risk. The Channel Tunnel was completed a long time ago and is fully operational. The financial stakeholders are still involved with the project. The Channel Tunnel was not delivered on time and suffered cost overruns, and it did not meet its demand and revenue projections. The financial stakeholders had the ability to shut down the project but kept it open despite making losses. Most financial stakeholders do not have a philanthropic interest in projects – they are interested in performance and returns.

On infrastructure projects in developing countries there is the influence of aid agencies and institutions like the World Bank. These organisations act as banker for projects. They also specify a variety of conditions on some projects relating to project performance, resourcing, equality, and so on. The World Bank has also started to look at the manner in which projects are procured. It is looking at the use of agents and facilitators as well as the manner in which projects are procured. The World Bank is particularly concerned about corrupt practices in the procurement of projects. It has threatened to blacklist companies if they are found to be involved in corrupt practices on projects. The European Union (EU) is also bringing this approach to their project practices. Organisations like the EU and the World Bank have some financial stakes in projects but also look at the social impact of projects.

18.6 Government

The government of any country holds a unique position as a stakeholder in that it is a major client of engineering projects as well as a regulator across all sectors of industry. The main concern of the government, as a client, is to deliver projects that will ultimately be for the benefit of society. Public sector projects are set up to be run along commercial lines with a greater recognition for 'value for money'. The growth of private sector financing on public sector projects is helping to establish a more commercial culture on public sector projects and within supporting agencies.

The government is the major procurer of engineering projects in every country. Projects are commissioned to support education, health, transport, defence, judiciary, prison services, welfare and so on. Although there may be separate departmental projects, their combined influence can have a considerable impact to shape the future of the industry. The American Corp of Engineers helped change perceptions by adopting partnering. This helped popularise partnered and collaborative approaches in the USA and subsequently was adopted in the UK. Similarly, the adoption of private finance initiative (PFI) by the government in the UK has reshaped the construction industry because only certain companies now meet the criteria for involvement on these projects. The volume of projects and the procurement policy of a government is a major influence on engineering projects.

The government is also a stakeholder as a regulator and legislator. Governments are expected to act in the best interests of the countries they govern. They legislate and regulate across industries and have an influence on projects. Government influence does not take place at one level. In the UK, there is the national, regional and local government influence. There is also the influence of the EU regulations on projects. All projects require approval at either a local or national level. The process of approval often requires consultation in a public arena. These activities take time and can have a knock-on effect on the project success. The Terminal Five facility at London Heathrow airport held a public enquiry that took more than 3 years to complete and had 11 local authorities and a variety of local, national and international interest groups taking an interest in the scheme. Although Terminal Five may be seen as an extreme case, all built projects require planning approval.

The government also acts as a stakeholder through its legislative agenda. New and existing laws have the ability to influence projects. Following the Piper Alpha disaster, the governments set up the Cullen enquiry and a key piece of legislation that came into being. The key change to the regulatory environment was the introduction to the Offshore Installations (Safety Case) Regulations, 1992 (amended 2005). They require the operator/owner of every fixed and mobile installation operating in UK waters to submit to the Health and Safety Executive (HSE) for their acceptance, a safety case. Safety cases are also required by the HSE at the design stage for fixed installations and when installations are involved in temporary drilling or well activities or are decommissioned. The safety case must give full details of the arrangements for managing health and safety and controlling major accident hazards on the installation. It must demonstrate, for example, that the company has safety management systems in place, has identified risks and reduced them to as low as reasonably practicable, has introduced management controls, provided a temporary safe refuge on the installation and has made provisions for safe evacuation and rescue. The entire project, from design through to operation, has to meet strictly regulatory guidelines. This is a case where the regulator has a major influence on the project.

On a national level the government can be an enabler. The government sets out policy for the development of a country. The choices that governments make influence various industrial sectors and projects. In the UK during the 1990s, the government moved away from the development of roads towards mass transit systems. The result was a number of light-rail projects being proposed. However, it became apparent that the road network needed to be upgraded for the economic development of the country. Government policy moved towards supporting and enabling road development. In turn, the government started to look at the mass transit systems more closely for value for money. Significantly, a number of transit schemes did not receive support as the government shifted its priority to roads. The government through its policy can enable or hinder projects.

The government is a major stakeholder in a number of ways and can influence projects significantly.

18.7 Community

Community influence appears very early in the life cycle of the project during the feasibility and approval stages. Terminal Five was a high-profile project that met a great deal of local resistance, ensuring that a comprehensive enquiry is undertaken. Communities are vocal stakeholders as they are the people that have to live with the projects.

Communities are also stakeholders through their political influence. In active communities, they have the ability to mobilise political support for their causes. Political mobilisation is a major force that can influence the outcomes or directions that projects take. Direct action campaigns against road transportation projects in the UK to customer boycotts of soft drinks in the south of India have had an impact on projects. If there is a mix of community activism with political influence, then this creates a powerful stakeholder influence. Politicians under threat are likely to support causes that may work against projects as well as putting their influence behind projects that communities want.

Communities are not always involved in large-scale projects; they often have a very local perspective on projects. It is often issues that relate to the quality of life, employment or disruption of the community that influences actions. Onshore and offshore wind farm projects in the UK have had a great deal of community interest. Onshore communities look at the impact of noise and visual intrusion, while in the case of offshore it is normally the impact on fishing grounds and wildlife that influence communities. The issue of compensation to fishermen was not generally considered a part of the project budgeting process. Community networks opposed to wind farm projects have sprung up across the UK and share information on how to oppose such projects. This is an indication of how communities can oppose the project and by sharing information become more powerful. They can also mobilise to support projects.

18.8 Interest groups

All projects may be susceptible to interest groups with a variety of 'stakes' in a particular project. There is no simple way to identify who these groups are and whether they may take an interest in a particular project. Many of these groups are considered to be NGOs. Larger interest groups are relatively easy to identify as they have clear causes or issues that they support. Many of the larger groups support issues such as the environment, wildlife, poverty relief, humanitarian effort and human rights. These large groups have the power to mobilise resources and opinion. In the 1970s Greenpeace was a very small organisation largely based in Europe; today Greenpeace has membership in over 150 countries. It has the ability to use its resources to prevent projects or mobilise opinion against projects. This was amply demonstrated during the decommissioning of the Brent Spar oil platform, during which Greenpeace took direct action as well as mobilising opinion worldwide against the plans. The owners of the Brent Spar had to change the project strategy for the decommissioning from sinking the platform at sea to its disposal on land. Brent Spar was a demonstration on the organising and media capability of larger interest groups.

Interest groups also take a 'stake' in projects away from their home base. Dam projects in developing countries have come under particular scrutiny. Dams are intended to provide cheap power to countries, but many Western NGOs and charities oppose them primarily because of the environmental impact and the displacement of people. Most of the activism does not happen at the place where the project is being executed but rather at the corporate home of the construction companies. Many UK construction firms find themselves under pressure to pull out of dam projects around the world because of pressure on the home base. Well-organised NGOs have a great deal of leverage and media power to influence the debate about projects.

18.9 Summary

Projects do not operate in isolation away from the realities and influences of the world. All projects have stakeholders who share an interest in the project. Either the project can influence their organisations and lives or they can influence the project. At the heart of the stakeholders are the project sponsors or clients. The project is central to their business strategies and success is critical for their corporate requirements. There are other stakeholders that also have an interest in the project. Identification of these stakeholders helps the project manager to identify potential interest and risks to the project. Stakeholder identification helps to prevent any surprises suddenly appearing during the project life cycle. No project manager can ignore the influence of stakeholders on projects.

Reference

Cleland, D. I. and King, W. R. (1998) *Project Management Handbook*, second edition, Van Nostrand Reinhold.

Further reading

Friedman, A, and Miles, S. (2006) *Stakeholders: Theory and Practice*, Oxford University Press.

Grayson, D. and Hodges, A. (2001) *Everybody's Business. Managing Risks and Opportunities in Today's Global Society*, Dorling Kindersley.

Robinson, S., Dixon, R., Preece, C. and Moodley, K. (2007) *Engineering, Business and Professional Ethics*, Butterworth-Heinemann.

Sommer, F., Bootland, J. and Hunt, M. (2004) *Engage. How to Deliver Socially Responsible Construction. A Client's Guide. CIRIA C627*, CIRIA.

Chapter 19
Project Management in Developing Countries

Civil engineering and the construction industry in developing countries are suf-ficiently different to warrant the inclusion of this chapter in a book on project management. The range of types and size of construction companies is differ-ent, the environment in which they operate is different, the resources that are employed may be different and the way projects are funded is different. This chapter first reviews some of the main issues that contribute to the distinctive nature of developing countries and how these affect projects.

19.1 What makes developing countries different?

Construction projects, and the construction industry, in developing countries are significantly different from those in the developed industrialised world. The main differences are related to climate, population and human resources, material resources, finance and economics, and socio-cultural factors. Due recognition of these differences is a prerequisite for the successful management of projects in developing countries.

Climate

Many poor developing countries experience climatic conditions quite different from those in the temperate north. The type of project that is required, the most appropriate technology to be applied and the way in which the project is managed can be influenced by these variations in climatic conditions. For example, communities living in hot climates have quite different requirements for power and water, giving rise to alternative approaches to the planning and design of the requisite infrastructure facilities. Climate will also affect the design and type of technology used: solar power may be a realistic alternative to thermal power generation; high temperatures and long hours of sunlight may indicate alternative forms of sewage treatment such as waste stabilisation ponds; and the design of buildings must be aimed at reducing glare from sunlight and ensuring heat is kept out (rather than in).

During construction, it may be necessary to take precautions not required in cooler climates, such as chilling or adding crushed ice to the water used in mixing concrete, and paying particular attention to the curing of concrete. Planning and scheduling of construction work can also be affected by the climate – particularly when constructing roads, bridges and hydraulic structures in areas affected by heavy seasonal monsoons. The project manager therefore needs to be fully aware of the climatic implications from the very earliest stages of the project.

Population and human resources

Of the 6 billion people currently living on this planet, less than 1 billion live in what the World Bank categorises as high-income countries, while some 3.5 billion live in low-income countries (over 2 billion of these in China and India). Nearly a half of the world's population have an income of less than US$2 per day, and 1.2 billion must survive on less than US$1 per day. Up to 25% of the children born in many low-income countries do not reach the age of 5 years, and the average life expectancy in the poorest countries is less than 40 years. Over a billion people do not have access to a safe water supply, a third of the world's population have inadequate sanitation facilities, and nearly 5 million deaths per year in developing countries are directly attributable to water-borne diseases and polluted air. In addition to this lack of basic water and sanitation facilities, many live in substandard housing, transportation and communication links are poor or non-existent and fuel for heating and cooking is in short supply.

The problem is compounded by exponential population growth and rapid migration from rural to urban areas worldwide. It is estimated that the world population will rise to 8.5 billion by 2025, of which over 7 billion will be in developing countries. By 2050, the total world population could increase to as much as 10 billion. In the past, much of the developing world's population was in rural areas, but this is now changing rapidly, and population in urban areas is growing much faster than in the rural areas. In 1994, about one billion people in the developing world lived in towns and cities; it is estimated that by 2025, this will have risen to around 4 billion.

These facts and figures point to a huge need for infrastructure development throughout the developing world – a need that can only be fulfilled through the implementation of well-managed construction projects.

Population and human resources affect not only the need for projects but also the way projects are implemented. The large pool of available and relatively cheap labour, much of which is unemployed or underemployed, points to a less mechanised approach to construction and a greater use of human labour. Labour-intensive construction requires an approach different from the planning, design and management of projects and these issues must be addressed at the earliest stages of the project. Although the labour force in developing countries may be plentiful, it is also likely to be relatively unskilled. The questions of training and technology transfer therefore need to be taken into consideration throughout the planning and implementation of the project.

Materials, equipment and plant

Many of the materials commonly used in construction projects are often not readily available in developing countries. Cement and steel may have to be imported and paid for with scarce foreign exchange. Delays in importation and difficulties in gaining passage for imported goods through customs are not uncommon and need to be allowed for. Even if materials are manufactured in the country of the project, supplies cannot always be guaranteed and the quality may be inferior to that normally expected in industrialised countries. Production capacity and quality should therefore be assessed before detailed design is done.

In some cases, it may be necessary to consider alternative materials such as stabilised soil, ferrocement, round-pole timber or pozzolana as a cement replacement. Many such alternatives are traditional indigenous building materials and may be more acceptable than steel or concrete. An assessment of what is available and appropriate needs to be made at an early stage and used in the design.

Imported mechanical equipment, whether for construction or for incorporation within the completed project, is expensive and requires maintenance. Trained maintenance technicians and a reliable supply of spare parts are the absolute minimum requirements if the equipment is to continue to function over its anticipated life. There is generally a shortage of reliable and operable construction plant in developing countries, and it is often not possible to hire plant, because plant-hire companies do not exist. Managers of projects, particularly those of large ones, will have to import plant for the project and then decide on whether to sell it or transport it back to the home country on completion of the project.

Finance and economics

Although there is a great need for new projects in developing countries, there is also a lack of funds from the normal sources expected in the developed countries. Many projects are funded externally from national aid agencies, international development banks or non-governmental organisations (NGOs) such as international charities. Project managers involved in the identification, preparation and appraisal stages of a funded project need to be fully aware of the requirements of the grant- or loan-awarding agency to whom they are making application for funding, as each has its own specific requirements.

Socio-cultural factors

The successful management of a project in a developing country requires an understanding of the ways society is organised and the indigenous cultural and religious traditions. In Muslim countries, time must be allowed for workers to participate in daily prayers and during the month of Ramadan fasting is mandatory during daylight hours; thus affecting productivity and the way work is organised. The respective roles of men, women, religious and community leaders, and landowners must be understood, particularly when managing projects in which the community is actively participating.

If socio-cultural factors are not taken into consideration, a project may not be successful even if it is successfully constructed. A new water supply may not be used if the community members feel they do not own it, or if existing traditional sources of water have a strong cultural significance; sanitation facilities might be underused or neglected if the orientation offends religious beliefs or if men's and women's toilet blocks are sited too close together. As with other factors already mentioned, knowledge of socio-cultural influences is therefore necessary at the earliest stages of a project because they may have a significant effect on project identification, appraisal and design, as well as on construction and operation.

Working with other professionals

Project managers often have to work with a variety of professionals – electrical and mechanical engineers, chemical engineers, heating and ventilating engineers, environmental scientists, architects, quantity surveyors and planners. But nowhere is the diversity of professionals so great as when the project is located in a developing country. In addition to the preceding list, the project manager working in a developing country may well have to work closely with agricultural scientists, health and community workers, educationalists and professional trainers, economists, community leaders, sociologists, ecologists, epidemiologists, local and national politicians and perhaps many others. Good written and oral communication skills and an ability to understand the views and perspectives of other professions are valuable qualities in any project manager, but they could be vital when the project is set in a developing country.

19.2 The construction industry in developing countries

Unlike the developed countries, many developing countries do not have a mature construction industry consisting of well-established contracting and consulting companies. Much, if not most, of the building and construction is done by the informal sector. This consists of individual builders and tradesmen, who are mainly concerned with building family shelters, and community and self-help groups, who may construct small irrigation works, community buildings, grain storage facilities, water wells and the like. These individuals and small groups rarely attract funding and the works are completed using labour-intensive methods and locally available materials. Small community groups may gain the attention of national and international NGOs, who are more inclined to fund and work with such groups than are formal aid agencies and development banks.

The formal sector consists of public or state-owned organisations and private domestic contractors. The proportion of work carried out by the public sector is usually much higher in the less developed countries than it is in the richer countries, but low salaries, lack of incentives and poor promotion prospects often result in highly demotivated professional and technical personnel.

Contracting is a risky business in any country, but in many poor developing countries the lack of access to financing, excessively complex contract documents, failure to ensure fair procurement practices, the high cost of importing equipment and the fluctuations of demand for construction often mean that the private sector of the construction industry has not had the opportunity to establish itself sufficiently to bid for major infrastructure projects. These are almost entirely carried out by international contractors and funded by national and international loans and grants. In some countries, up to 80% of major building and civil engineering is executed in this way.

Because of the importance of the construction industry to the development of the overall economy, the growth of an indigenous construction industry needs to be encouraged. In an attempt to do this, the World Bank and other agencies have encouraged the 'slicing' and 'packaging' of contracts – breaking up large contracts into smaller ones that local contractors would have the resources and capability to bid for. Other initiatives have included a greater use of labour-intensive construction, help with finance for smaller construction companies, the encouragement of plant-hire companies and management training for principals and staff of construction companies. Technology transfer and training, particularly management training, play an important role in this.

19.3 Finance and funding

Developing countries are, by definition, poor. Funding for projects will therefore be scarce, loan finance difficult to obtain and resources scarce. In many cases the only source of finance will be from development banks, aid agencies or charitable non-governmental agencies, many of whom obtain at least part of their funding from national aid agencies.

The major development banks include the World Bank, the Asian Development Bank (ADB), the African Development Bank (AfDB) and the European Bank for Reconstruction and Development (EBRD). These are all multilateral funding agencies, drawing their funds from several different countries. They operate as commercial banks, loaning money at agreed rates of interest. The loans have to be repaid, albeit the loan conditions are often more favourable than commercial banks, and they may allow a period of grace before repayments commence.

Most industrialised countries have their own government bilateral aid agencies, such as the Department for International Development (DfID), UK. These agencies fund projects in developing countries through loans and grants, and also direct some of their allocated funds to those development banks of which they are members. Aid awarded directly by these agencies is often 'tied', that is, the grant or loan is conditional upon some of the goods and services needed for the project being procured from the donor country.

Loan finance for construction companies to expand, to buy equipment or simply to maintain adequate balances and ease cash-flow difficulties is extremely difficult to obtain from commercial banks in developing countries. Contractors may therefore be forced to borrow from other sources at inflated rates of interest. However, some countries have development finance companies, which act as intermediaries to channel funding from external agencies such as the World Bank to the construction industry and other developers.

19.4 Appropriate technology

The distinctive nature of the construction industry in developing countries suggests alternative approaches to the design, construction and management of projects. The application of appropriate technology is one approach that has been promoted as a way to overcome some of the problems associated with the implementation and long-term sustainability of development projects in the Third World. Appropriate technology should be able to satisfy the requirements for fitness for purpose in the particular environment in which it is to be used. It should also be maintainable using local resources, and it should be affordable. Many would argue that all technology should be appropriate, and perhaps, therefore, it is intermediate technology that we should be focusing our attention upon.

The concept of intermediate technology was first developed by E. F. Shumacher who defined it in terms of the equipment cost per workplace. He suggested that the traditional indigenous technology of the Third World could be represented as a £1 technology, while that of the industrialised world was a £1000 technology. An example, from the agricultural sector, is the traditional hand or garden hoe as a £1 technology, compared with a modern tractor and plough as the £1000 technology. He pointed out that, throughout the world, the equipment cost of a workplace was approximately equal to the average annual income, and that any budding entrepreneur could save sufficient money over a 10–15 year period to purchase the equipment necessary to start a small business. However, if the budding entrepreneur is a resident of a developing country, where salaries are a fraction of that in the industrialised world, it would take him or her over 100 years to purchase equipment of the advanced £1000 technology type, and this was clearly impossible. Schumacher therefore advocated an intermediate technology – a £100 technology (the animal-drawn plough) – which would be affordable to people in the Third World and would improve efficiency, reduce drudgery and help to develop and improve the economy.

In the context of construction technology, concrete can be mixed slowly and inefficiently by hand, using a flat wooden board and a shovel (£1 technology). Alternatively, a mechanised concrete batching and mixing plant can be used (£1000 technology). The labour-intensive method is very slow and the quality of the concrete is likely to be inferior. The mechanical mixer is not only expensive but may be difficult to maintain because of lack of local skilled mechanics and the difficulty and cost of obtaining spare parts. Supplies of electrical power or fuel for the mixer may also be unreliable. An intermediate technology concrete mixer, developed in Ghana, is illustrated in Figure 19.1. It consists of a box lined with thin galvanised metal and fitted with a hinged top, mounted on simple wooden wheels and fitted with a handle.

Sand, aggregate, cement and water are placed in the box, and the lid is fastened and the device is then simply pushed around the site until the concrete is mixed. The device is cheap, simple to maintain and repair, requires no source of power apart from human labour and produces quite good quality concrete – laboratory tests indicated cube strengths of approximately 90% of those obtained from a mechanical mixer.

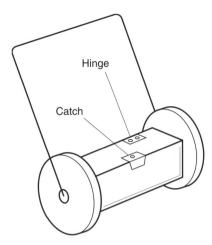

Figure 19.1   Intermediate-technology concrete mixer developed in Ghana.

There are critics of the use of intermediate technology, including many from developing countries themselves, who point to the rapidly expanding economies of Korea, Malaysia and Taiwan as examples, and argue that, without access to the most advanced technology available, the Third World will never catch up with the industrialised world. On the other hand, there are numerous examples illustrating the unsustainability of advanced technology – the factories operating at a fraction of their design capacity because of inadequate distribution networks or due to lack of maintenance staff or spare parts; the water and wastewater treatment plants that are not functioning because of lack of funds for consumables and spare parts; and the impassable sealed roads that have not been resurfaced because of lack of suitable plant or materials.

The project manager must decide on what is or is not appropriate in any given context. He or she needs to address the questions of fitness of purpose, maintainability, cost and sustainability. Whether 'local indigenous', 'intermediate' or 'advanced' technology is the most appropriate will depend on the physical, social, cultural and economic environment of the particular developing country in which the project is set.

19.5 Labour-intensive construction

One facet of appropriate construction technology is the use of less plant and equipment, and more labour. This is termed *labour-based* or *labour-intensive construction*. Much of the infrastructure – sewers, aqueducts, canals and railways – in industrialised countries was built by our ancestors using labour-based methods, and it is only fairly recently that extensive use has been made of heavy machinery for construction and building. With the change to machine-based construction, engineers and project managers, in both industrialised and developing countries, became less familiar with labour-based methods. Training for civil engineers has become based on equipment-intensive methods, and this has become the norm throughout most of the world. This situation was encouraged in many developing countries through the provision of aid, particularly bilateral tied aid, by funding agencies, expecting or requiring trade and business as a condition for loans and grants.

In the 1970s, it became clear that the plight of the world's rural poor was not improving, partly because the small and isolated infrastructure projects required for these rural areas were not attractive to either funding agencies or to large equipment-based contractors. Research into the potential for labour-based construction revealed that there was a reluctance to adopt such methods because it was felt that the costs could not be accurately predicted, the labour force was unreliable and that it would be more expensive and more prone to delays than equipment-based construction. In addition to this were the many problems associated with the welfare of large numbers of labourers.

However, the reliance on plant-intensive construction has a number of drawbacks for developing countries, the main one being the high expenditure of foreign exchange, something that few Third World countries can afford. In contrast, labour-based construction actually generates employment and produces an income for those engaged on the project.

It has now been demonstrated that, in certain circumstances, labour-intensive methods can compete with plant-based methods both in terms of technical quality and cost. This can be made possible with good management. Good management of labour-based construction entails ensuring that the morale and motivation of workers remains high at all times by offering incentive payments, providing good training, attending to the welfare of the workers and ensuring that materials and tools are always available and in good condition. Because labour-based construction projects can often entail work being carried out at a number of small, dispersed and remote sites, good communications between the various sites and good planning and coordination are essential.

Most labour-based construction projects will in fact use a combination of various forms of motive power – human, animal and machine. The key to efficient construction is selecting the optimum combination for a particular project. Long-distance haulage is probably best done by truck; animals are better and cheaper for shorter distances over steep or rough terrain; and even shorter haulage distances might be most suited to labourers with wheelbarrows or head baskets. In many construction operations, a mix of labour and machine will provide the optimum solution – earthworks may use labour for excavation and a truck for haulage, for example. Balancing these resources to ensure full productivity is one of the tasks of the manager.

Some activities are always more suited to machine-based construction, while others are better carried out by labour-based methods. For example, although a concrete pipe culvert might be installed using machines, brick arch culverts can be constructed only using labour. An asphalt road surface requires equipment and it would be totally impractical to consider labour-based methods for this form of construction. Earthworks, excavation and quarrying are all well suited to labour-based methods.

The implicit choice of equipment-based construction methods has an influence on the planning and design of the works, and a fair comparison between the financial costs of labour-based and equipment-based methods may be difficult once the design is finalised. Therefore, if labour-based methods are to be given an unprejudiced assessment, alternative designs suited to labour-based construction should be considered. This is sometimes referred to as *design equivalence*. For an even more comprehensive comparison, a full economic appraisal should be carried out, whereby the economic benefits of enhanced employment and the potential for growth and development are included as benefits of the project.

Finally, it is important to note that the availability of labour in rural agricultural areas can vary considerably with the seasons: during planting and harvesting time, full employment and higher wages can be obtained on the farms. It may be necessary to raise project wages at these times to ensure continuity of work, although this may have the disadvantage of workers refusing to return to normal rates of pay later. It may also have the added drawback to the community of decreased agricultural output. In some cases, it may be necessary to employ migrant workers from another area and this will entail providing suitable accommodation, normally in camps, which must be provided with water, sanitation and other essential services. In any case, there is a need to collect data on availability of labour prior to making the decision on whether or not to adopt a labour-intensive approach to the project.

19.6 Community participation

In industrialised countries, the general public may be involved in the sanction and approval stage of projects through public enquiries or public protest, but they are unlikely to be involved in the planning and design of projects, and it is even less conceivable that they would participate in their construction or operation. For projects in developing countries, particularly small projects in rural areas, the concept of 'community' or 'beneficiary' participation is now accepted as being expedient, if not essential, for success.

The participation of the community for whom the project was intended was initially employed to provide assistance with the construction of rural projects such as the development of water, sanitation and irrigation schemes. This was primarily to reduce costs and to ensure that the local community would have sufficient expertise to enable them to operate and maintain the schemes once they had been constructed. The idea has since been developed further to encompass identification, planning and even design of projects, the argument being that the community have a much better local knowledge, they know what they want and, of course, they are aware of all the social, cultural and religious factors that may affect the design.

For project managers, this can entail a great deal of additional work and it can delay the start and completion of a project. The reported advantages are that it will help to ensure that the project is used and that it can and will be successfully operated and maintained.

In some cases, such as the improvement of water quality, rural communities may not be aware of the advantages of a project, and community training and education may be necessary. The construction of pilot schemes may be necessary to convince villagers of the benefits to be derived from a project and to obtain their views on the design and possible improvements. All this places an extra burden on the project manager and highlights the importance of communication skills and the ability to work with a variety of different people and organisations.

19.7 Technology transfer

The Third World needs new construction technologies and management expertise in order to develop – but only if those technologies and techniques are appropriate. The long-term benefits acquired when new technology is introduced and used for one short contract will be negligible, and may possibly even be detrimental to the achievement of long-term goals. However, if there is a real commitment to the ideal of technology transfer in terms of the acquisition of knowledge, skills and equipment, the benefits may be considerable.

Effective technology transfer is more than just education and training. Although training is essential, a much greater understanding of techniques, processes and machinery will be acquired from their actual use than from merely observing or learning about their use. Even more control and mastery will be obtained by owning and being responsible for them. This can be achieved through direct purchase of equipment, entering into a joint venture, becoming a licensee or franchisee for an established process or entering into some other form of contractual arrangement with an experienced company or organisation in the developed world.

Direct purchase of equipment involves the least amount of risk on the part of the vendor and the greatest risk for the purchaser. However, provided that good training, reliable after-sales support, established channels for the supply of spare parts and sound warranties are included in the package, this can be a quick and effective method of transferring some technologies.

Becoming a licensee may be seen as purchasing intellectual knowledge in addition to purchasing or leasing equipment. It is usually in the interests of the licensor to provide a higher level of support than might be expected from a vendor because the terms of the licence would normally entail payment of a percentage of the value of the work carried out. Hence, some of the risk of such a venture will be taken by the licensor.

Joint ventures between an established contractor in an industrialised country and a contractor in a developing country can be a very effective way to transfer technology and encourage the development of the indigenous construction company. There are a variety of different forms of joint venture, all of which require some investment from both the partner in the developed country and the one in the developing country. The investment from the developed country partner is in the form of either cash or equipment, together with technical and management expertise. The developed country partner may invest cash and/or premises, labour and locally available equipment. Joint ventures will be an effective way of transferring technology only if both parties are firmly committed to the idea. Very often joint ventures are formed merely to satisfy a local requirement for overseas contractors to enter into joint venture with a local company to bid for work in that country. In such cases, there is often little or no commitment on either side, the overseas company being involved out of necessity and the local company allowing their foreign partner to make the decisions while they take a back seat, and their share of the profits! Joint ventures set up for short one-off projects are likely to be a less effective means of technology transfer than long-term ventures.

Contractors working in developing countries can help the process by providing training to local staff and subcontractors, and most funding agencies allow for training within the loans or grants awarded for projects. Agencies also provide finance specifically for training and technical assistance, which can be provided through an arrangement known as *twinning*. Twinning is a formal professional relationship between an organisation in a developing country and a similar but more mature and experienced organisation in another country. Unlike training programmes, these are often long-term arrangements and may involve lengthy visits from key personnel in both organisations to their counterparts in the twinned institution.

19.8 Corruption

Although corruption does occur in industrialised countries, it would appear to be more widespread in developing countries, and any project manager working in such areas needs to be aware of the problems and of their own professional codes of conduct and ethics. Corruption can affect a project at a number of stages in the project cycle, such as the marketing of construction services, obtaining planning permissions and building consents, pre-qualification, tender evaluation and award, construction supervision and quality control, getting materials and equipment through customs, issuing payment certificates, and making decisions on claims and completion. It is not unusual to read of buildings and other structures collapsing as a consequence of poor-quality construction and building-code infringements, which have been traced to bribery and corruption.

It takes two parties to make bribery effective – the party that requests the payment and the party that provides the payment. In this regard, international companies have been as culpable as indigenous government officials in some cases. It has not been unusual for international construction companies to employ local 'agents' to facilitate business dealings in a foreign country. Such agents may have a perfectly legitimate function, but payments or commissions to agents can also be an effective way of concealing illicit payments to corrupt officials so as to secure a contract or to facilitate completion of the work. The World Bank has suggested that an average of around 10% of contract costs are siphoned off in bribes and other illicit payments, with up to 80% being lost in this way in extreme cases.

Until fairly recently a blind eye was turned to this problem, and some economists and business people argued that such practices actually stimulated economies by allowing more wealth to enter and be distributed throughout the country. It was further argued that bribes could cut through overly bureaucratic red tape and allow construction projects to meet important deadlines, thus saving costs in the long term.

These arguments have since been thoroughly discredited, and major international institutions such as the World Bank, OECD, United Nations and International Federation of Consulting Engineers (FIDIC) have all taken a firm stand on the issue. The World Bank points out that apart from money finding its way into corrupt individuals' bank accounts and thus being less available for other development projects, bribery often means that the wrong projects are implemented, leaving a graveyard of white-elephant infrastructure facilities that may never be used to their full potential.

FIDIC is quite clear on the issue and modified their code of ethics in 1996, by adding the following.

The consulting engineer shall:

- neither offer nor accept remuneration of any kind which in perception or in effect either (a) seeks to influence the process of selection or compensation of consulting engineers and/or their clients or (b) seeks to affect the consulting engineer's impartial judgement;
- co-operate fully with any legitimately constituted investigative body which makes inquiry into the administration of any contract for services or construction.

In their policy statement on corruption, FIDIC make it clear that the term *bribe* includes indirect payments, using mechanisms such as scholarships, currency exchange facilities and the actions of agents, as well as direct cash payments. Not only FIDIC, but many other powerful international organisations too are now prepared to act firmly by sanctioning their members, terminating contracts or taking other disciplinary action against any individual or firm that is found to be involved in corrupt practices.

In 1993, Transparency International initiated a major crusade against corruption, and further details of their work can be found at the website listed in the Further Reading section at the end of the chapter.

19.9 Summary

The world's population is still growing at an alarming rate, and much of this growth is in the less developed world. The world urban population is growing even faster, and once again it is in the developing world where the largest and fastest growing cities are to be found. Although infrastructure development is taking place, it is barely keeping pace with the ever-expanding population, which requires the basic necessities of food, shelter, water and education. The need is overwhelming. Without civil engineers and people who can effectively manage civil engineering projects, this need will never be met. But successful projects in developing countries require managers who recognise the needs and are knowledgeable of the clear and distinctive differences, difficulties, peculiarities and rewards of managing projects in these countries.

Reference

FIDIC, Code of Ethics: http://www.fidic.org/about/ethics.asp.

Further reading

Cooper, L. (1984) The twinning of institutions – its use as a technical assistance delivery system, World Bank Technical Paper 23.

Côté-Freeman, S. (1999) *False Economies, Developments, Fourth Quarter*, DFID.

Coukis, B. and World Bank (1983) *Labor-based Construction Programs*, Oxford University Press.

Institution of Civil Engineers (1981) *Appropriate Technology in Civil Engineering*, Thomas Telford.

Smillie, I. (1991) *Mastering the Machine – Poverty, Aid, and Technology*, Intermediate Technology Publications.

Transparency International: http://www.transparency.org/

USAID, Anticorruption Web site: http://www.usaid.gov/democracy/anticorruption/

World Bank (1984) *The Construction Industry – Issues and Strategies in Developing Countries*, The World Bank.

World Bank (1992) *World Development Report 1992 – Development and the Environment*, Oxford University Press.

World Bank (1994) *World Development Report 1994 – Infrastructure for Development*, Oxford University Press.

World Bank, Data and Maps: http://www.worldbank.org/data/

World Resources Institute (1994) *World Resources 1994-95*, Oxford University Press.

Chapter 20
Projects in Controlled Environments 2:
PRINCE2™ *

This chapter provides an introduction to PRojects IN Controlled Environments 2 (PRINCE2™), a project management methodology that has been embraced by public and private sector organisations in the UK and internationally. The methodology provides a 'start-to-finish' approach to managing and controlling a project throughout a project's life cycle (specifying, designing, developing, testing and change over) and covers the necessary prerequisites. The structure, contents and functions used within PRINCE2™ are described in this chapter to provide an introductory understanding of the methodology. Within this text capital letters are used to denote PRINCE2™ key words. Readers who are interested in learning more about PRINCE2™ are referred to the references, further reading and websites at the end of the chapter.

20.1 Introduction

An all too common occurrence in project management has been projects exceeding their forecast cost and due date and not meeting their intended specifications. While causes for this phenomenon are complex, the problems can frequently be attributed to an inadequate business case, poor planning of tasks and resources, lack of control, lack of stakeholder consultation and lack of quality control. The PRINCE2™ project management methodology has been developed to combat these project management problems.

PRINCE2™ was developed from PROMPTII (Project Resource Organisation Management Planning Technique) a project management method created in 1975; PROMPTI was for managing programmes and PROMPTIII for managing operations; however, neither PROMPTI nor PROMPTIII ever got off the ground. PROMPTII, whose foundations were in the information technology (IT) industry, had five major components covering strategic planning, systems development, operations and maintenance and enhancements, quality assurance and software tools support. While PROMPTII provided a first-stage management methodology for projects, it was deficient in several areas: critically it used the

---

* PRINCE2™ is a Trade Mark of the Office of Government Commerce.

term stage manager and did not mention project managers or critical path analysis and assumed a pre-defined life cycle. In 1979 the Central Computing and Telecommunications Agency (CCTA), now the Office of [as per website] Government and Commerce (OGC), adopted PROMPTII as their project management standard for IT projects; and in 1989 PRINCE** superseded PROMPTII (OGC, 2005). The PRINCE methodology underwent further development by the CCTA and was launched as PRINCE2™ in 1989.

PRINCE2™ provides an 'off-the-shelf' project management methodology that is free for use with no licence fee. The methodology can be scaled and adapted to small- and large-scale projects in any business sector. This is clearly demonstrated by some of the corporate users of PRINCE2™ including Racal, Smiths Engineering, Ericsson, Reading Borough Council, National Health Service, Camelot, British Telecommunications, Tesco, Cheshire County Council and many others. Ideally it is applied to a project at its earliest stage, that is, once a project mandate has been written. However, if the project mandate is incomplete, the first process of PRINCE2™, Starting Up a Project (SU), does ensure that all information required is available; see the section titled 'Processes'.

20.2 Benefits and limitations

A common misconception is that PRINCE2™ is too bureaucratic. However, this is a result of over-application of the methodology within an organisation. The methodology provides a process approach to managing projects that can be easily tailored by an organisation by scaling down the methodology to suit the project and the way in which the organisation prefers to manage their projects effectively. For instance, formal project meetings are necessary but not at all stages within the project. PRINCE2™ provides a best practice for managing projects and not another level of bureaucracy; providing a readily available project management methodology for an organisation and saving significant development expenditure. The methodology offers a common understanding for all those that work on the project. According to the OGC (2005), PRINCE2™ provides an organisation with:

> controlled management of change, active involvement of stakeholders throughout the process, controllable use of resources, a proven best practice in project management and encourages formal recognition of responsibilities within a project.

PRINCE2™ is based on the concept of disaggregating the project into stages and then setting objectives, planning, seeking authorisation, reporting, reviewing and controlling each stage. Importantly, while work is being carried out on one stage, the next stage is being planned. This allows the PRINCE2™ user to look ahead and plan the next stage of work, update the business case and assess whether the project still has a viable business case.

** PRINCE © is a Registered Trade Mark and a Registered Community Trade Mark of the Office of Government Commerce, and is Registered in the U.S. Patent and Trade Mark Office.

At the project level PRINCE2™ provides projects with a controlled start and finish, regular reviews against the plan and business case, flexible decision points, automatic management control from deviations, involvement of managers and stakeholders, transparent communication channels and agreement on product quality at the outset (OGC, 2005). The methodology although robust does not include procurement or contractual processes, technical methods for developing cost estimates or revenue forecasts for the business case or risk analysis, budgetary control or earned value analysis, team leadership and motivation, all fundamental to delivering a project successfully. PRINCE2™ recognises an organisation will have its own preferred techniques that are used for projects; however, these will frequently integrate with the methodology.

20.3 Project management standards and methodologies

PRINCE2™ provides a project management methodology. It provides a 'start-to-finish' approach showing 'how to plan, manage and control a project'. Whereas project management 'bodies of knowledge', representing 'things to know' in project management are promoted by the Project Management Institute (PMI), the Association for Project Management (APM) and British Standards (BS 6079-1:2002). The only other methodology available in the public domain that comes close to this approach is project cycle management (PCM) methodology which is widely used for social/development projects; whereas PRINCE2™ is more for commercial project management. While each of these methodologies has its merits, it is PRINCE2™ popularity that has grown exponentially and it is easy to see why this is the case.

20.4 Structure and contents

PRINCE2™ comprises eight components, eight processes and three techniques that combine to produce products for managing and controlling the project and delivering the project, as shown in Figure 20.1. The eight components are the Business Case, Organisation, Plans, Controls, Management of Risk, Quality, Configuration Management and Change Control. Components are essentially groups of management products, of which there are 33, which provide critical information to the project board for managerial decision making.

The eight processes are Starting Up a Project (SU), Initiating a Project (IP), Controlling a Stage (CS), Managing Product Delivery (MPD), Managing Stage Boundaries (SB), Planning (PL), Directing a Project (DP) and Closing a Project (CP). Management and technical products are created, updated or output from processes. PRINCE2™ techniques include Product-Based Planning (PBP) used with PL, Change-Control Approach used with CS and Quality Review Technique that can be invoked at any point in the project but has links with PL, MP and CS. Techniques are optional with the exception of the PBP.

PRINCE2™		
Components:	**Processes:**	**Techniques:**
• Business Case	• Starting Up a Project (SU)	• Product-based Planning
• Organisation	• Initiating a Project (IP)	• Change Control Approach
• Plans	• Directing a Project (DP)	• Quality Review Technique
• Controls	• Planning a Project (PL)	
• Management of Risk	• Controlling a Stage (CS)	
• Quality in a Project Environment	• Managing Product Delivery (MPD)	
• Configuration Management	• Managing Stage Boundaries (SB)	
• Change Control	• Closing a Project (CP)	
Management and Technical Products		

Figure 20.1 PRINCE2™'s components, processes, techniques, and products.

Components

Business Case

The Business Case, a management product, is central to the PRINCE2™ methodology. It provides the justification for the project and is central to decision making and directing of a project by the project board. It is continually updated as the project develops to ensure the project is viable. The development and continuation of the business case is the prime responsibility of the executive – see Organisation. The business case forms part of the Project Initiation Document (PID) and comprises: reasons for the project, description of options considered, forecast benefits expected, key project risks, forecast costs and an investment appraisal. The PID contains static and dynamic information (expected to be updated as the project develops). Static information includes the Background, Project Definition, Project Organisation Structure, Communication Plan, Project Quality Plan and Project Controls. Dynamic information includes the initial Business Case, initial Project Plan and initial Risk Register.

Organisation

The organisational structure in PRINCE2™ provides a standardised format for the roles and responsibilities of a project management team. The project organisation structure comprises the corporate or programme management responsible for instigating the project and defining the overall constraints (i.e., time, cost and quality) and a project management team. The project management team structure comprise a project board responsible for project assurance, decision making and project directing, project managers and team managers.

In PRINCE2™ three project interests must be consistently represented on the project board: the Business Need, the User and the Supplier. Therefore, to reflect these interests the project board comprises an executive, senior user(s) and senior supplier(s). The executive is responsible for the business need and the customer's interests. The senior user is responsible for ensuring the user's needs are specified correctly and are met by suppliers. The senior supplier is accountable for all products delivered by supplier(s) and for providing resources to deliver the project.

The project manager is responsible for the day-to-day running of the project and is the interface between the project board and the team manager(s). Specialist project skills (estimating, forecasting and risk analysis) and configuration management (administration, documentation and version control) are provided to the project manager and team managers by a project support function. Each project board member has a project assurance function within its role. The project board may decide to delegate some of all of its project assurance responsibilities, but they will still be accountable for this activity. PRINCE2™ states any delegation must be independent of the project manager. The project board's and the project manager's decision-making responsibilities cannot be delegated.

Plans

PRINCE2™ uses Product-Based Planning (PBP) to establish plans; see Techniques. This requires a Product Breakdown Structure (PBS) to be created to identify products to be delivered, and a Product Flow Diagram (PFD) illustrating the relationships of the products to each other and activities identified as needed to create the products. The durations and costs of activities will then be estimated and represented in a Gantt chart, comprising key deliverables and control points for stage boundaries. Once the PID has been approved by the project board, the plan is baselined and updated at the end of each stage to provide the project board with the latest position on project progress, changes and revised costs and forecasts.

In PRINCE2™ three levels of project plans exist: a Project Plan used by the Project Board, a Stage Plan used by the Project Manager and a Team Plan used the Project Team. The project plan is mandatory in PRINCE2™ and is used by the project board to monitor and control the overall project. The stage plan is used by the project manager to control and monitor the day-to-day running of the project. It is produced by the project manager for the next stage as the current stage approaches completion. Team plans are optional in PRINCE2™, and represent a further level of detail represented in the stage plan for the work package. If a plan at any of the three levels is forecast to exceed its tolerances an Exception Plan may be requested by the next higher level of management to replace the plan.

Controls

Control and decision making are exercised at the project board level and the project manager level. Project control requires information on project progress to allow progress to be monitored, comparisons to be made against plans and review of future options. Once a stage plan has been approved by the project board, the project manager has the responsibility for delivering the stage. The project manager keeps the project board informed of the progress by event-driven trigger events such as assessment at the end of stage-driven or time-driven reports such as Highlight reports providing progress during a stage: the frequency of the latter is advised by the project board. The concept of 'management by exception' is used in PRINCE2™ whereby, if there are deviations from a plan against tolerance, the project board needs to be informed, thereby avoiding frequent project meetings. Equally, the project manager exercises control over the project teams on a day-to-day basis within a stage. A project team will report to the project manager on the progress of a Work Package typically through a Checkpoint report, a time-driven report. Other reports include Product Descriptions, Project Issues and Risk Log – see the section on Products.

Management of risk

The identification, evaluation and identification of countermeasures, and monitoring and controlling risk, are fundamental to PRINCE2™. This is facilitated through creating and maintaining a project Risk Log. The Risk Log comprises risk identification, risk description, risk category, impact (time, quality, benefit and people/resources), probability of occurrence, proximity of risk to occurring, countermeasure, owner, author, date identified, date last updated and current status of risk. The project manager has responsibility for risk management with PRINCE2™; however, ownership is with the executive.

Quality in a project environment

Quality in projects is about identifying customer's needs, identifying these aspects of quality with a product and delivering these to satisfy the customer's expectations. In PRINCE2™ quality management comprises quality system, quality assurance, quality planning and quality control. The quality system covers organisational structure, processes and procedures to implement quality within projects, importantly forming part of the corporate quality system and not a replacement. Quality assurance ensures quality standards are set and monitored and are being used effectively on the project. Quality planning ensures quality is planned with objectives and criteria set during the start up of a project, and is specified in detail within each stage plan setting quality criteria with the product descriptions. Quality control assesses whether the products produced are meeting the required standards set within the product descriptions utilising quality reviews.

Configuration management

This provides a means of identifying, capturing and tracking a project's assets – in this case the Management and Technical products produced by the project. Configuration management enables the project team to know where products are and their version. Configuration management is mandatory in PRINCE2™. The five basic functions for configuration management are Planning, Identification, Control, Status Accounting and Verification at the end of each stage.

Change control

Specification changes are likely on projects. A change to a project specification will have a 'knock-on' effect to the project plan and budget, and there needs to be a controlled mechanism to facilitate these changes. A Project Issue can capture a change, which is then dealt with in one of two ways: Request for Change or by raising an Off-Specification Report when the product has failed or likely to fail to meet criteria. These procedures ensure changes to the baseline project are made in a controlled manner and documented.

Processes

PRINCE2™'s eight processes are either one-off or repeatable. Each process receives information as an input and transforms this information into one of the components and provides an output. The three one-off processes include Starting Up a Project (SU) to ensure prerequisites are in place before initiating a project; Initiating a Project (IP) to ensure the project's Business Case is sufficiently robust and resources are in place to commence the project, and Closing a Project (CP) to close the project in a controlled manner. The five repeatable processes include: Planning (PL) to develop robust plans and risks for the project; Directing a Project (DP) a Project Board management and decision-making process running from SU to CP; Managing Stage Boundaries (SB) a process to provide management information to the project board to decide whether to continue with the project; Controlling a Stage (CS) a process to ensure work packages are maintained and controlled by the project manager; and Managing Product Delivery (MP) a process to ensure planned products are created and delivered.

Techniques

Three Techniques are explicitly provided within PRINCE2™: Product-Based Planning, Change Control and Quality Review Technique. Each of these techniques creates or updates management products.

Product-Based Planning

PRINCE2™'s planning process utilises Product-Based Planning (PBP), which is mandatory. PBP is a technique to establish the products required, the sequence in which products are needed, the structure and contents and the quality of the product to be produced. The three main stages with PBP are identification of the Product Breakdown Structure (PBS), creation of Product Descriptions and production of a Product Flow Diagram (PFD). The Product Description is a Management product and comprises an identifier for the product, title, purpose, composition, derivation, format and presentation, allocation, quality criteria, quality methods and quality skills required.

A PBS comprises products produced (represented by a rectangle) by the project and those external (represented by an ellipse) to the project. The PFD provides the dependency of products on the project from which the activities are generated to deliver these products. The PFD is utilised to estimate activity durations and resources required which are then represented using a Gantt chart or network diagram.

Change-control approach

Change Control in PRINCE2™ is solely for Technical products rather than Management products. It is driven by the identification of a project issue that triggers a Request for Change or Off-Specification Change. The project manger will try to solve an off-specification change within planned project tolerances. Where this is not feasible the issue is escalated to the project board as part of Controlling Stage (CS) process and requesting Ad hoc Direction, part of the Directing Project (DP) process. Where a project issue deviates beyond stage or project tolerances the project manager must follow this procedure. The decision by the Project Board might be a request to raise an Exception Plan.

Quality Review technique

Quality Reviews are aimed at assessing whether the product is 'fit for purpose' and meets its quality criteria specified in the product's description. A quality review can be called at any time within PRINCE2™, but it is most commonly associated with the following processes: PL, MP, and CS (authorising a work package and assessing progress). The key stages required to deliver a quality review are preparation, conduct review meeting and follow-up actions. To facilitate a quality review four roles have been identified: Chair Person, the Producer (of the product), the Reviewers and the Scribe (administrative support). The project team is involved in the quality review process; the project manger plans the review and the team manager planning the review in further detail with project assurance policing the process.

Products

Two types of products exist in PRINCE2™: Management products used to collect and gather information and Technical products produced by the project. Table 20.1 lists PRINCE2™'s Management products, referred to within this section, that are input and output by processes. Three frequently cited

Table 20.1 PRINCE2™'s management products (OGC, 2005).

Product group	Management products	Purpose
	Business case	Provides document for justification of the project.
	Project approach	Project delivery method.
Reports	Checkpoint report	Time-driven report on status of work for each member of the team.
	End project report	To report on performance against PID plan and cost.
	End-stage report	Summary of progress to date, used to decide on action to take, approve next stage, amend project scope or stop project.
	Exception report	Inform Project Board of adverse.
		Stage forecast to exceed tolerance levels.
	Follow-on action recommendations	Pass on unfinished project work to operational group responsible.
	Highlight report	Stage status provided at frequent intervals to project board to monitor stage progress.
Controls	Lessons learned log	Lessons learnt from project to apply to others.
	Lessons learned report	Final report on project lessons learnt.
	Daily log	Project Manager's daily diary.
	Risk log	Risks identified, analysed and managed on project.
	Product breakdown structure	Shows all products to be developed and quality controlled.
	Product flow diagram	Sequence or products to be delivered and dependencies.
	Project brief	Project information used to initiate projects.
	Project initiation document	Contains critical project information to control and manage project.
	Project issue	Issues raised on project.

	Project mandate	Project information to trigger and start up project
	Work package	Products to be produced suppliers agreed to by project manager and team manager
	Organisational structure	Project management team structure and responsibilities
	Project board approval	Approval for a project stage by the project board
Plans	Communication plan	Define how and when to communicate with interested parties
	Post-project review plan	Plan for Executive showing when and how to measure expected project benefits
	Stage plan (or exception plan)	Plan containing products to be produced, objectives to be achieved, control points, activities, resources, tolerances and deliverables
	Project plan	Shows how and when the project objectives are to be achieved
Quality products	Acceptance criteria	Defines in measurable terms what must be done for the final product to be accepted by customer
	Product checklist	To monitor all products to be produced in a stage plan
	Product description	Description of products, quality and production method
	Project quality plan	Defines quality techniques to be applied to products
	Quality log	Quality checks planned and conducted on project
	Issue log	Project issues and status
	Configuration item record	Information about project's status and versions
	Configuration management plan	How and who controls project's products
	Off-specification	Produced when product is forecast to fail to meet specification covering errors or omissions in work
	Request for change	Modifications to products for whatever reason

management products in PRINCE2™ are Project Mandate, Project Brief and Project Initiation Document (PID).

The Project Mandate is typically provided by corporate or programme management and exists as a result of a strategic study or identified business need. PRINCE2™ defines the Project Mandate as 'whatever information comes to trigger the project, be it a feasibility study or back of an envelope' (OGC, 2005). The project mandate provides the project's terms of reference that triggers the start of the PRINCE2™ methodology. PRINCE2™ suggests this document should contain the authority responsible, background, project objectives, scope, constraints, interfaces, quality expectations, outline business case (reasons), reference to associated products or documents, an indication of the executive and project manger, the customer(s), user(s) and known interested parties.

The Project Brief created from the project mandate provides the basis on which to initiate the project. The suggested contents of the project brief are project background, project definition (objectives, scope, outline deliverables, exclusions, constraints and interfaces), outline business case, customer's quality expectations, acceptance criteria for products and any known risks. The PID contains critical project information on which to commence delivery of a project. It is used to convey project information to the project board to assess viability of the project, control and monitor the project as it moves through its project life cycle.

20.5 Structured walk-through

In this section the eight Components, eight Processes and the three Techniques described in Section 20.4 are brought together to provide a simplified structured walk-through of PRINCE2™. To assist with understanding the interaction between Processes, Components and the Project Organisation, Figure 20.2 will be used. The walk-through is essentially in two sections 'Planning Work': covering SU and IP processes and 'Doing Work' covering CS, MP, SB and CP processes; DP and PL processes are utlilised within each of these processes.

Planning work

Process SU and IP 'Planning Work' on the project in preparation for processes that are 'Doing work' on the project. A Project Mandate (Table 20.1) external to the methodology triggers the PRINCE2™ methodology that commences with sub-processes SU1 through to SU6. The SU process designs and appoints the project management team (SU2 and SU3); prepares the project brief and risk log (SU4); prepares the project approach in outline terms and customer's quality expectations (SU5) and designs a project plan (PL1). When the draft initiation stage plan has been completed (SU6) the DP process is invoked.

DP1 sub-process seeks to approve the appointment of the project management team, reviews the project brief and draft initiation stage plan and obtains commitment to the project's resources by the project board. Once approval

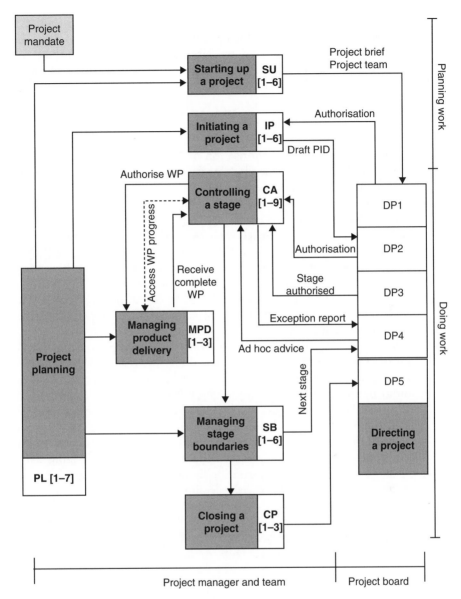

Figure 20.2   An overview of PRINCE2™.

has been granted by the project board, DP1 triggers IP, a process to produce a draft PID. The sub-processes to implement this include Planning Quality (IP1) to develop a quality plan and log, planning a project (IP2) to develop a risk log and project plan, refining the business case and risks (IP3) to develop the business case, setting up project files (IP4) to develop a communication plan and project controls and assembling the PID (IP6) to produce a draft PID and next stage plan.

The PID comprises the project definition and background, the Project Approach, an acceptable Business Case, the Project Plan, a Project Quality Plan, the Project Organisation Structure, an updated Risk Log and a note of the project tolerances. When the IP6 sub-process is complete, DP2 is triggered that seeks to authorise the next stage plan; once authorised, CS is invoked.

Doing work

Process CS starts with authorisation of Work Packages (CS1) and Managing Product Delivery (MP) whereby a project team accepts a work package (MP1), executes it (MP2) and delivers it (MP3).

CS continues to assess the progress of the work package (CS2), capturing (CS3) and examining project Issues (CS4), reviewing a stage status (CS5), reporting highlights (CS6), taking corrective action (CS7) and receiving completed work package delivered by the project team to the project manager (CS9). Project issues (CS8) can be escalated to the project board using an Exception report, by invoking SB (SB6) process to produce the exception plan, if tolerances are exceeded to obtain ad hoc advice (DP4) from the project board. The project board in response may request the project manager to invoke the SB (SB6) process to produce the Exception plan.

Managing Stage Boundaries (SB) process is used to plan a stage (SB1), update a project plan (SB2), the business case (SB3), the risk log (SB4) and report performance at the end of a stage (SB5) to the project board using DP3 that seeks authorisation for a stage or an Exception Plan. SB6 is used to produce an exception plan. The 'Doing Work' cycle continues between processes CS, MP and SB, triggering PL and DP processes until the final planned stage has been completed; CS5 that then triggers CP. CP is triggered by CS and requires confirmation from the project board DP5 for project closure. Clearly, before commencing project closure the stage needs authorisation by the project board DP3.

Closing a Project (CP) processes comprises three sub-processes and involves decommissioning a project (CP1), identifying follow-on actions (CP2) and project evaluation review (CP3). CP can be triggered by three processes: DP3 or DP4 for premature project closure or CS5 reviewing stage status that results in the notification of a project end.

20.6 Summary

PRINCE2™ offers a robust best practice in the form of a methodology whereby project problems are raised early providing decision makers with more control allowing them to have an opportunity to consider whether the project is still viable. While PRINCE2™ is a major step forward in the eradication of poor technical project management practices, it will not guarantee the success in planning and managing projects. This is simply due to the complexity of factors in particular the 'softer' aspects that are critical to project success.

Reference

Office of Government Commerce (2005) Managing Successful Projects with PRINCE2™, twelfth impression, HMSO, London.

Further reading

APM Group website: www.apmgroup.co.uk
APM web site: www.apm.or.uk
Managing Successful Projects with PRINCE2, OGC, twelfth impression, 2005.
British Standards Project Management (BS: 6079-1:2002).
Project Cycle Management Handbook, European Commission.
PRINCE2 a Practical Handbook, second edition, 2002, Colin Bentley, Butterworth Heinemann.

Chapter 21
The Future of Engineering Project Management

The discipline of project management continues to evolve. The recognition of project management as a profession has spread across all business sectors, from construction to information technology (IT), from 'time'-based systems to 'performance and quality'-based systems, from discrete operations to managing by projects.

Over the last few years changes in project management have been reflected in increased documentation, manuals of good practice and professional qualification standards. Notable among these are, in the UK, BS 6079 and the Association for Project Management Body of Knowledge and, in the US, the revised Project Management Institute Body of Knowledge and the International Project Management Association's Platinum Standard in Certification.

Industry has greater expectations of successful project management which together with growing commercial pressure to adopt business practices has resulted in more collaborative forms of procurement. The increasingly sophisticated and complex implementation of collaborative procurement forms including joint ventures, consortia and management contracts, private finance concessions, partnering, alliancing, PRIME Contracting, framework contracts and public–private partnerships has resulted in placing further duties and obligations onto the project manager. To fulfil these requirements new skills and knowledge are necessary, and this in turn requires new thinking about project management.

In many countries there has been academic research into developing a way forward for project management. In the UK the Engineering and Physical Science Research Council funded a major research network aimed at 'rethinking project management'. The work has been largely multidisciplinary involving cooperation between university business schools, engineering departments and project management units. Periodic reviews of what has come to be regarded as current practice (or the traditional or conventional approach to project management) is always useful in facilitating the most effective applications. It is important to remember that there are no prescribed 'silver bullet' project management solutions but only better-trained project managers with a greater project understanding and with the background to deal with increasing complexity with more success.

In the past, engineering project management was dominated by the single-project paradigm. Developments in collaborative procurement, as described earlier, have led to projects being undertaken in a multi-project milieu. Programme and portfolio management are examples of these multi-project concepts that seek to bridge the gap between organisation and project strategies. Realisation that the challenges of managing the whole multi-project package is not necessarily the sum of managing a number of individual projects has led to an increasing demand for research and best practice. The emphasis on multi-project management is on the need for soft skills rather than more tools and techniques.

21.1 Key roles in project management

The promoter

The ultimate responsibility for a project lies squarely with the promoter. Consequently, any project organisation structure should ensure that this ultimate responsibility can be effected. This is not to say that the promoter must be involved in the detailed project management but it does mean that machinery must be in place for the promoter to make critical decisions affecting the investment promptly whenever necessary.

The project manager may be an in-house member of the promoter's organisation or more often a contracted service for all or part of the project life cycle. Irrespective of the basis of the appointment, the project manager must ensure that the promoter's organisation supports the project team with direction, decision and drive, and that it regularly reviews both objectives and performance.

The project manager

The primary role of the project manager is to control the evolution and execution of the project on behalf of the promoter. This role requires a degree of executive authority so as to coordinate activities effectively and to take responsibility for progress. It is necessary to define the extent of such delegated authority and the means by which instructions will be received with regard to those decisions the project manager is empowered to make.

Ideally the project manager should be involved in the determination of the project objectives and subsequently in the evaluation of the contract strategy. The project manager must therefore drive the project forward and think ahead, delegate routine functions and concentrate on problem areas.

If the project manager is to fulfil the task of the effective realisation of the project on behalf of the promoter, decisions taken on engineering matters cannot be divorced from all other factors affecting the investment. Control may only be achieved by regular reappraisal of the project as a whole so that the current situation in the design office, on fabrication, on the supply of materials and on site may be related to the latest market predictions. If this is done, the advantage to be gained, say, from early access to the site may be equated with any additional costs in full knowledge of the value of early or timely completion. The continual updating of a simple 'time and money' model of the project originally compiled for appraisal will greatly facilitate effective control during the engineering phase of the project.

21.2 Guidelines for project management

The research papers published in 2006 in the UK by Winter and Morris, as part of the 'rethinking project management' programme identified five key areas – project complexity, social process, value creation, project conceptualisation and practitioner development – as an agenda 'to inform and structure future project management research'. This work has been one of the most detailed and most comprehensive studies of project management practice and research and has resulted in a common base for further work, albeit there may be some variations in the views of project management researchers regarding the understanding and domain of the five areas.

Project complexity is increasingly cited as a factor to be recognised and tackled in project management procedures. At the simplest level it is interpreted as 'interconnectedness'; in other words, it is not size or extent of a project that is most important but the degree to which elements of a project can influence each other and/or form self-regulating links. Complexity has been recognised as a barrier to successful project management by traditional methods and a range of responses have been instigated from theoretical investigations of complexity science to the establishment of the joint US/UK/Australian Complex Project Management College for the defence sector. In the future, it is proposed that managers of major defence projects will have to be graduates of the college who will have studied complex project management and undergone a similar accreditation programme to the IPMA (International Project Management Association) Platinum Standard.

The realisation that 'political' support, in its widest sense, was one of the essential prerequisites for project success occurred many years ago. Nevertheless, textbooks and project management training have been slow to recognise the importance of the interactions of people and groups in an array of social practices. Business has been much quicker to recognise the significance of human actions and interactions and the roles of culture, politics and power. The fact that project management is a non-project-specific methodology does not mean that it is apolitical. International politics, corporate politics and personal politics all affect each and every project undertaken to some extent. Project management must be placed in the context of its social processes so as to function effectively.

The old saying 'An engineer produces for a shilling (5p) what anyone can produce for a pound' mirrors part of the traditional project management thinking about the cost-effectiveness of a project manager. It is also true that today value for money is important, but the focus is now on value creation and revenue generation rather than on cost saving. This again reflects the reality of many projects as business investments, with the recognition of the rapidly changing business markets, opportunities and objectives.

The term 'project conceptualisation' is not always easily understood. The traditional view of projects is that they have, as far as is practicable, clear, transparent, well-defined and largely fixed objectives. To reflect the reality in which many projects are actually being executed a broader, less specific concept of a project is required. Today projects may be subject to renegotiation or change at all stages of the project life cycle. Despite this being diametrically opposed to the traditional view that ambiguity over project objectives can be a critical source of project uncertainty, it is more realistic. It often means that a number of permeable outcomes and objectives could be identified and projects managed as multi-purpose.

In increasingly chaotic conditions it is no longer sufficient to apply standard project management procedures in a prescribed methodology. More pressure is placed on the project manager to be able to learn from the project and respond creatively to difficult and changing aspects of the project environment. The competency of project managers must therefore go beyond technical capability and include attributes such as: achievement orientation, initiative, focus on client's needs, assertiveness, leadership, analytical and conceptual thinking and, above all, flexibility.

Project management is concerned with the achievement of realistic objectives for a project. This will demand effort, it will not happen as a matter of course, and it will require the dedication and motivation of people. The provision and training of an adequate project management team is therefore an essential prerequisite for a successful job, for it is their drive and judgement, their ability to persuade and lead, which will ensure that the project objectives are achieved.

21.3 Project management: The way ahead

It is always difficult to prejudge the evolution and development of existing systems, particularly in the more subjective management areas. The further into the future the target the less that can be seen but attempts to manage the future have to be made if progress is to be achieved.

As time goes on new methods and techniques will be developed and promoted with varying degrees of success. In the last 20 years total quality management and business process re-engineering could be viewed in this context but could also be viewed as part of the overall process of project management in their own right. It is almost certain that further novel and innovative processes will be derived and incorporated into best practice for the delivery of world-class project management. Project management relies on good management practice but has the overall goal of the project completion in terms of the project objectives as its prime aim. Therefore, it is likely that project management will, over the next few years, subsume business objectives as a part of the investment project. This is likely to involve some changes in approach and terminology but will result in improved project management procedures integrating discrete business functions to enhance the effectiveness of decision making.

Project management will be required until such time all projects can be delivered to meet all the customer's requirements, and any other project constraints, first time and every time. Judging by current performance, there is likely to be a need for project management in the foreseeable future.

Further reading

CIOB (2202) *Code of Practice for Project Management for Construction and Development*, third edition, Blackwell Publishing.

Pryke, S. and Smyth, H. (2006) *The Management of Complex Projects: A Relationship Approach*, Blackwell Publishing.

Special Edition (2006) Rethinking project management, *International Journal of Project Management*, **24**.

Index

Note: Page numbers in *italics* refer to figures and those in **bold** to tables